Liquid Metal Batteries

This book offers a broad overview of the current knowledge and future challenges in liquid metal batteries. It details the fundamental science of liquid metal batteries, materials for cathode and anode, various synthesis methods, and approaches to modify their properties to provide high energy and power densities, along with long life. It offers a broad overview of current research and future challenges in this exciting field.

- Covers materials, chemistry, and technologies for liquid metal batteries.
- Discusses the state of the art and challenges.
- Explains the fundamentals of electrochemical behavior of liquid metal batteries.
- Offers insight into design and architectural aspects.
- Provides new direction to scientists, researchers, and students to better understand the chemistry, electrochemical properties, and technologies of liquid metal batteries.

With chapters written by experts in energy from around the globe, this book serves as a textbook for advanced students and offers new guidelines to researchers and industries working in energy and materials engineering.

Ram K. Gupta is Professor of Chemistry at Pittsburg State University and Associate Vice President for Research and Support at Pittsburg State University, Pittsburg, KS, USA. Dr. Gupta's research focuses on green energy production and storage using conducting polymers and composites, electrocatalysts for fuel cells, nanomaterials, optoelectronic and photovoltaic devices, organic-inorganic hetero-junctions for sensors, nanomagnetism, bio-based polymers, bio-compatible nanofibers for tissue regeneration, scaffold and antibacterial applications, and bio-degradable metallic implants. Dr. Gupta has published over 380 peer-reviewed articles, made over 500 national/international/regional presentations, chaired many sessions at national/international meetings, and edited/written several books/chapters for the American Chemical Society, Royal Society of Chemistry, CRC, Elsevier, Springer, and Wiley. He has received several million dollars from external agencies for research and educational activities. He serves as an associate editor, guest editor, and editorial board member for various journals.

Liquid Metal Batteries

Chemistry, Advancements, and Challenges

Edited by Ram K. Gupta

CRC Press
Taylor & Francis Group
Boca Raton London New York

CRC Press is an imprint of the
Taylor & Francis Group, an **informa** business

Designed cover image: © 2026 Shutterstock

First edition published 2026
by CRC Press
2385 NW Executive Center Drive, Suite 320, Boca Raton FL 33431

and by CRC Press
4 Park Square, Milton Park, Abingdon, Oxon, OX14 4RN

CRC Press is an imprint of Taylor & Francis Group, LLC

ISBN: 978-1-032-78912-5 (hbk)
ISBN: 978-1-032-78915-6 (pbk)
ISBN: 978-1-003-48972-6 (ebk)

DOI: 10.1201/9781003489726

Typeset in Times
by Apex CoVantage, LLC

Contents

About the Editor

Dr. Ram K. Gupta is Associate Vice President for Research and Support and Professor of Chemistry at Pittsburg State University. Stanford University has recently named Gupta among the top 2% of research scientists worldwide. Before joining Pittsburg State University, he worked as Assistant Research Professor at Missouri State University, Springfield, MO, and then as Senior Research Scientist at North Carolina A&T State University, Greensboro, NC. Dr. Gupta's research spans a range of subjects critical to current and future societal needs, including semiconducting materials and devices; biopolymers; flame-retardant polymers; green energy production and storage using nanostructured materials and conducting polymers; electrocatalysts; optoelectronics and photovoltaics devices; organic-inorganic heterojunctions for sensors; nanomagnetism; biocompatible nanofibers for tissue regeneration; scaffold and antibacterial applications; and bio-degradable metallic implants. Dr. Gupta has mentored ten PhD/postdoc scholars, 76 MS students, and 58 undergraduate/high school students. He has published over 380 peer-reviewed journal articles (14,000+ citations, 65 h-index, 298 i10-index), made over 500 national/international/regional presentations, chaired/organized many sessions at national/international meetings, wrote several book chapters (120+), worked as Editor for many books (50+) for American Chemical Society, CRC, Springer, Elsevier, etc., and received several million dollars for research and educational activities from external agencies. He is also serving as Editor, Associate Editor, Guest Editor, and editorial board member for various journals.

Preface

With the rapid advancements in modern lifestyle, the pervasive use of electronics, automobiles, and the relentless pace of industrialization, there arises an urgent need for high-performance energy devices. Since Alessandro Volta's invention of the first battery nearly two centuries ago, batteries have made remarkable strides through diverse approaches and materials. Among these, lithium-ion batteries have demonstrated immense potential across consumer and industrial applications; yet they face significant challenges in long-term performance improvement.

A critical issue affecting the cycling performance of lithium metal anodes is the persistent problem of dendritic growth, volumetric expansion, and continuous electrolyte decomposition during stripping deposition. In response, liquid metal batteries have emerged as a promising alternative. These innovative batteries offer a solution to the intrinsic dendrite problem associated with lithium and other alkali metals. They are characterized by scalability, cost-effectiveness, and extended cycle life.

This book aims to provide comprehensive insights into the current state-of-the-art knowledge and future challenges in liquid metal batteries. It covers fundamental principles, materials for cathodes and anodes, various synthesis methodologies, and approaches to enhancing their properties for achieving high energy and power densities, as well as stable electrochemical characteristics over extended operational lifetimes. Each chapter is authored by experts in the field of energy from around the globe, making this text not only suitable for students but also a valuable resource for researchers and industries engaged in energy, materials science, and nanotechnology.

Contributors

Alaa Bedair
Department of Analytical Chemistry
University of Sadat City
Sadat City, Egypt

Mayankkumar L. Chaudhary
National Institute for Materials Advancement
Pittsburg State University
Kansas, United States

Ana K. Cuentas-Gallegos
Centro de Nanociencias y Nanotecnología
Universidad Nacional Autónoma de México
Baja California, México

Mercy Thelakkattu Devassy
Energy Systems Group
Vikram Sarabhai Space Centre
Kerala, India

Chunxia Gao
School of Chemistry and Chemical Engineering
Yangzhou University
Jiangsu, China

Paola N. García-Hernández
Unidad de Energía Renovable
Centro de Investigación Científica de Yucatán A. C.
Yucatán, México

Ram K. Gupta
Department of Chemistry
National Institute for Materials Advancement
Pittsburg State University
Kansas, United States

Yamini Gupta
Department of Physics
Goswami Ganesh Dutta Sanatan Dharma College
Chandigarh, India

Mahmoud Hamed
Pharmaceutical Chemistry Department
Misr International University
Cairo, Egypt

Mohammad Harun-Ur-Rashid
Department of Chemistry
International University of Business Agriculture and Technology
Dhaka, Bangladesh

Bing Joe Hwang
Department of Chemical Engineering
National Taiwan University of Science and Technology
Taipei, Taiwan

Abu Bin Imran
Department of Chemistry
Bangladesh University of Engineering and Technology
Dhaka, Bangladesh

Md. Hafizul Islam
Department of Chemistry
International University of Business Agriculture and Technology
Dhaka, Bangladesh

Saiful Islam
Department of Chemistry
International University of Business Agriculture and Technology
Dhaka, Bangladesh

Suresh Jayakumar
Department of Physics and Nanotechnology
SRM Institute of Science and Technology
Tamilnadu, India

Mengjin Jiang
College of Polymer Science and Engineering
Sichuan University
Sichuan, China

Bibin John
Energy Systems Development Division
Vikram Sarabhai Space Centre
Kerala, India

Phuoc-Anh Le
Institute of Chemistry
Vietnam Academy of Science and Technology
Hanoi, Vietnam

Tianyi Liu
Brewer Science Inc.
Missouri, United States

Ntakadzeni Madima
Advanced Materials Division
Mintek
Randburg, South Africa

Fotouh R. Mansour
Medicinal Chemistry Department
King Salman International University
South Sinai, Egypt

Siphelo Ngqoloda
Advanced Materials Division
Mintek
Randburg, South Africa

Bulelwa Ntsendwana
Advanced Materials Division
Mintek
Randburg, South Africa

Daniella E. Pacheco-Catalán
Unidad de Energía Renovable
Centro de Investigación Científica de Yucatán A. C.
Yucatán, México

Rutu Patel
National Institute for Materials Advancement
Pittsburg State University
Kansas, United States

Perla J. Pérez-Díaz
Unidad de Energía Renovable
Centro de Investigación Científica de Yucatán A. C.
Yucatán, México

Akhilash Mohanan Pillai
Energy Systems Development Division
Vikram Sarabhai Space Centre
Kerala, India

A. V. Radhamani
Department of Physics and Nanotechnology
SRM Institute of Science and Technology
Tamilnadu, India

Aminur Rahman
Department of Chemistry
Bangladesh University of Engineering and Technology
Dhaka, Bangladesh

Mpfunzeni Raphulu
Advanced Materials Division
Mintek
Randburg, South Africa

Patteth S. Salini
Energy Systems Development Division
Vikram Sarabhai Space Centre
Kerala, India

P. Chinnappan Santhosh
Department of Physics and Nanotechnology
SRM Institute of Science and Technology
Tamilnadu, India

Montree Sawangphruk
Department of Chemical and Biomolecular Engineering
Vidyasirimedhi Institute of Science and Technology
Rayong, Thailand

Kriti Sharma
Department of Physics
Goswami Ganesh Dutta Sanatan Dharma College
Chandigarh, India

Kassie Nigus Shitaw
Department of Chemical Engineering
National Taiwan University of Science and Technology
Taipei, Taiwan

Rajendra Kumar Singh
Institute of Science
Banaras Hindu University
Varanasi, India

Poonam Siwatch
Department of Physics
Arya Kanya Mahavidyalaya
Haryana, India

Wei-Nien Su
Graduate Institute of Applied Science and Technology
National Taiwan University of Science and Technology
Taipei, Taiwan

Md. Abu Bin Hasan Susan
Department of Chemistry
University of Dhaka
Dhaka, Bangladesh

S. K. Tripathi
Department of Physics
Panjab University
Chandigarh, India

Vikas Yadav
Institute of Science
Banaras Hindu University
Varanasi, India

Jiadeng Zhu
Brewer Science Inc.
Missouri, United States

1 Liquid Metal Batteries
An Introduction

Suresh Jayakumar, P. Chinnappan Santhosh, and A. V. Radhamani

1.1 NEED FOR LMBs

To achieve carbon neutrality by 2060, increasing the use of non-fossil energy sources and expanding the use of renewable green technologies is essential. This includes establishing a circular economy and ensuring the widespread adoption of smart grid architecture. Transitioning from fossil-fuel-powered plants to renewable sources, such as solar, hydro, and wind power, has the potential to significantly reduce carbon emissions in key industrial sectors. However, the intermittent nature and uneven geographical distribution of these renewable sources limit their widespread market dominance [1]. Thus, to mitigate grid inflexibility and maintain a stable energy supply, using robust energy storage systems is crucial. Grid-scale energy storage requires batteries with a prolonged operational lifetime of 20 to 30 years, exceptional safety, and low cost. Traditional batteries, such as Li-ion, Na-ion, lead-acid, and Zn-ion, experience capacity reduction because they employ solid electrodes and acidic or alkaline electrolytes. This leads to phase transformation, dendrite growth, and side reactions, which pose challenges in meeting the durability standards required for grid-level energy storage. These limitations necessitate the exploration of alternative technologies to provide more reliable and sustainable energy storage solutions [2]. Liquid metal batteries (LMBs) present a novel approach to grid-level energy storage by utilizing liquid metals as both positive and negative electrodes. Metals such as antimony (Sb), bismuth (Bi), tin (Sn), lead (Pb), and tellurium (Te) are used for positive electrodes, while lithium (Li), sodium (Na), potassium (K), calcium (Ca), and magnesium (Mg) function as negative electrodes [3]. The integration of liquid metal electrodes with the stable, non-decomposing behavior of inorganic molten salts as electrolytes enables LMBs to overcome degradation processes, theoretically granting them an endless cycle life. Additionally, the absence of thermal effects contributes to the high safety level of LMBs, making them suitable for grid-scale applications [4]. Further, the global liquid metal battery market is expected to grow significantly, with the market size estimated to reach $1299.8 million by 2029, up from $816 million in 2022, with a compound annual growth rate (CAGR) of 6.9%. The COVID-19 pandemic has impacted market dynamics, but the demand for innovative energy storage solutions continues to drive growth. Companies like Ambri, Eos Energy Storage, and Form Energy are at the forefront of developing and commercializing LMB technology. These batteries promise to revolutionize power grids

DOI: 10.1201/9781003489726-1

by enhancing renewable energy integration, extending battery life, and reducing costs, thereby addressing some of the most pressing energy challenges. Interestingly, market dynamics for LMBs vary by region. In North America, particularly the USA, the market is expected to grow due to strong industrial automation, high consumer spending, and technological advancements. The region's focus on increasing efficiency through the growth of the Internet of Things (IoT) will facilitate this growth. In Europe, both Western and Eastern regions show promise. Western Europe benefits from robust economic conditions, continuous innovation, and infrastructure development, while Eastern Europe, despite challenges such as the Russia-Ukraine conflict, is experiencing growth in manufacturing and technological hubs. Asia, led by China, Japan, South Korea, and India, will continue to dominate global manufacturing, with a shift toward higher efficiency and quality standards. The Middle East is also investing in diversifying its economy away from fossil fuels, with significant plans like Saudi Vision 2030 driving growth. Africa is poised for the largest growth, supported by investments in agriculture, mining, manufacturing, and other key sectors, along with improvements in political stability and initiatives like the African Continental Free Trade Area (AfCFTA) facilitating easier movement of goods and services [5]. In general, the growth of the LMB market is now closely linked with a country's broader economic growth and development plans.

Overall, the development of LMBs represents an innovative solution to the challenges of grid-scale energy storage. By improving the integration of renewable resources into power grids, these batteries can enhance the stability and reliability of energy supply, contributing to the broader goals of decarbonization and sustainable development. As the technology matures and becomes more widely adopted, it holds the potential to transform energy storage and distribution on a global scale, addressing both current and future energy needs.

1.2 HISTORICAL EVOLUTION OF LMBs

The evolution of batteries dates to the late eighteenth century with the invention of the voltaic pile by Alessandro Volta, which marked the beginning of electrochemical energy storage. Over the centuries, advancements in battery technology have been driven by the need for more efficient, durable, and powerful energy storage solutions. From the lead-acid batteries of the nineteenth century to the nickel-cadmium and nickel-metal hydride batteries of the twentieth century, each development aimed to enhance energy density, reduce costs, and improve cycle life. The introduction of lithium-ion batteries in the late twentieth century revolutionized the industry, providing high energy density and lightweight solutions for portable electronics and electric vehicles. However, the limitations of lithium-ion technology, such as safety concerns and resource scarcity, prompted the search for alternative energy storage systems. The development of LMBs represents a significant leap in battery technology, addressing some of the key challenges faced by previous generations of batteries.

LMBs have an extensive and dynamic history that can be traced back to the early twentieth century. The initial development of LMBs was inspired by the rapid advancements in classical electrometallurgy, particularly the invention of the three-liquid-layer Hoopes cell for the electrolytic production of high-purity aluminum in the 1920s. This cell featured three distinct layers of liquid metals, which self-segregated based on their densities and immiscibility. Further, in 1960, the first patent on the thermally regenerative bimetallic cell was filed by Bernard Agruss at General Motors. This patent essentially describes a liquid metal battery that is thermally "recharged." The early twenty-first century witnessed a renewed interest in LMBs, largely driven by the growing demand for grid-scale energy storage solutions [6]. Early all-LMBs typically consisted of two kinds of high-melting-point liquid metals as electrodes and a molten salt (e.g., halide salt) electrolyte, forming three distinct layers and operating at temperatures generally above 350°C. Despite their high operating temperatures, these batteries offered several advantages such as low cost due to inexpensive inorganic salt electrolytes, high-rate capacity, long lifespan, and ease of assembly and scaling up [7]. Sadoway et al. demonstrated the potential of high-temperature LMBs with a highly reversible magnesium-antimony (Mg-Sb) cell. Antimony (Sb) emerged as a more cost-effective option at $7/kg and was more abundant compared to bismuth (Bi) at $24/kg and tellurium (Te) at $150/kg. On a per-mole basis, Sb ($0.74/mol) is significantly cheaper than Bi ($4.40/mol) and Te ($19.19/mol). This study marked the first use of Sb in a liquid metal battery. Magnesium was chosen as the negative electrode due to its low cost ($5.15/kg, $0.125/mol), abundance, low electronegativity, and compatibility with Sb and the electrolyte. The selected electrolyte, a blend of $MgCl_2$: NaCl (50:30:20 mol%), featured a low melting point (396°C) and greater electrochemical stability. Operating at 0.5 V, the Mg-Sb cell achieved a low cost of $230 per kWh, stimulating further research and development in this area [8]. Subsequent studies explored the use of post-transition and semimetals (e.g., Bi, Sb) as materials for positive electrodes and alkaline metals (e.g., Li, Na, Ca) for negative electrodes, achieving rechargeable potentials surpassing 1.0 V. However, operating these batteries at temperatures above 350°C posed challenges, including the need for rigorous thermal management and high corrosion rates of active cell components. Concurrently, another type of LMB emerged, utilizing solid electrolytes instead of molten salts. Notable examples include Na-S and ZEBRA batteries, developed since the 1960s, which employed molten sodium as the anode and an Na^+ selective ceramic conductor, β/β″-alumina, as the solid-state electrolyte. These batteries could operate at temperatures above 120°C due to the relatively low melting points of sodium and sulfur, but maintaining optimal ion conductivity and wettability required operation at 300–350°C. Since the 2000s, research has also focused on Li-based LMBs, which use Li^+ solid electrolytes to circumvent high-melting-point lithium salts. Garnet-type solid electrolytes have garnered significant attention due to their high Li^+ conductivity, stability with Li metal, and wide electrochemical window. Using molten Li anodes and low-melting-point Sn-Pb and Bi-Pb alloy cathodes, these LMBs can operate at intermediate temperatures around 240°C, paving the way for the development of medium-temperature LMBs for grid energy storage. Generally, LMBs that operate at high temperatures around 300°C face significant challenges such as thermal corrosion, sealing problems, and increased maintenance costs [9]. More recently, the emphasis on reducing the operating temperature of LMBs has prompted researchers to explore fusible

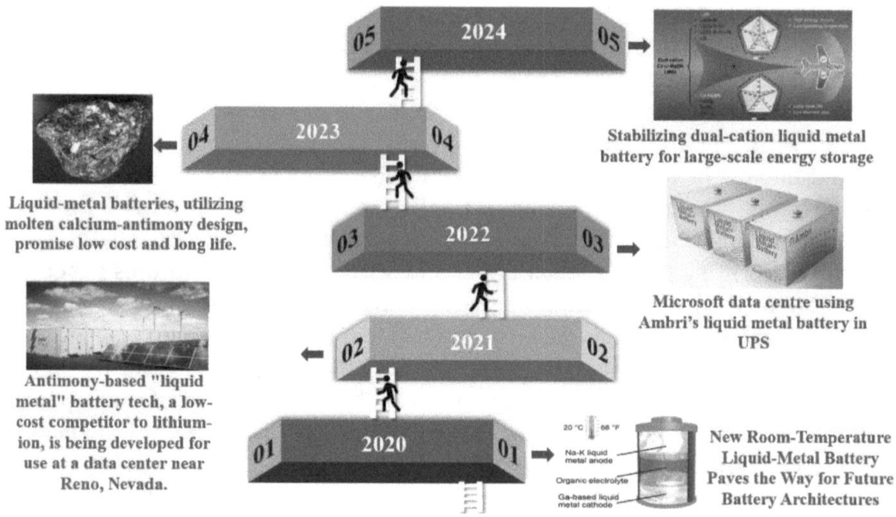

FIGURE 1.1 The evolution of liquid metal batteries from 2020 to 2024 [11–13].

materials such as Ga metal, Ga-based alloys, and Na-K liquid alloys, which are liquid at or near room temperature (0–40°C). Among these, Ga-based liquid metals are particularly promising due to their safety and moderate potential window. Room-temperature Na-K liquid alloys, distinct from Ga, act as charge carrier reservoirs and could potentially replace Li anodes. Despite their potential, room-temperature LMBs still face challenges related to the reactivity of liquid metals with electrolytes, which affects electrochemical reversibility [10]. Further, Figure 1.1 shows the evolution of LMBs of the recent era.

1.3 FUNDAMENTALS OF LMBs

LMBs are a significant innovation in energy storage technology, particularly for applications requiring high capacity, extended life, and scalability, such as grid-scale energy storage. The liquid metal battery consists of three different layers, and the separation of the three liquid layers is based on density, which is essential for maintaining their distinct positions during charge/discharge [14]. As mentioned before, the design involves a three-layer configuration: the low-density with lower electronegativity liquid metal anode on top, molten salt electrolyte in the middle, and high-density with higher electronegativity liquid metal cathode at the bottom [15]. In the LMB electrochemical cell reactions, the negative liquid metal (A) undergoes oxidation, releasing A^{z+} ions and electrons. These A^{z+} ions move through the electrolyte and reduce the positive liquid metal (B) to the liquid metal (A). This leads to the formation of a liquid alloy A-B during the discharge process [16]. The deposition of the liquid metal (A) on the liquid metal (B) increases the thickness of the liquid metal (B). These processes are reversed during charging.

The half-cell reaction in the negative liquid metal (A) and positive liquid metal (B) is as follows [17]:

$$\text{For negative} \qquad A(1) = A^{z+} + ze^{-}\,; \tag{1.1}$$

$$\text{For positive} \qquad A^{z+} + ze^{-} = A(in\,B)\,; \tag{1.2}$$

$$\text{The overall cell reaction is} \quad A(1) = A(in\,B)\,. \tag{1.3}$$

The driving force behind thermodynamic processes is linked to changes in the partial molar Gibbs free energy,

$$\Delta\bar{G}_{Ce_{ll}} = \bar{G}_{A(in\,B)} - \bar{G}_{A(1)}\,. \tag{1.4}$$

For each component, this energy, denoted as $\bar{G}i$, is calculated using the formula,

$$\bar{G}_{A(in\,B)} = G^{0}_{A(1)} + RT\,ln\,a_{A(in\,B)}\,, \tag{1.5}$$

$$\bar{G}_{A(1)} = G^{0}_{A(1)} + RT\,ln\,a_{A(1)} \qquad \left(a_{A(1)} = 1\right). \tag{1.6}$$

In this equation, a_i represents the activity of the component, $G^{0}_{A(1)}$ represents the standard chemical potential, R is the gas constant, and T represents the temperature [18].

A negative electrode in LMBs typically consists of low-density metals such as lithium, sodium, or magnesium, chosen for their high electrochemical potential, ensuring effective electron donation during discharge. The selection of the negative electrode material considers its ability to form a liquid alloy that remains less dense than the electrolyte, as well as the incorporation of noble elements to prevent undesirable reactions and preserve the battery's integrity and performance. The oxidation of the negative electrode during discharge generates cations that migrate through the electrolyte, contributing to the overall energy output of the battery. Conversely, the positive electrode is made of higher-density liquid metals, such as antimony, lead, or bismuth, which are selected for their suitable melting points and capacity to form stable alloys with negative electrode materials. During discharge, as the negative electrode oxidizes and cations move across the electrolyte, the positive electrode grows by incorporating these cations into a liquid A-B alloy, an essential aspect linked to electrochemical reduction processes. The electrolyte serves as the medium for ion transport between the negative and positive electrodes, typically a molten salt mixture chosen for its low melting point, high ionic conductivity, and chemical stability at elevated temperatures [17]. The criteria for material selection for LMBs require the components to remain in a liquid state at practical temperatures, exhibit high electrical conductivity, and be non-radioactive. The batteries have the unique ability to self-heal due to the fluid nature of the components, correcting physical imperfections or damage and extending the operational lifespan. The natural separation of the liquid layers based on density stratification ensures that the anode, electrolyte, and cathode remain in their respective layers, preventing mixing and potential short circuits, contributing to reliability and safety, particularly in large-scale energy storage applications. In design, LMBs are inherently scalable, enabling the creation of larger battery packs by stacking and connecting individual cells, a feature advantageous in integrating renewable energy sources into the power grid. This modularity facilitates easy maintenance and replacement of individual cells, further enhancing the overall reliability and lifespan of the battery system [18]. The current research and development in liquid metal battery technology are focused on improving the materials and design to enhance performance and reduce costs. For example, magnesium-antimony (Mg-Sb) cells have been identified as promising candidates due to their affordability and high reversibility. Specifically, antimony provides a cost advantage over other materials such as bismuth and tellurium, making it a more attractive option for large-scale deployment. The use of antimony in LMBs, in combination with magnesium as the negative electrode, demonstrates the potential of this technology to deliver a low-cost, high-performance energy storage solution.

1.4 RECENT ADVANCEMENTS IN LMBs

The three-liquid-layer configuration of LMBs is gaining attention as a solution for large-scale energy storage due to its simple manufacturing process, affordability, and long cycle life. However, a major challenge is the formation of solid discharge products at the interface between the electrolyte and electrode, which slows down electrode reactions and leads to increased polarization voltages, thus reducing the efficiency of LMBs. To tackle this issue, Xie et al. have introduced an innovative positive electrode made from a bismuth-gallium (Bi-Ga) alloy with a 70:30 mol% ratio. The alloy utilizes bismuth's ability to lower operating temperatures and enhance energy density, along with gallium's low melting point and density to promote the formation of a Ga-rich phase. This phase coexists with the solid discharge product, Li_3Bi, creating an efficient pathway for lithium diffusion, which accelerates electrode reactions. The resulting Li||Bi-Ga system has shown impressive performance, achieving a discharge voltage of 0.67 V at 200 mA/cm^2 and an energy efficiency of 45% at 1200 mA/cm^2, almost double the efficiency of the conventional Li||Bi system. Furthermore, it demonstrates remarkable stability, with a minimal capacity fade rate of 0.08% per cycle over 300 cycles, as shown in Figure 1.2. Furthermore, the lithium diffusion coefficient (D_{Li}) of the Bi-Ga alloy electrode is measured to be 8.90×10^{-5} cm^2/s, which is an order of magnitude higher than that of the Bi electrode (8.61×10^{-6} cm^2/s). The galvanostatic intermittent titration technique (GITT) was used to study lithium diffusion in the positive electrode throughout the entire discharge process. In line with the results obtained from electrochemical impedance spectroscopy (EIS), the Bi-Ga alloy electrode exhibits significantly greater D_{Li} compared to the pure Bi electrode throughout the discharge process. These results suggest that the Bi-Ga alloy could be a highly effective positive electrode material for LMBs, providing a promising approach to overcoming the limitations of electrode reaction kinetics and improving the overall electrochemical performance of LMBs for large-scale energy storage [19]. Cui et al. conducted a comprehensive study on the corrosion behavior of SS304 stainless steel in an Sb-Sn alloy under both static and cell-cycling conditions. The study focused on understanding the mechanisms and their impact on battery performance. Through electrochemical measurements and ex-situ analyses, the researchers discovered that Fe and Ni in SS304 stainless steel dissolve more readily in the liquid

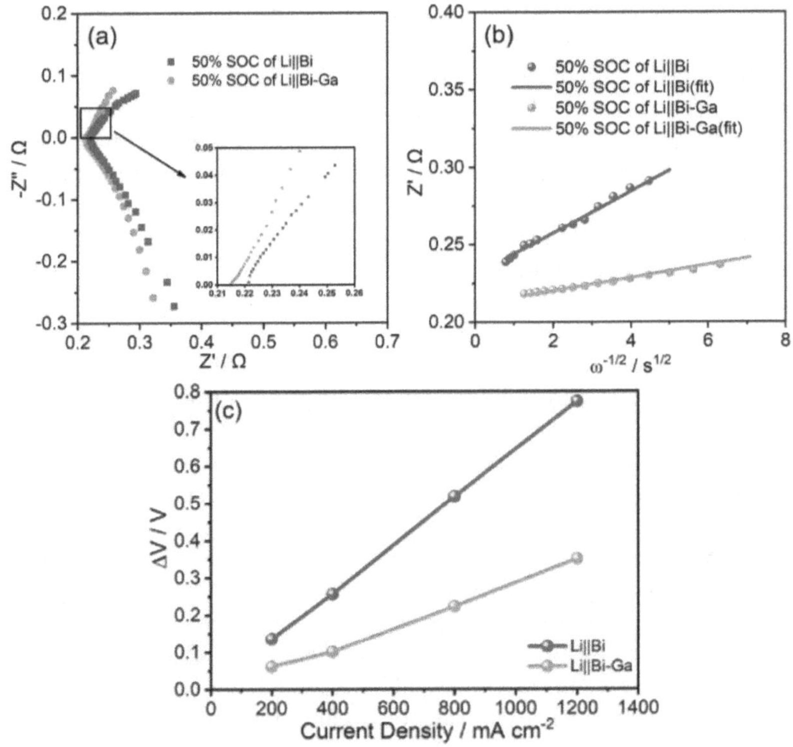

FIGURE 1.2 (a) Nyquist plots of Bi-Ga alloy and pure Bi electrode at 50% DOD (depth of discharge), (b) Linear relationship of Z_0 versus $\omega^{-1/2}$, (c) Polarization voltage (ΔV) versus current density of Li||Bi-Ga and Li||Bi cells.

Source: Adapted with permission [19], Copyright (2020), Elsevier.

Sb-Sn alloy than Cr. This leads to the formation of Fe-Ni-Sb-Sn corrosion products, significantly altering the composition of the positive electrode, which in turn results in abnormal discharge voltage plateau and capacity fading. This low discharge voltage plateau is mainly caused by the reaction between Fe-Sb and Li. Additionally, the presence of Sn in the positive electrode accelerates the corrosion process by promoting the diffusion of dissolved elements. This, combined with the high operating temperatures, exacerbates the degradation of the current collector. The study concludes that Fe- and Ni-based alloys, such as stainless steel, are unsuitable as positive current collectors in Li||Sb-Sn LMB due to their susceptibility to corrosion. Instead, the study suggests developing a conductive and corrosion-resistant protective coating or optimizing the positive electrode composition as more effective strategies. The investigation reveals that the corrosion of SS304 in the Sb-Sn alloy is primarily driven by the dissolution of Fe, which is more soluble in the alloy than Cr or Ni. As cycling progresses, the corrosion products evolve into the active positive electrode material, leading to a new low discharge voltage plateau due to the lithiation of Fe-Sb and capacity loss associated with the formation of Ni-Sb. These findings emphasize the need for protective coatings for the current collector rather than relying on stainless steel, which is prone to corrosion in this high-temperature environment. In order to resolve these challenges and improve the wettability of the electrodes [20], Yan et al. have taken a novel approach in Li||Sb-Sn batteries by incorporating additives in the electrode. The absence of Ti additives leads to poor contact between the Sb-Sn electrodes and the graphite-based collector due to the high surface tension of the liquid alloy. This poor wettability increases the internal resistance, decreasing the battery's efficiency. However, adding Ti to the Li||Sb-Sn system improves the wettability of the liquid alloy to the graphite substrate due to the lower surface tension of the multicomponent alloy. This enhancement ensures better contact throughout the charging and discharge processes. The introduction of Ti also leads to the formation of two layers in the positive electrode after the reaction between Li and Sb, further contributing to the battery performance and stability. Adding 1.5 mol% Ti into the Sb-Sn positive electrode results in a Li||Sb-Sn-Ti battery achieving a voltage efficiency of 88.4% at a 0.2C rate, representing a 2.8% increase compared to batteries without the Ti additive [21]. In Li||Bi LMB system, Cui et al. have introduced a new chemistry for LMBs that uses a LiCl–KCl electrolyte combined with Sb–Bi–Sn and Sb–Bi–Pb positive electrodes. This combination significantly lowers the operating temperature of Li-based LMBs while maintaining high energy density. Their research reveals that the low-melting-point LiCl–KCl molten salt is compatible with lithium and that the displacement reaction between them reduces the KCl content. By adding a small amount of potassium (K) to lithium, this reaction is inhibited, stabilizing the molten salt composition and markedly enhancing the battery's cycling stability at a reduced operating temperature of 400°C. The LiK|LiCl–KCl||$Sb_{30}Bi_{40}Sn_{30}$ system delivers an impressive energy density of approximately 241 Wh/kg with a low material cost of about $68.8 kW/h, while the LiK|LiCl–KCl||$Sb_{30}Bi_{40}Pb_{30}$ system achieves an energy density of around 194 Wh/kg at an even lower material cost of $62.8 kW/h. The study demonstrates that the low-melting-point Sb-Bi-Sn and Sb-Bi-Pb alloys, which primarily govern the battery voltage plateaus, can be effectively matched with the LiCl-KCl electrolyte to achieve stable operation and high energy density at significantly lower temperatures, over 100°C lower than most other reported Li-based LMBs [22]. Yan et al.'s study presents a new Zn-based positive electrode that uses a displacement reaction to enhance the voltage and reduce the cost of LMBs. The compatibility between LiCl-KCl and Zn was analyzed, and it was found that a complete displacement reaction could reduce the active components. To address this, a small amount of Bi was added to Zn, which improved stability and cycling performance. The reaction mechanism of the Zn-Bi positive electrode is illustrated in Figure 1.3. Initially, the positive electrode consists of two liquid-phase layers (Figure 1.3a). During the charging process, Zn in the positive electrode reacts with LiCl in the electrolyte to form Li_2ZnCl_4 (Figure 1.3b). Due to the limited solubility of Li_2ZnCl_4 in the LiCl-KCl electrolyte and its higher density ($\rho = 2.54$ g cm^{-3}) compared to the LiCl-KCl eutectic salt ($\rho = 1.6$ g cm^{-3}), Li_2ZnCl_4 is unevenly distributed within the electrolyte, creating a Li_2ZnCl_4-rich layer near the positive electrode and a Li_2ZnCl_4-poor layer near the negative electrode (Figure 1.3c). This distribution ensures the continued reaction of Li and Li_2ZnCl_4 during the discharge process (Figure 1.3d). In the initial charging phase (Stage II), Li-based compounds are formed due to potential side reactions. In the subsequent

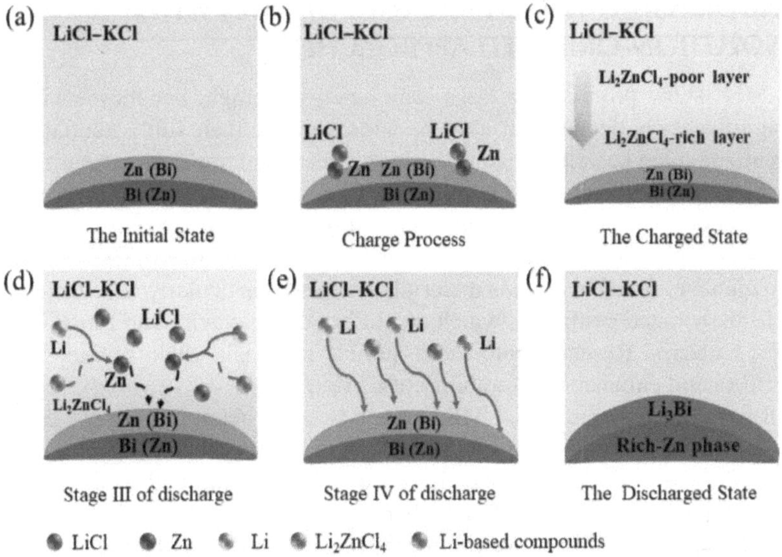

FIGURE 1.3 Illustrates the schematic diagram of the Li‖ Zn–Bi system reaction mechanism. (a) Initial state, (b) the charging process, (c) the charged state, (d) the beginning of discharge, (e) the reaction between Li, Li-based compounds, and Bi, and (f) the discharged state.

Source: Adapted with permission [23], Copyright (2023), American Chemical Society.

TABLE 1.1
Shows the Performance and Cost Metrics of Various LMB Systems

LMB system	Operating Temperature (°C)	Discharge Voltage (V)	Electrolyte	Energy density (Wh/kg)	Material cost ($ kW/h)	Ref
$Ca_{60}Li_{30}Mg_{10}$‖Bi	550	0.6	$CaCl_2$-LiCl	54.68	139.44	[24]
Bi_9Sb‖Na	450	0.95	LiCl-NaCl-KCl	–	30.99	[4]
Sb-Sn‖Li	530	0.6	LiF-LiCl-LiBr	–	57.83	[25]
Sb‖Li	550	0.8	LiF-LiCl-LiBr	421.6	42.4	[26]
Sb-Sn‖Li	500	0.79	LiF-LiCl-LiBr	200.4	71.58	[27]
Zn‖Na	560	1.1	NaCl-$CaCl_2$	–	4.3	[28]
Te-Sn‖Li	500	1.6	LiF-LiCl-LiBr	495	–	[29]
Sb‖Li	550	0.754	LiF-LiCl-LiBr	135	42.4	[30]

discharge phase (Stage IV), in addition to the Li-Bi reaction, these Li-based compounds further react with Bi to form Li_3Bi, continuing until the discharge process is complete. The Li‖LiCl-KCl‖Bi_3Zn_7 battery demonstrates a high discharge voltage of 0.93 V at a current density of 100 mA cm^{-2} and achieves an energy density of 202.04 Wh/kg, with a material cost as low as $50.29 kW/h [23]. Table 1.1 shows the performance and cost metrics of various LMB systems.

1.5 ADDRESSING THE CHALLENGES OF LMBs WITH SOLUTION-ORIENTED APPROACHES

- LMBs show promise for large-scale energy storage, but they encounter significant challenges that must be addressed for their full potential to be realized. One key challenge is the high operating temperature required for LMBs, which increases energy consumption and complexity in the overall system. Developing new materials with lower melting points could help reduce high-temperature operation, making LMBs more efficient and easier to manage. Another issue is material corrosion, particularly concerning the electrolyte and container, which can affect the longevity and reliability of the batteries. Research into more durable materials could mitigate these effects and enhance the durability of LMBs.
- In terms of performance, LMBs currently suffer from low energy density and inferior rate capability compared to other battery technologies. To overcome these challenges, researchers are exploring alternative liquid metal combinations that could offer higher energy density and optimizing electrode materials to improve rate performance. Additionally, the formation of multiple solid intermetallic phases during operation can lead to performance degradation. Investigating new alloys and methods to control these phase formations is essential for maintaining the efficiency of LMBs.
- The high cost of raw materials is a concern that could limit the widespread adoption of LMBs. To counter this, there is a need to seek out cost-effective and abundant materials that can reduce overall costs.
- Finally, the environmental impact of metals used in LMBs is an important consideration. Developing recycling processes and more environmentally friendly materials will be crucial in ensuring that LMBs are a sustainable energy storage solution.

Table 1.2 presents an overview of the challenges faced by LMBs and their corresponding solutions, emphasizing the key areas of focus for researchers and industry

TABLE 1.2
The Challenges of LMBs and Proposed Solutions to Address Them

Category	Challenges	Solutions
Material	High operating temperature	Develop new materials with lower melting points to reduce operating temperatures.
Material	Material corrosion	Research more durable electrolyte and container materials to reduce the impact of corrosion.
Performance	Low energy density	Explore different combinations of liquid metals to enhance energy density.
Performance	Inferior rate capability	Optimize electrode materials and electrolyte composition to enhance rate performance.

TABLE 1.2 *(Continued)*
The Challenges of LMBs and Proposed Solutions to Address Them

Category	Challenges	Solutions
Material	Multiple solid intermetallic phases formation	Explore new alloys and develop strategies to manage or reduce the formation of intermetallic phases.
Economic	High cost of raw materials	Reduce overall cost of LMBs by using cost-effective and abundant materials.
Environmental	Environmental impact of metals	Develop new recycling processes to maintain sustainability.

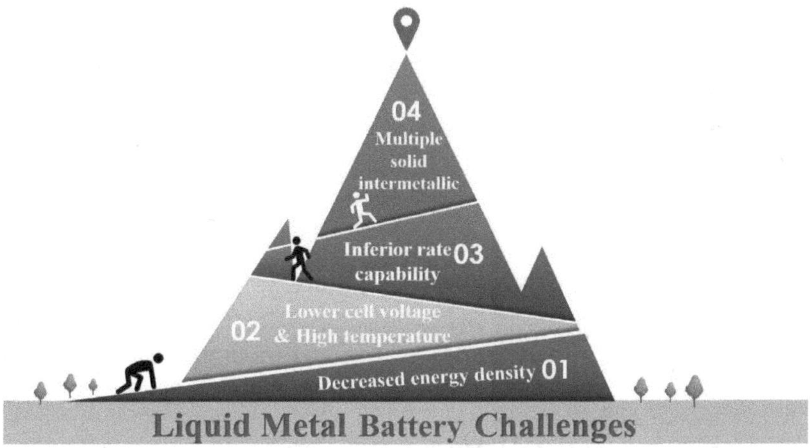

FIGURE 1.4 Illustrates the significant challenges associated with liquid metal batteries.

stakeholders. It is crucial to address these challenges to advance LMB technology and facilitate its widespread adoption in the energy storage sector. Figure 1.4 shows the major challenges of LMBs.

1.6 FUTURE PERSPECTIVES FOR LMBs

- **Integration with Smart Grids:** LMBs could play a pivotal role in smart grid systems by providing reliable, long-duration energy storage that helps balance intermittent renewable energy sources. Their ability to rapidly respond to grid fluctuations and store excess energy makes them well-suited for this integration.
- **Development of Low-Temperature LMBs:** Research is moving toward creating LMBs that operate at lower temperatures. Innovations in low-melting-point metals and alloys could expand the applicability of LMBs,

making them more suitable for a wider range of energy storage applications and reducing the cost of thermal management.

- **Enhanced Performance Through Advanced Electrolytes:** Future developments may focus on improving the performance of LMBs by exploring advanced electrolyte compositions that offer higher ionic conductivity, greater stability, and lower melting points. This could lead to the development of batteries with enhanced efficiency and longer operational life.
- **Exploration of Hybrid LMB Technologies:** Combining LMBs with other energy storage technologies, such as solid-state or flow batteries, could offer synergistic benefits. Hybrid systems could leverage the strengths of different technologies to achieve superior performance, safety, and cost-effectiveness.
- **Customization for Specific Applications:** Future research could lead to the development of specialized LMBs tailored for specific applications, such as grid stabilization, renewable energy integration, or electric vehicle charging. Customization could optimize performance and cost-effectiveness for these targeted uses.
- **Energy Density and Storage Capacity:** Improving the energy density of LMBs will be crucial for expanding their use. Research into new electrode materials and configurations could significantly increase the amount of energy that LMBs can store and deliver, making them more competitive with other storage solutions.
- **Safety and Stability Enhancements:** Ongoing efforts will likely focus on further improving the safety and stability of LMBs. This includes developing new materials that resist degradation and minimize risks associated with high temperatures and reactive components, enhancing the overall reliability of LMBs.
- **Economic Feasibility and Market Penetration:** The future success of LMBs will depend on achieving economic feasibility. This involves lowering production costs, improving the efficiency of large-scale manufacturing processes, and demonstrating the economic advantages of LMBs compared to existing technologies.
- **Regulatory and Standardization Developments:** As LMBs move closer to commercial deployment, establishing regulatory standards and industry benchmarks will be crucial. This will ensure safety, performance, and compatibility with existing energy infrastructure, facilitating broader adoption and market acceptance.

REFERENCES

[1] T. Mu, Z. Wang, N. Yao, M. Zhang, M. Bai, Z. Wang, X. Wang, X. Cai, Y. Ma, Technological penetration and carbon-neutral evaluation of rechargeable battery systems for large-scale energy storage, *J. Energy Storage*. 69 (2023) 107917.

[2] J. Kim, D. Shin, Y. Jung, S.M. Hwang, T. Song, Y. Kim, U. Paik, LiCl-LiI molten salt electrolyte with bismuth-lead positive electrode for liquid metal battery, *J. Power Sources*. 377 (2018) 87–92.

[3] N. Weber, C. Duczek, G.M. Horstmann, S. Landgraf, M. Nimtz, P. Personnettaz, T. Weier, D.R. Sadoway, Cell voltage model for Li-Bi liquid metal batteries, *Appl. Energy.* 309 (2022) 118331.

[4] H. Zhou, H. Li, Q. Gong, S. Yan, X. Zhou, S. Liang, W. Ding, Y. He, K. Jiang, K. Wang, A sodium liquid metal battery based on the multi-cationic electrolyte for grid energy storage, *Energy Storage Mater.* 50 (2022) 572–579.

[5] https://reports.valuates.com/market-reports/QYRE-Auto-34G16037/global-liquid-metal-battery-energy-storage-system.

[6] S. Wu, X. Zhang, R. Wang, T. Li, Progress and perspectives of liquid metal batteries, *Energy Storage Mater.* 57 (2023) 205–227.

[7] Y. Ding, X. Guo, G. Yu, Next-generation liquid metal batteries based on the chemistry of fusible alloys, *ACS Cent. Sci.* 6 (2020) 1355–1366.

[8] D.J. Bradwell, H. Kim, A.H.C. Sirk, D.R. Sadoway, Magnesium–antimony liquid metal battery for stationary, *J. Am. Chem. Soc.* (2012) 2011–2013.

[9] Y. Ding, X. Guo, Y. Qian, L. Xue, A. Dolocan, G. Yu, Room-temperature all-liquid-metal batteries based on fusible alloys with regulated interfacial chemistry and wetting, *Adv. Mater.* 32 (2020) 2002577.

[10] S. Zhang, Y. Liu, Q. Fan, C. Zhang, T. Zhou, K. Kalantar-Zadeh, Z. Guo, Liquid metal batteries for future energy storage, *Energy Environ. Sci.* 14 (2021) 4177–4202.

[11] https://eepower.com/news/new-room-temperature-liquid-metal-battery-paves-the-way-for-future-battery-architectures.

[12] https://spectrum.ieee.org/liquid-metal-battery.

[13] https://thestlouisgroup.com/antimony-battery-storage.

[14] T. Ouchi, D.R. Sadoway, Positive current collector for Li||Sb-Pb liquid metal battery, *J. Power Sources.* 357 (2017) 158–163.

[15] L. Fu, Q.C. Le, X.B. Wang, X. Liu, W.T. Jia, Overview on the liquid metal battery for grid-level large-scale energy storage, *Key Eng. Mater.* 723 KEM (2017) 572–578.

[16] J. Xia, Q. Shi, H. Li, M. Zhou, K. Jiang, K. Wang, A novel sorting method for liquid metal batteries based on deep learning and sequential features, *J. Energy Storage.* 64 (2023) 107093.

[17] H. Li, H. Yin, K. Wang, S. Cheng, K. Jiang, D.R. Sadoway, Liquid metal electrodes for energy storage batteries, *Adv. Energy Mater.* 6 (2016) 1600483.

[18] H. Kim, D.A. Boysen, J.M. Newhouse, B.L. Spatocco, B. Chung, P.J. Burke, D.J. Bradwell, K. Jiang, A.A. Tomaszowska, K. Wang, W. Wei, L.A. Ortiz, S.A. Barriga, S.M. Poizeau, D.R. Sadoway, Liquid metal batteries: Past, present, and future, *Chem. Rev.* 113 (2013) 2075–2099.

[19] H. Xie, H. Zhao, J. Wang, P. Chu, Z. Yang, C. Han, Y. Zhang, High-performance bismuth-gallium positive electrode for liquid metal battery, *J. Power Sources.* 472 (2020) 228634.

[20] K. Cui, F. An, W. Zhao, P. Li, S. Li, C. Liu, X. Qu, Feasibility research of SS304 serving as the positive current collector of Li||Sb-Sn liquid metal batteries, *J. Phys. Chem. C.* 125 (2021) 237–245.

[21] S. Yan, H. Li, Z. Li, W. Chen, X. Zhou, H. Zhou, W. Zhang, Y. He, K. Jiang, K. Wang, In situ transition layer design based on Ti additive enabling high-performance liquid metal batteries, *ACS Appl. Mater. Interfaces.* 15 (2023) 5265–5272.

[22] K. Cui, W. Zhao, S. Li, D. Zhou, C. Liu, X. Qu, P. Li, Low-Temperature and high-energy-density Li-based liquid metal batteries based on LiCl-KCl molten salt electrolyte, *ACS Sustain. Chem. Eng.* 10 (2022) 1871–1879.

[23] S. Yan, L. Fan, H. Li, X. Zhou, H. Zhou, W. Zhang, Z. Li, Y. He, B. Li, K. Jiang, K. Wang, Novel high-voltage Zn-based electrode based on displacement reaction for liquid metal batteries, *ACS Sustain. Chem. Eng.* 11 (2023) 11693–11699.

[24] H. Zhou, L. Huang, M. Yu, X. Ning, Stabilizing dual-cation liquid metal battery for large-scale energy storage: A comprehensive hybrid design approach, *Energy Storage Mater.* 70 (2024) 103540.

[25] S. Yan, H. Li, W. Chen, H. Zhou, X. Zhou, W. Zhang, Z. Li, Y. He, K. Wang, K. Jiang, Full-lifetime recycling and reutilization of key materials of low-cost and sustainable liquid metal batteries, *ACS Sustain. Chem. Eng.* 11 (2023) 17038–17045.

[26] S. Yan, X. Zhou, H. Li, Y. Shen, Y. He, H. Zhou, K. Wang, K. Jiang, Utilizing in situ alloying reaction to achieve the self-healing, high energy density and cost-effective Li||Sb liquid metal battery, *J. Power Sources.* 514 (2021) 230578. https://doi.org/10.1016/j.jpowsour.2021.230578.

[27] H. Li, K. Wang, S. Cheng, K. Jiang, High performance liquid metal battery with environmentally friendly antimony-tin positive electrode, *ACS Appl. Mater. Interfaces.* 8 (2016) 12830–12835.

[28] J. Xu, O.S. Kjos, K.S. Osen, A.M. Martinez, O.E. Kongstein, G.M. Haarberg, Na-Zn liquid metal battery, *J. Power Sources.* 332 (2016) 274–280.

[29] H. Li, K. Wang, H. Zhou, X. Guo, S. Cheng, K. Jiang, Tellurium-tin based electrodes enabling liquid metal batteries for high specific energy storage applications, *Energy Storage Mater.* 14 (2018) 267–271.

[30] X. Zhou, H. Zhou, S. Yan, Y. He, W. Zhang, H. Li, K. Wang, K. Jiang, Increasing the actual energy density of Sb-based liquid metal battery, *J. Power Sources.* 534 (2022) 231428.

2 Fundamentals of Liquid Metal Batteries

Tianyi Liu, Chunxia Gao, Mengjin Jiang, and Jiadeng Zhu

2.1 BASIC PRINCIPLES

Different from the traditional rechargeable batteries, liquid metal batteries (LMBs), also known as molten salt batteries, employ liquid metal for both their anodes and their cathodes, separated by a molten salt electrolyte, requiring no separators [1–7]. LMBs are usually secondary batteries, and the structure of the battery is typically vertical, with the cathode, electrolyte, and anode from bottom to top. Normally, the density of the cathode is greater than that of the electrolyte, and the density of the electrolyte is greater than that of the anode [8]. Also, these three materials should be immiscible, so the battery naturally forms a three-layer structure. The battery configuration is shown in Figure 2.1, where A and B represent the anode and cathode, respectively. During discharge, the anode metal is oxidized, and the resulting metal ions travel through the electrolyte to the cathode, forming an alloy at the cathode (Figure 2.1a). While charging, A is reduced from the A-B alloy and migrates back to the anode (Figure 2.1b). The whole process can be expressed as follows:

Discharging:	Anode	$A—ze^- \rightarrow A^{z+}$;
	Cathode	$B + A^{z+} + ze^- \rightarrow AB$.
Charging:	Anode	$A^{z+} + ze^- \rightarrow A$;
	Cathode	$AB \rightarrow B + A^{z+} + ze^-$.

At present, the most common anode and cathode are Li and Pb-Sb, respectively, with a stability of ~10 years and a Coulombic efficiency of ~73% [8]. This combination is paired with an electrolyte of eutectic LiF-LiCl-LiI or LiCl-KCl. There are also some studies focusing on other metal systems to pursue higher energy density, lower cost, and lower operating temperatures [9,10]. For example, Mg||Sb battery can reach a capacity of more than 1000 mAh/cm^2 at 700°C, with a cost of 375 $kW/h (Li-based LMBs are lower than 100 $kW/h) [11]. Besides, Ga has a melting point as low as 29.87°C and can form low-melting-point alloys with other metals. Thus, Wang et al. developed LMBs using Ga-Sn alloy, which has a melting point of 21°C. The resultant battery could be cycled at 35°C with a capacity of 409 mAh/g [12].

To fabricate an optimal cell, it is crucial to understand battery design, fluid mechanism, and features of LMBs, along with the materials selection and performance

DOI: 10.1201/9781003489726-2

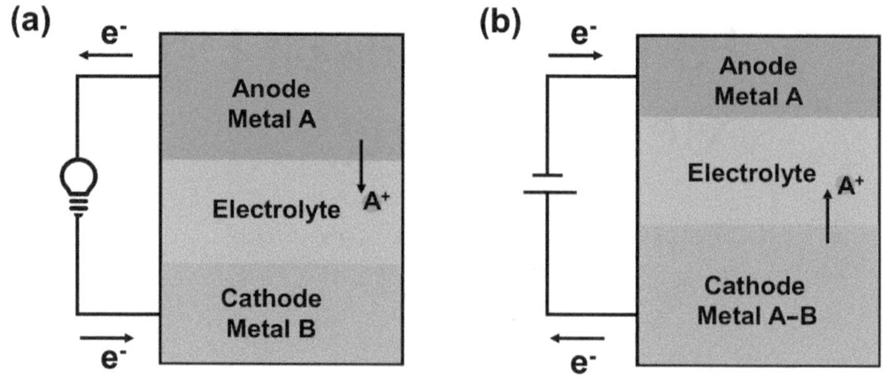

FIGURE 2.1 Schematic diagram of a liquid metal battery upon (a) discharging and (b) charging.

characterization, which will be discussed in detail in the following section. This chapter presents the configuration of batteries and fluid dynamics within the cells. Furthermore, a thorough examination of the materials used in LMBs will be given. This includes the discussion of the anode and cathode materials, electrolytes, and other components, as well as how those factors influence the battery's performance and longevity. Finally, parameters for evaluating the effectiveness of the overall battery's performance will also be discussed.

2.2 BATTERY DESIGN, FLUID MECHANISM, AND FEATURES OF LMBs

This section presents the evolution of battery design and introduces the battery's operation by illustrating the cell's structure. Moreover, the section covers fluid dynamics and key features of LMBs.

2.2.1 Battery Design

In 1925, Hoopes and his coworker patented an industrial electrolytic cell that was used to refine aluminum to ultra-high purity levels [13]. This cell is considered a prototype of LMBs. The Hoopes cell is designed with three distinct layers, each having different densities and functions, as shown in Figure 2.2a. The bottom layer is an impure aluminum layer, which acts as the anode. It contains aluminum that needs to be refined. The middle layer is an electrolyte layer, typically composed of a molten mixture of cryolite (Na_3AlF_6), along with additives like BaF_2. The top layer is the pure aluminum, which serves as the cathode. Pure aluminum is deposited here as the electrolysis process progresses. A cooled water wall helps stabilize these layers by preventing convective currents or disturbances caused by temperature fluctuations. Also, several graphite collectors are used in Hoopes cells. This system produces high-purity aluminum but can be seen as a rechargeable battery, even though the

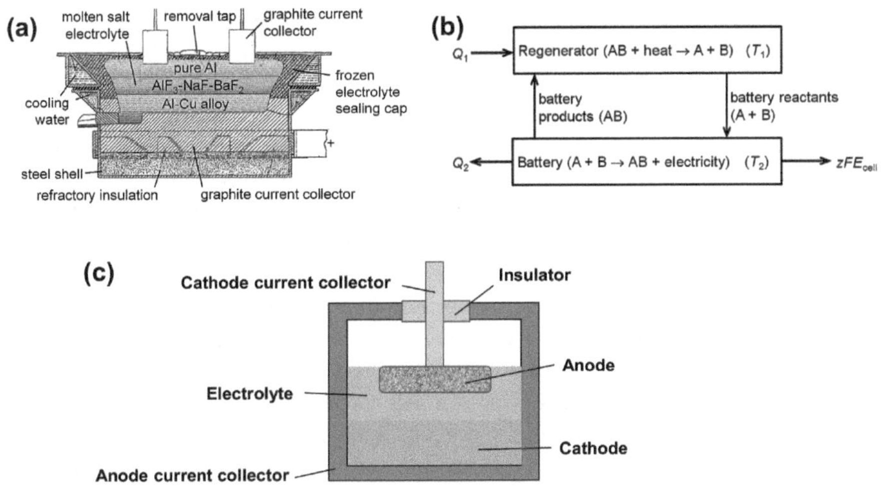

FIGURE 2.2 (a) Scheme of a Hoopes cell. (b) Scheme of Yeager's original 1958 concept of a thermally regenerative battery. (c) Cell schematic of a typical liquid metal battery.

Source: (a) Adapted with permission [8]. Copyright The Authors, some rights reserved; exclusive license American Chemical Society. Distributed under a Creative Commons Attribution License 4.0 (CC BY) https://creativecommons.org/licenses/by/4.0/. (b) Adapted with permission [8]. Copyright The Authors, some rights reserved; exclusive license American Chemical Society. Distributed under a Creative Commons Attribution License 4.0 (CC BY) https://creativecommons.org/licenses/by/4.0/.

theoretical voltage is lower than 30 mV. The electrolytic process is a "charging" process, which is capable of producing aluminum with a purity of 99.99%. Although this device is simple, it has operated reliably for over 20 years.

In 1958, a thermally regenerative closed-cycle or thermally regenerative battery was developed, considered the next-generation configuration of LMBs. This type of energy storage system was first proposed by Yeager et al. [14] and is especially attractive for applications where waste heat is abundant, such as in industrial processes or solar thermal power plants. The corresponding illustration is shown in Figure 2.2b. Similar to conventional batteries, the thermally regenerative battery consists of electrodes and electrolytes. However, one particular requirement for the electrodes is that they are made of materials that can undergo reversible redox reactions driven by thermal energy. During the discharging process, the reaction AB → A+B occurs at temperature T_2, generating electricity and thermal energy Q_2. During the charging process, unlike traditional batteries, which are recharged by applying an electrical current, thermally regenerative batteries are recharged by applying heat. The thermal energy Q_1 reverses the electrochemical reaction (A+B → AB), restoring the battery to its original state and preparing it for another cycle of electricity generation. Nevertheless, this system is restricted by Carnot cycle efficiency, which limits the application of a thermally regenerative battery.

Modern LMBs usually use ceramic-lined steel crucibles or other high-temperature-resistant materials as the container [12]. This container also works as a cathode current collector. The bottom layer is molten cathode material. The electrolyte is added over the cathode. Due to its lower density, the electrolyte will naturally float on the molten cathode. The anode material is generally encased in a retainer, which is usually made of Fe-Ni foam. The retainer is connected to the anode current collector and insulated from the cathode current collector. This design ensures that the anode does not contact the cathode directly, and it is more reliable than the initial design, allowing the material to float under its own gravity (self-segregation). The system should be well-insulated to maintain high temperatures and minimize heat loss. Figure 2.2c demonstrates an example of a recent LMB. The cathode material was placed in a container made of graphite or stainless steel, which also acted as the current collector. Then, the molten salt electrolyte was added, which fully covered the cathode. The anode was made of alkali metal, which was absorbed and retained by Fe-Ni foam. The foam was connected to the negative current collector made of stainless steel and merged into the electrolyte and away from the cathode. Current collectors are electrically isolated by employing an alumina insulator. In addition to this method of retaining anode, there is another approach to limit the flow of the electrolyte—"paste electrolyte" [15,16], in which electrolytes are mixed with ceramic powders. In this way, the mobility of the electrolyte is restricted even at high temperatures, which prevents the cathode and the anode from contacting each other through the electrolyte. However, this method will cause the conductivity of the electrolyte to decrease due to the insulating properties of ceramics. Whether by retraining the electrodes or the electrolytes, an additional mechanical method is necessary to prevent the cathode and anode from contacting, since relying solely on gravity is insufficient to separate them.

As room-temperature liquid metals have drawn more and more attention, they have been used in traditional batteries, such as Ga and Ga-based alloys [17]. Additionally, in some recent studies, "self-healing batteries," in which only part of the materials are liquid (cathode or anode), are also classified as LMBs [18,19]. The structure and the preparation method of these batteries are the same as traditional batteries—coin or pouch formats can be applied.

2.2.2 Fluid Mechanism

The fluid mechanism inside LMBs highly influences the battery performance. It can lead to interior and surface instability, especially for the electrolyte and the cathode. On the other hand, fluid flow also has benefits. Fluid flow can cause intermediates to be rapidly destroyed, which is beneficial because these intermediates are usually accompanied by significant volume changes. Fluid flow caused by gravity, temperature gradient, and volume expansion is affected by component viscosity, battery sizes, magnetic field, and thermal diffusion. So far, thermal convection, compositional convection, Marangoni flow, and electro-vortex flow are the most studied categories for today's LMBs [20–24]. However, it should be noted that most of the research results are based on simulations, calculations, or a single metal liquid layer. Due to the limitations of experimental conditions, it is almost impossible to investigate real LMBs. We look forward to more research in the future that can break through existing limitations and solve these problems, which is essential for LMB design to improve battery efficiency and thermal management.

2.2.3 Features of LMBs

LMBs have attracted tremendous attention due to their unique advantages that make them promising for large-scale energy storage applications:

(1) **High-Rate Capability.** Unlike solid-electrode batteries, which struggle with rapid charging, LMBs are capable of fast charging and discharging. Since the electrodes are in the liquid state, charge carrier diffusion is faster than that of solid-state electrodes [2, 25]. Additionally, the absence of a solid-liquid interface barrier between electrodes and electrolytes enables faster ion mobility. Moreover, the higher temperature also increases the reaction kinetics at the electrodes [4]. The liquid metal layers allow for a uniform current distribution across the electrodes, preventing localized hotspots and ensuring the battery can handle high current loads without performance loss. Some studies indicate that fast charging has less impact on the battery than fast discharging because intermetallic compounds, which are usually accompanied by large volume changes, are formed on cathodes during discharging, while charging is not [24].

(2) **Long Lifetime.** This is the most attractive advantage of LMBs. The liquid nature of the electrodes and electrolytes helps in maintaining consistent contact and reduces the formation of the dendrite and passivation layer [26]. This self-healing property ensures the battery maintains high performance even under fast cycling conditions. Additionally, as there is no solid structure to crack or degrade, the battery's lifespan is significantly extended. The stability of the electrolyte also contributes to the overall longevity of the batteries. The liquid electrolyte in these batteries is typically made from stable, high-temperature molten salts, which are less prone to decomposition or side reactions compared to the liquid electrolytes used in other battery types, such as lithium-ion batteries.

(3) **Low Cost.** Most candidates for electrode materials are relatively common metals, so the price of the entire battery can be very low [27].

(4) **Scalability.** Being able to respond rapidly to changes in electricity demand, LMBs are applied to grid-scale energy storage (0.1–1.0 GWh). They can also assist in frequency regulation, voltage support, and other grid services [28,29].

Even though LMBs offer several advantages, some disadvantages and challenges still need to be addressed with the final goal of fabricating a robust and high-performance cell.

(1) **High Operating Temperature.** LMBs typically operate at temperatures higher than 240°C. Maintaining stability in such a high-temperature

environment is critical for safe operation. Otherwise, the overall efficiency of the system will be reduced. Also, since the materials are all liquid and the whole battery construction is based on their densities, the battery cannot be moved freely.

(2) **Material Constraints.** The requirements of melting points, density, and immiscibility limit the choice of electrodes and electrolytes. Also, the high operating temperature limits the materials for containment and insulation, as they must withstand extreme heat without degrading or reacting with the battery components.

(3) **Low Voltage and Low Energy Density.** Due to the limitation of material combinations, the voltage of LMBs is relatively low. Typically, the voltage of LMBs is between 0.5 and 2.0 V (i.e., the Mg||Sb system is ~ 1.2 to 1.5 V, the Na||Pb system is ~ 0.9 to 1.2 volts, the Li||Pb-Sb is ~ 1–1.2 V, etc.). Compared to the lithium-ion battery, which produces a voltage of ~3.2–4.2 V, the voltage produced by LMBs is not satisfactory. The low voltage will result in low energy density.

(4) **Safety Concerns.** The use of molten metals and salts introduces potential safety risks, particularly in breaking or corrosion in the containment system. A leak could result in high-temperature molten materials escaping, which could be hazardous.

In addition, the relatively long warm-up time, critical sealing requirement, and extraordinary maintenance might limit the extensive applications of LMBs.

In summary, LMBs have undergone a long evolution and have various structural forms. Modern LMBs take advantage of the self-segregation of the battery materials to maintain their structures at high temperatures. Fluid mechanisms within LMBs, such as thermal and compositional convection, significantly impact battery performance. LMBs offer significant advantages, including high-rate capability, long lifespan, low cost, and scalability, making them a promising choice for large-scale energy storage. However, they also present challenges, such as high operating temperatures, material constraints, low voltage, low energy density, and safety concerns, which limit their widespread application. The features learned above will allow us to utilize and enhance LMBs in a targeted way more effectively.

2.3 MATERIALS

The choice of materials is crucial to battery performance. It is not only the individual properties of the materials that matter but also how well they work together. This section introduces materials applied to LMBs based on their functions.

2.3.1 Electrodes

The anodes for LMBs are usually alkali metals, such as Li, Na, Mg, or Ca, as they have strong electronegativity. During discharging, these metals are oxidized ($M - xe^- \rightarrow M^{x+}$). In contrast, the cathodes for LMBs are usually made of post-transition or semi-metals, which are less electronegative, such as Bi, Sb, or Pb. During the discharging process, these metals form alloys with the anode metals. Among the metals listed above, Bi is a promising electrode candidate because it is a white, silver-pink, crystalline metal with a density of 9.78 g/cm^3, denser than most metals. Besides, it has a relatively low melting point of 271.5°C (520.7°F) and good alloying behavior with alkali metals. However, the cost of Bi is high. In comparison, Sb is a good complement due to its low cost. It has a density of 6.68 g/cm^3 and can form stable alloys with alkali metals, especially Mg and Li. However, its melting point (630.6°C) is relatively higher than that of other cathode candidates, so Pb is often alloyed with Sb to be used as a cathode. Pb has a bluish-gray color with a density of 11.34 g/cm^3. The melting point of Pb is 327.5°C, which is in the middle among the melting points of all commonly used metals. Recently, Ga has attracted attention, as its melting point is close to room temperature (29.76°C). Additionally, it has an extremely high theoretical capacity of 769 mAh/g [19]. Different cathode materials combined with the corresponding anode materials will produce various voltages and capacities. The compatibility between electrodes and electrolytes and operating temperatures will also influence the performance of electrodes [30].

With the development of battery technology, single metal electrodes are less common, and more binary and ternary alloy materials are utilized. Alloying strategies, which have been applied to both cathodes and anodes, overcome the shortcomings of single metals. Widely studied alloys include Pb-Sb, Sb-Sn, Bi-Pb, and Te-Sn. By alloying with cost-effective metals, the cost is greatly reduced, thus enabling more practical applications. Besides, the melting temperature of the alloy can be lower than every individual metal it is made up of. This can be achieved by tuning the ratios of these metals by modifying the metal crystal structure and charge density [31].

Alloying could also change the ion diffusion pathway and the intermediate compounds during the charging and discharging processes, which may affect the battery's efficiency. Moreover, the formation of new compounds by anode metals with alloy metals could cause volume change. In the Sb-Bi-Sn system, Li^+ forms Li_3Sb with Sb first and then forms Li_3Bi with Bi during the discharge process. The large volume change and uneven deposition of Li_3Bi cause Li_3Sb to crack and allow the electrolyte to thoroughly come in contact with the cathode material, as shown in Figure 2.3 [32]. The prepared Li||Sb-Bi-Sn system could achieve an energy density of 260 Wh/kg at 500°C. It should be noted that when introducing an inactive compound, the voltage and the battery energy density may be negatively affected. Therefore, it is essential to carefully balance the positive and negative impacts.

2.3.2 Electrolytes

Electrolytes play a critical role in battery performance. Like other batteries, the electrolytes for LMBs should have i) high ionic conductivity; ii) a stability window within the operating voltage range; and iii) low solubility of the electrodes. Besides those, there are some special requirements for the LMB electrolyte: i) low molten temperature; ii) proper density, since the structure of the battery is maintained by gravity; and iii) immiscibility with the positive and negative electrodes at high temperatures.

So far, the most studied and widely used electrolytes for LMBs are a mixture of halide salts (such as Cl, F, or Br) that remain in a liquid state at the operating

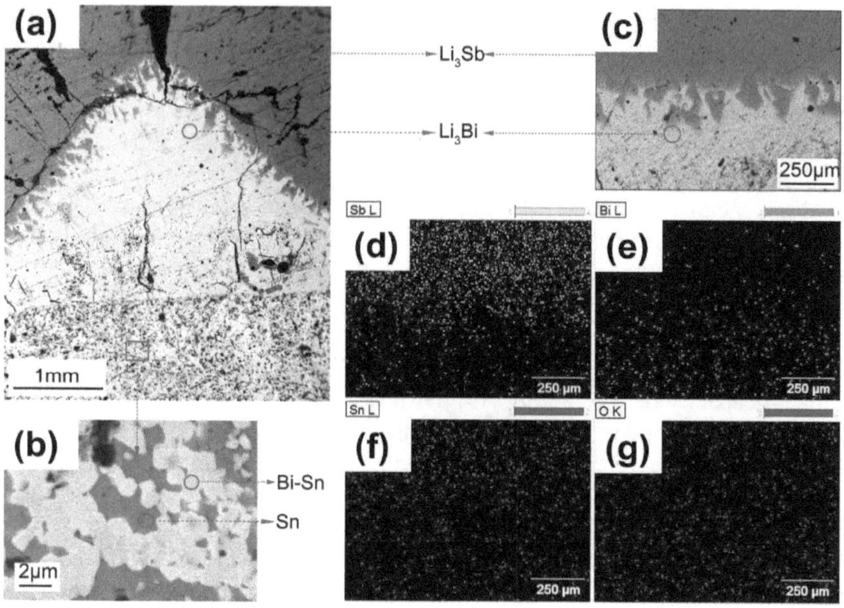

FIGURE 2.3 (a) A cross-section scanning electron microscopy (SEM) image of Sb-Bi-Sn positive electrode at 75% discharge state. Enlarged images of the alloy at the bottom (b) and top (c). (d), (e), (f), and (g) energy-dispersive X-ray spectroscopy (EDS) element mapping images of the intermediate compound phase interface.

Source: Adapted with permission [32].

temperature of the battery [33,34]. One example is **LiCl-KCl eutectic,** a common mixture used in lithium-based LMBs [30,35]. The melting point is 353°C with a mole ratio of 59.2:40.8. And the solubility of Li^+ in this system is only about 0.17% [36]. However, when the Li anode comes into contact with KCl, the displacement reaction occurs (Li + KCl → K + LiCl). With the addition of K to the Li anode, the cycling stability of the battery can be improved, as shown in Figure 2.4 [30]. Another example is $NaCl\text{-}CaCl_2$, which has a low melting point and high ionic conductivity [37]. $NaCl\text{-}CaCl_2$ was used for Na-Zn LMBs, and the battery was assembled at a fully discharged state. While charging, the following reaction occurs at the anode: $Zn—2e^- \rightarrow Zn^{2+}$.

At the cathode, unlike other electrolytes that do not participate in the redox reaction, Na^+ is deposited from the electrolyte, forming the anode [37]. Actually, the reactions on the anode are complicated. Due to the small electrodeposition potential gap between Na^+ and Ca^{2+}, both cations in the electrolyte will be deposited on the anode side. Additionally, Ca will naturally alloy with Zn. Therefore, Ca-Zn is found on the anode, and a small amount of Zn migrates to the anode. All these compounds at the anode side provide capacity when discharged. Therefore, multiple successive discharge voltage plateaus are observed in the Na||$NaCl\text{-}CaCl_2$||Zn battery voltage profile [26].

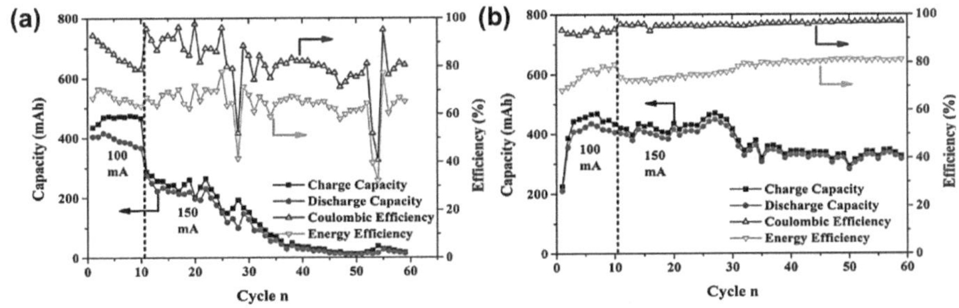

FIGURE 2.4 The charge/discharge capacity, Coulombic efficiency, and energy efficiency of cycle number for a (a) Li||$Sb_{30}Bi_{40}Sn_{30}$, and (b) LiK||$Sb_{30}Bi_{40}Sn_{30}$ battery tested at 400 °C (100 mA for the initial 10 cycles and 150 mA for the subsequent cycles).

Source: Adapted with permission [30]. Copyright 2022, American Chemical Society.

One unique electrolyte is a β-alumina solid electrolyte (BASE) (Na-β″-Al_2O_3), which is a commercialized electrolyte used in Na-S and sodium-metal halide (ZEBRA) batteries. BASE possesses an ionic conductivity of 0.2 S/cm at 300 °C and good electrochemical properties. However, the greatest challenge of BASE in Na batteries is that Na has a poor wetting ability with the BASE anode as a result of the oxidation layer on the Na surface. The wetting issue can be alleviated by coating [36] or adding oxygen getters, such as Ti or Al [37]. Alloying Na with Cs can also improve Na's wetting behavior on the BASE surface. Moreover, the operating temperature of the whole battery system can be decreased to 95 °C [38].

2.3.3 Insulators

Insulators are critical for maintaining electrical isolation between the different components of the battery while allowing ionic conduction through the electrolyte. The main insulators used in LMBs are typically materials that can withstand high temperatures, resist corrosion, and maintain their insulating properties under the battery's operating conditions. Some common insulators used in LMBs are ceramics such as alumina (Al_2O_3), magnesia (MgO), and zirconia (ZrO_2) [39,40]. These materials are excellent insulators with high thermal stability, making them suitable for high-temperature environments. It should be pointed out that the compatibility of the insulator is essential. Al_2O_3 and BeO demonstrate resistance to corrosion when exposed to molten sodium and Na-Pb, Na-Bi, and Na-Sn alloys, while chemically compatible insulator materials for molten lithium include BeO, $CaZrO_3$, Y_2O_3, $MgAl_2O_4$, and AlN [8]. Insulators are closely related to hermetic sealing, which is crucial in ensuring the battery's long-term stability. There are two common sealing methods: compressive and adhesive sealing, with the latter being more widely used [8].

The choice of materials is vital for LMBs to perform effectively. Anode metals typically include alkali metals like Li, Na, Mg, or Ca, while cathodes often consist of less electronegative metals like Bi, Sb, Pb, or their alloys. Electrolytes in LMBs need to have high ionic conductivity, low melting temperature, and immiscibility with electrodes, with mixtures like LiCl-KCl or NaCl-CaCl2 being commonly used. Additionally, solid electrolytes like β-alumina are used despite challenges like poor wetting with Na anodes. Insulators, often ceramics like alumina or zirconia, are crucial for electrical isolation and thermal stability and maintaining battery integrity under high-temperature conditions. Materials need to be carefully chosen as it is crucial to consider individual material properties, how these materials interact, and how they are assembled.

2.4 BATTERY PERFORMANCE

Evaluating battery performance involves analyzing a range of parameters that collectively determine how well a battery can perform in its intended application. These parameters provide insights into the battery's efficiency, longevity, safety, and suitability for different uses [41,42]. In this section, key parameters utilized to evaluate battery performance are introduced:

Capacity. It represents the amount of electric charge a battery can store and deliver. If not specified, the capacity usually refers to the discharge capacity. It is typically measured in ampere-hours (Ah) or milliampere-hours (mAh). As for LMBs, capacity is expressed in Ah because LMBs are usually used for large-scale applications. The capacity is determined by the amount of active material (cathode or anode, depending on which one is the limiting factor) in the battery and the efficiency of the chemical reactions.

Energy Density. It refers to the amount of energy a battery can store relative to its volume (Wh/cm^3), area (Wh/cm^2), or mass (Wh/kg). Higher energy density means the battery can store more energy in a given space or weight.

Power Density. It indicates how quickly energy can be delivered from the battery, which is important for applications requiring high power output over short periods. The units for power density are Wh/cm^3, Wh/cm^2, or Wh/kg.

Battery Life. The number of charge and discharge cycles that a secondary battery can undergo before its capacity degrades is an important parameter for evaluating its lifespan. Normally, 80% of its original capacity is regarded as the degradation point.

Coulombic Efficiency (Charge/Discharge Efficiency). This is the ratio of discharge capacity to charge capacity, expressed as a percentage (%). A high Coulombic efficiency indicates minimal energy loss during charging and discharging, which is crucial for maximizing battery efficiency and longevity. For LMBs, because the electrode material dissolves in the molten salt electrolyte more extensively than in conventional batteries, self-discharge is more severe, leading to a lower Coulombic efficiency compared to other batteries.

Internal Resistance. It refers to the resistance of the flow of charge carriers within the battery, which causes voltage drops and heat generation. Lower internal resistance is preferable as it results in higher efficiency, less heat generation, and better power delivery, especially under high-load conditions. Due to the degradation of materials, increased contact resistance (such as the passivation layer), and changes in structure, the increase in the battery's internal resistance is inevitable. The battery's internal resistance is usually evaluated quantitatively and measured by electrochemical impedance spectroscopy, and the results are plotted in Nyquist plots. One could make a comparison of Li‖ Sn battery cycled between 0.5 V and 1.2 V at different charged states after the 100th cycle. The internal resistance of the battery can be roughly reflected by the horizontal axis value of the endpoint of the semicircles. Even though determining the specific electron transfer resistance value would require model fitting, it can be concluded that the battery in the discharged state has a smaller resistance than in the charged state.

Operating Temperature. It refers to the range of temperatures within which the battery can safely and effectively operate. The operating temperature can highly affect the battery's capacity, rate capability, and cycle life. Batteries must operate reliably across the temperatures they are exposed to in their intended application. Extreme temperatures will not only affect battery performance, safety, and longevity but also be more demanding for sealing and container materials. For a typical LMB, the operating temperature is usually above 240 °C. Some recent studies focus on room-temperature LMBs, which are promising for next-generation energy storage applications [10]. It should be noted that the battery regeneration temperature is not the operating temperature. The regeneration temperature is usually higher than the operating temperature.

Table 2.1 summarizes several metals commonly used as electrodes, their corresponding operating temperatures, and the batteries' normal voltage.

TABLE 2.1
Summary of Typical LMBs

Anode	Cathode	Electrolyte	Operating temperature	Average discharge voltage	Discharge capacity
Li	Bi[43]	LiCl-LiF	550°C	0.6V	190 mAh/cm^2
	Sb[44]	LiF-LiCl-LiBr	550°C	0.8 V	527 mAh/g
	Sb-Pb[28]	LiF-LiCl-LiI	500°C	0.69V	600 mAh/cm^2
	Sb-Sn[45]	LiF-LiC	550°C	0.75V	1.2–2.5 Ah/cm^2
	Sn-Ga[46]	ethylene carbonate/ dimethyl carbonate	35°C	0.7V	409 mAh/g
	Bi-Pb[47]	LiCl-LiI	410°C	0.6V	140 mAh/g
	Sn-Bi-Pb[30]	LiCl-KCl	400°C	0.84 – 0.70 V	252–371 mAh/g
	Sn-Bi-Sb[32]	LiF-LiCl-LiBr	500°C	0.72V	361 mAh/g
	Sb-Te[48]	LiF-LiCl-LiBr	550°C	1.02 V	450 mAh/g

(Continued)

TABLE 2.1 ***(Continued)***
Summary of Typical LMBs

Anode	Cathode	Electrolyte	Operating temperature	Average discharge voltage	Discharge capacity
Na	Zn[37,49]	$NaCl\text{-}CaCl_2\text{-}ZnCl_2$	556°C	1.4 V–1.8 V	N/A
NaK	Ga-In[50]	Typical organic electrolyte (dimethoxyethane/ fluoroethylene carbonate, etc.)	Room temperature	0.5V	290 mAh/cm^3
K	Hg[16,51]	KOH-KBr-KI	500°C to 700°C	0.1 – 0.5 V	N/A
Mg	Sb[52]	$MgCl_2\text{-}KCl\text{-}NaCl$	700°C	0.2–0.4V	1000–1400 mAh/cm^2
Ca	Sb[53]	$LiCl\text{-}NaCl\text{-}CaCl_{2igh}$	650°C and 700°C	0.8–1.02	0.124–0.136 Ah
	Bi[54]	$LiCl\text{-}NaCl\text{-}CaCl_2$	600°C	1V	N/A

2.5 SUMMARY

This chapter provides a comprehensive overview of the key aspects involved in the development of LMBs. It has delved into the intricacies of battery design and explored the factors that influence battery performance, such as fluid flow, electrode, electrolyte composition, and insulator selection. The battery mechanism based on materials is also presented. Furthermore, parameters that are used to evaluate LMBs are explained. Performance characteristics for commonly studied LMBs using various materials are listed. Challenges and opportunities remain in advancing LMB technology and identifying areas for potential improvement and innovation. By understanding the critical elements of battery design, battery working mechanisms, fluid mechanics, materials selection, and performance characterization, researchers and engineers can work toward developing more efficient, durable, and safer LMBs.

REFERENCES

[1] Y. Jiang, C. Yang, Y. Yu, Y. Zhou, Z. Shang, S. Zhang, P. Liu, J. Zhu, M. Jiang, Aromatic polyaroxydiazole pseudocapacitive anode materials with tunable electrochemical performance through side group engineering, *J. Mater. Chem. A.* 12 (2023) 364–374.

[2] S. Wu, X. Zhang, R. Wang, T. Li, Progress and perspectives of liquid metal batteries, *Energy Storage Mater.* 57 (2023) 205–227.

[3] J. Zhu, E. Yildirim, K. Aly, J. Shen, C. Chen, Y. Lu, M. Jiang, D. Kim, A.E. Tonelli, M.A. Pasquinelli, P.D. Bradford, X. Zhang, Hierarchical multi-component nanofiber separators for lithium polysulfide capture in lithium–sulfur batteries: An experimental and molecular modeling study, *J. Mater. Chem. A.* 4 (2016) 13572–13581.

[4] D. Agarwal, R. Potnuru, C. Kaushik, V.R. Darla, K. Kulkarni, A. Garg, R.K. Gupta, N. Tiwari, K.S. Nalwa, Recent advances in the modeling of fundamental processes in liquid metal batteries, *Renew. Sustain. Energy Rev.* 158 (2022) 112167.

[5] C. Yang, X. Liu, J. Zhu, Z. Sun, Y. Yu, M. Jiang, A unique redox active ion-electron conductor enabled 5C fast-charging graphite anode for lithium-ion batteries, *J. Energy Storage.* 76 (2024).
[6] Y. Yu, J. Zhu, Y. Li, Q. Xu, Y. Jiang, C. Yang, L. Shi, L. Chen, P. Liu, J. Zhang, M. Jiang, Mixed ion-electron conductive binders coupling superior stiffness and toughness establish dual crosslinking stable silicon anodes, *Chem. Eng. J.* 479 (2024) 147807.
[7] J. Zhu, C. Yan, G. Li, H. Cheng, Y. Li, T. Liu, Q. Mao, H. Cho, Q. Gao, C. Gao, M. Jiang, X. Dong, X. Zhang, Recent developments of electrospun nanofibers for electrochemical energy storage and conversion, *Energy Storage Mater.* 65 (2024).
[8] H. Kim, D.A. Boysen, J.M. Newhouse, B.L. Spatocco, B. Chung, P.J. Burke, D.J. Bradwell, K. Jiang, A.A. Tomaszowska, K. Wang, W. Wei, L.A. Ortiz, S.A. Barriga, S.M. Poizeau, D.R. Sadoway, Liquid metal batteries: Past, present, and future, *Chem. Rev.* 113 (2013) 2075–2099.
[9] T. Daeneke, K. Khoshmanesh, N. Mahmood, I.A. De Castro, D. Esrafilzadeh, S.J. Barrow, M.D. Dickey, K. Kalantar-Zadeh, Liquid metals: Fundamentals and applications in chemistry, *Chem. Soc. Rev.* 47 (2018) 4073–4111.
[10] Y. Ding, X. Guo, G. Yu, Next-generation liquid metal batteries based on the chemistry of fusible alloys, *ACS Cent. Sci.* 6 (2020) 1355–1366.
[11] H. Li, K. Wang, S. Cheng, K. Jiang, High performance liquid metal battery with environmentally friendly antimony-tin positive electrode, *ACS Appl. Mater. Interfaces.* 8 (2016) 12830–12835.
[12] K. Wang, J. Hu, T. Chen, J. Tang, Z. Wang, N. Fan, W. Zhang, K. Wang, A high-performance room-temperature Li||Ga–Sn liquid metal battery for grid energy storage, *Energy Technol.* 9 (2021) 2100330.
[13] W. Hoopes, *Electrolytically-refined aluminum and articles made therefrom*, 1922. https://patents.google.com/patent/US1534315A/en
[14] E. Yeager, Fuel cells: Basic considerations, in power sources division, *Proceedings of 12th Annual Battery Research and Development*, Fort Monmouth, 1958.
[15] H. Shimotake, G.L. Rogers, E.J. Cairns, Secondary cells with lithium anodes and immobilized fused-salt electrolytes, *Ind. Eng. Chem. Process Des. Dev.* 8 (1969) 51–56.
[16] H. Li, H. Yin, K. Wang, S. Cheng, K. Jiang, D.R. Sadoway, Liquid metal electrodes for energy storage batteries, *Adv. Energy Mater.* 6 (2016) 1600483.
[17] S. Zhang, Y. Liu, Q. Fan, C. Zhang, T. Zhou, K. Kalantar-Zadeh, Z. Guo, Liquid metal batteries for future energy storage, *Energy Environ. Sci.* 14 (2021) 4177–4202.
[18] Y. Wu, L. Huang, X. Huang, X. Guo, D. Liu, D. Zheng, X. Zhang, R. Ren, D. Qu, J. Chen, A room-temperature liquid metal-based self-healing anode for lithium-ion batteries with an ultra-long cycle life, *Energy Environ. Sci.* 10 (2017) 1854–1861.
[19] R.D. Deshpande, J. Li, Y.-T. Cheng, M.W. Verbrugge, Liquid metal alloys as self-healing negative electrodes for lithium ion batteries, *J. Electrochem. Soc.* 158 (2011) A845.
[20] N. Weber, V. Galindo, J. Priede, F. Stefani, T. Weier, The influence of current collectors on Tayler instability and electro-vortex flows in liquid metal batteries, *Phys. Fluids.* 27 (2015).
[21] N. Weber, P. Beckstein, W. Herreman, G.M. Horstmann, C. Nore, F. Stefani, T. Weier, Sloshing instability and electrolyte layer rupture in liquid metal batteries, *Phys. Fluids.* 29 (2017).
[22] Y. Shen, O. Zikanov, Thermal convection in a liquid metal battery, *Theor. Comput. Fluid Dyn.* 30 (2016) 275–294.
[23] L. Xiang, O. Zikanov, Subcritical convection in an internally heated layer, *Phys. Rev. Fluids.* 2 (2017) 063501.

[24] D.H. Kelley, T. Weier, Fluid mechanics of liquid metal batteries, *Appl. Mech. Rev.* 70 (2018).

[25] T. Liu, Z. Du, X. Wu, M.M. Rahman, D. Nordlund, K. Zhao, M.D. Schulz, F. Lin, D.L. Wood, I. Belharouak, Bulk and surface structural changes in high nickel cathodes subjected to fast charging conditions, *Chem. Commun.* 56 (2020) 6973–6976.

[26] F. Zhang, J. Jin, J. Xu, Z. Shi, Anode reaction mechanisms of Na|NaCl-CaCl2|Zn liquid metal battery, *J. Energy Chem.* 72 (2022) 81–87.

[27] J.G. Simpson, G. Hanrahan, E. Loth, G.M. Koenig, D.R. Sadoway, Liquid metal battery storage in an offshore wind turbine: Concept and economic analysis, *Renew. Sustain. Energy Rev.* 149 (2021) 111387.

[28] K. Wang, K. Jiang, B. Chung, T. Ouchi, P.J. Burke, D.A. Boysen, D.J. Bradwell, H. Kim, U. Muecke, D.R. Sadoway, Lithium–antimony–lead liquid metal battery for grid-level energy storage, *Nature.* 514 (2014) 348–350.

[29] H. Zhou, H. Li, Q. Gong, S. Yan, X. Zhou, S. Liang, W. Ding, Y. He, K. Jiang, K. Wang, A sodium liquid metal battery based on the multi-cationic electrolyte for grid energy storage, *Energy Storage Mater.* 50 (2022) 572–579.

[30] K. Cui, W. Zhao, S. Li, D. Zhou, C. Liu, X. Qu, P. Li, Low-temperature and high-energy-density Li-based liquid metal batteries based on LiCl-KCl molten salt electrolyte, *ACS Sustain. Chem. Eng.* 10 (2022) 1871–1879.

[31] H. Ohtani, K. Okuda, K. Ishida, Thermodynamic study of phase equilibria in the Pb-Sn-Sb system, *J. Phase Equilibria.* 16 (1995) 416–429.

[32] W. Zhao, P. Li, Z. Liu, D. He, K. Han, H. Zhao, X. Qu, High-performance antimony-bismuth-tin positive electrode for liquid metal battery, *Chem. Mater.* 30 (2018) 8739–8746.

[33] Q. Zeng, Z. Lv, S. Li, B. Yang, J. He, J. Song, Electrolytes for liquid metal batteries, *Mater. Res. Bull.* 170 (2024) 112586.

[34] X. Zhou, S. Yan, X. He, H. Zhou, J. Ning, H. Li, K. Wang, K. Jiang, Low-temperature, high cycling stability, and high Coulombic efficiency liquid metal batteries enabled by lithium halide-potassium halide molten salt electrolytes, *Energy Storage Mater.* 61 (2023) 102889.

[35] J. Serp, R.J.M. Konings, R. Malmbeck, J. Rebizant, C. Scheppler, J.P. Glatz, Electrochemical behaviour of plutonium ion in LiCl–KCl eutectic melts, *J. Electroanal. Chem.* 561 (2004) 143–148.

[36] J.R. Selman, D.K. DeNuccio, C.J. Sy, R.K. Steunenberg, EMF studies of lithium-rich lithium-aluminum alloys for high-energy secondary batteries, *J. Electrochem. Soc.* 124 (1977) 1160–1164.

[37] J. Xu, O.S. Kjos, K.S. Osen, A.M. Martinez, O.E. Kongstein, G.M. Haarberg, Na-Zn liquid metal battery, *J. Power Sources.* 332 (2016) 274–280.

[38] X. Lu, G. Li, J.Y. Kim, D. Mei, J.P. Lemmon, V.L. Sprenkle, J. Liu, Liquid-metal electrode to enable ultra-low temperature sodium–beta alumina batteries for renewable energy storage, *Nat. Commun.* 5 (2014) 1–8.

[39] H. Li, H. Yin, K. Wang, S. Cheng, K. Jiang, D.R. Sadoway, liquid metal electrodes for energy storage batteries, *Adv. Energy Mater.* 6 (2016) 1600483.

[40] Y. Jiang, D. Zhang, Y. Shi, R. Luo, N. Cai, Lithium transport and intermetallic generation in Li–Bi liquid metal batteries, *Electrochim. Acta.* 405 (2022) 139779.

[41] M. Armand, J.-M. Tarascon, Building better batteries, *Nature.* 451 (2008) 652–657. https://doi.org/10.1038/451652a.

[42] J.S. Yeo, J.H. Lee, E.J. Yoo, Electrochemical properties of environment-friendly lithium-tin liquid metal battery, *Electrochim. Acta.* 290 (2018) 228–235.

[43] X. Ning, S. Phadke, B. Chung, H. Yin, P. Burke, D.R. Sadoway, Self-healing Li–Bi liquid metal battery for grid-scale energy storage, *J. Power Sources.* 275 (2015) 370–376.

[44] S. Yan, X. Zhou, H. Li, Y. Shen, Y. He, H. Zhou, K. Wang, K. Jiang, Utilizing in situ alloying reaction to achieve the self-healing, high energy density and cost-effective Li||Sb liquid metal battery, *J. Power Sources.* 514 (2021) 230578.

[45] X. Zhou, H. Zhou, S. Yan, Y. He, W. Zhang, H. Li, K. Wang, K. Jiang, Increasing the actual energy density of Sb-based liquid metal battery, *J. Power Sources.* 534 (2022) 231428.

[46] K. Wang, J. Hu, T. Chen, J. Tang, Z. Wang, N. Fan, W. Zhang, K. Wang, A high-performance room-temperature Li||Ga–Sn liquid metal battery for grid energy storage, *Energy Technol.* 9 (2021) 2100330.

[47] J. Kim, D. Shin, Y. Jung, S.M. Hwang, T. Song, Y. Kim, U. Paik, LiCl-LiI molten salt electrolyte with bismuth-lead positive electrode for liquid metal battery, *J. Power Sources.* 377 (2018) 87–92.

[48] H. Li, K. Wang, H. Zhou, X. Guo, S. Cheng, K. Jiang, Tellurium-tin based electrodes enabling liquid metal batteries for high specific energy storage applications, *Energy Storage Mater.* 14 (2018) 267–271.

[49] J. Xu, A.M. Martinez, K.S. Osen, O.S. Kjos, O.E. Kongstein, G.M. Haarberg, Electrode behaviors of Na-Zn liquid metal battery, *J. Electrochem. Soc.* 164 (2017) A2335–A2340.

[50] Y. Ding, X. Guo, Y. Qian, L. Xue, A. Dolocan, G. Yu, Room-temperature all-liquid-metal batteries based on fusible alloys with regulated interfacial chemistry and wetting, *Adv. Mater.* 32 (2020) 2002577.

[51] C.E. Crouthamel, H.L. Recht, *Regenerative EMF cells: A symposium co-sponsored by the division of industrial and engineering chemistry and the division of fuel chemistry at the 149th meeting of the American chemical society, Detroit, Mich., April 8–9, 1965.* 64 (1967) 38. https://library.kaist.ac.kr/search/ctlgSearch/posesn/view.do?bibctrlno=638386&se=b0&ty=B

[52] D.J. Bradwell, H. Kim, A.H.C. Sirk, D.R. Sadoway, Bantimony liquid metal battery for stationary energy storage, *J. Am. Chem. Soc.* 134 (2012) 1895–1897.

[53] T. Ouchi, H. Kim, X. Ning, D.R. Sadoway, *Calcium-antimony alloys as electrodes for liquid metal batteries, J. Electrochem. Soc.* 161 (2014) A1898–A1904.

[54] H. Kim, D.A. Boysen, T. Ouchi, D.R. Sadoway, Calcium–bismuth electrodes for large-scale energy storage (liquid metal batteries), *J. Power Sources.* 241 (2013) 239–248.

3 Chemistries of Liquid Metal Batteries

Mayankkumar L. Chaudhary, Rutu Patel, and Ram K. Gupta

3.1 INTRODUCTION

The shift from fossil fuels to renewable and sustainable energy sources is being hastened by the growing push for a low-carbon economy, which has a profound impact on energy consumption patterns throughout the world [1]. Here, batteries and other electrochemical energy storage devices are crucial for storing intermittent renewable energy, such as that which comes from wind, solar, and geothermal power [2]. Among these technologies, lithium-ion batteries (LiBs) have been the go-to for portable electronics and are slowly but surely making their way into new industries such as electric vehicles and fixed energy storage systems. By 2027, the worldwide LIB market is expected to reach around \$200 billion [3]. Current LIB technologies still have a ways to go before they can meet the targets for cell-level energy density (~350 Wh kg^{-1}), volumetric energy density (~750 Wh L^{-1}), and cost (less than A\$170 kW h^{-1}), even though they are widely used [4].

Researchers have turned their attention to alternate battery chemistries, namely systems based on metal electrodes, which have been studied since the 1970s, to meet the demand for lower-cost energy densities (~500 Wh kg^{-1}) [5]. The electropositive potential (–3.04 V vs. SHE) and high specific capacity (3860 mA h g^{-1}) of metal anodes, including Li, provide substantial benefits. Dendrite generation, large volume variations, and continual electrolyte breakdown are important issues that solid metal anodes encounter, which compromise their safety and cycle life [6]. Since solid-to-solid chemistries have their limitations, researchers are looking into liquid-to-liquid electrochemistry as a possible replacement. In comparison to their solid-state equivalents, liquid systems offer a number of benefits, such as (i) higher voltage efficiencies and quicker kinetics at liquid-liquid interfaces, (ii) resistance to microstructural deformation and dendrite growth, (iii) the possibility of creating flexible energy storage components, and (iv) affordability owing to the availability of cheap active materials [7,8]. Liquid metal batteries (LMBs) are attractive for grid-level systems and other large-scale energy storage applications because of these advantages. The electrowinning of aluminum in 1886 introduced the idea of LMBs, and by the turn of the twentieth century, there had been substantial advancements, such as the three-liquid-layer Hoopes cell enabling the synthesis of highly pure aluminum [8]. In the 1960s, three-liquid-layer cells made a comeback as thermally regenerating bimetallic cells; they were the subject of a great deal of study until they were supplanted by

DOI: 10.1201/9781003489726-3

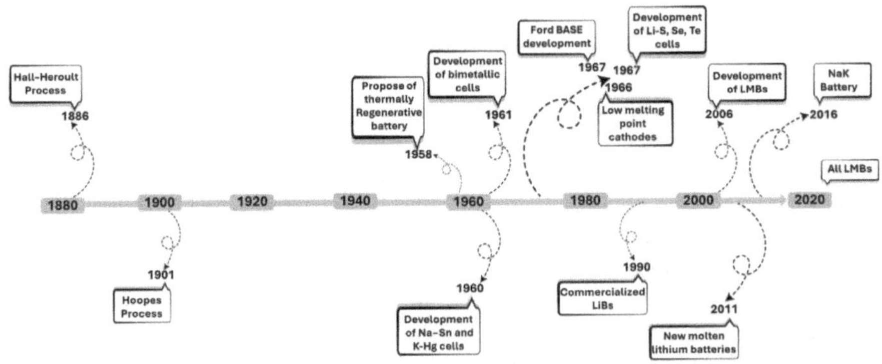

FIGURE 3.1 Timeline of LMBs.

portable high-energy-density batteries. Nevertheless, LMBs have once again come into the spotlight due to the revived interest in grid-scale energy storage that occurred around the turn of the millennium. In the early days of LMBs, the electrolyte was a molten salt (such as halides) and the electrodes were two high-melting-point liquid metals. The density differences and mutual immiscibility caused the electrodes to self-segregate into separate layers. The systems' limited practical applicability was caused by the high operating temperatures necessary to maintain their all-liquid condition [9]. Figure 3.1 represents the journey of LMBs [7,10].

Ga metal, Ga-based alloys, and Na-K liquid alloys are fusible materials that stay in the liquid phase at or near room temperature (0–40 °C) and have been used in recent attempts to lower LMB working temperatures [11]. The moderate potential window and safety features of Ga-based liquid metals make them an attractive candidate. For instance, during charge and discharge cycles, Ga anodes in LiBs display a phase shift from liquid to solid to liquid, showcasing their promise in advanced energy storage systems [12]. Just like traditional Li anodes, RT Na-K alloys have their own set of benefits when it comes to storing charge carriers. The inherent reactivity of liquid metals with electrolytes makes it challenging to achieve high cell voltages and adequate electrochemical reversibility in RT-LMBs, notwithstanding these developments [13]. To overcome these obstacles, a deeper understanding of the intricate multiphysical processes occurring in LMBs, including electric potential distribution, electro-vortex flow, thermal convection, mass transport, and heat transfer, along with advancements in battery chemistry, is required [14]. Improving performance, optimizing cell design, and allowing large-scale deployment of LMBs all depend on developing models that connect electrochemistry and thermophysics. While research into HM-LMBs has advanced, more work is required to bring chemistry's theoretical advances into line with the needs of actual applications.

3.2 FUNDAMENTAL PRINCIPLES OF LMBs

A novel energy storage technology, LMBs are renowned for their exceptional performance, distinct operating principles, and straightforward construction. A metal anode, a salt electrolyte, and a metal cathode make up the three layers that allow these batteries to work. Stable stratification is achieved throughout the operation by virtue of the immiscibility of these layers, which is a consequence of their density disparities. Anodes composed of molten metal are usually Na, Mg, or Li, which are low-density metals with high energy densities and low melting temperatures. Interposed between the anode and cathode is the molten salt electrolyte, which is a combination of salts (usually chlorides, fluorides, or bromides) with low melting temperatures. These salts provide high thermal stability and ionic conductivity. The molten cathode, meanwhile, is made of heavy metals such as Pb, Bi, or Sb. Their electrochemical qualities, such as compatibility with the electrolyte and anode and high voltage potential, are the deciding factors in their selection.

There is a simple yet effective process that allows LMBs to function. The anode metal oxidizes during discharge, releasing ions and electrons. An alloy or intermetallic complex is formed when ions travel through the molten salt electrolyte and are reduced at the cathode. At the same time, an electric current is generated when electrons move via an external circuit. When the battery is charged, the process is turned around, and the metal ions go back to the anode as the alloy breaks down. To maintain optimal functioning, LMBs keep all components in a liquid condition and function at temperatures exceeding 300 °C. The necessary thermal energy is usually maintained while cycling by means of exothermic processes. Simplifying the design and boosting the efficiency of the battery, the immiscibility of the layers lowers the chance of cross-contamination. Their natural self-segregation assures constant and dependable functioning without the need for separators [7,15].

When it comes to the practicality, effectiveness, and general performance of LMBs, the rules of electrochemistry and thermodynamics are what matter. Thermodynamically speaking, the voltage of an LMB is dictated by the Gibbs free energy (ΔG) of the cell reaction, which is written as $\Delta G = -nFE$, where n is the quantity of electrons transported, F is the Faraday constant, and E is the cell voltage. Ionic diffusion activation energy reduction and electrochemical reaction kinetics enhancement are both significantly improved at elevated operating temperatures. To keep parasite losses to a minimum and maximize system efficiency, it is crucial to carefully control the amount of thermal energy needed to keep certain temperatures.

The functionality of LMBs is also critically dependent on electrochemical principles. The battery's open-circuit voltage (OCV) is determined by the chemical potential difference between the anode and the cathode materials. The molten salt electrolyte allows for little resistance to ion transport due to its strong ionic conductivity. The electrodes' liquid condition improves reaction kinetics, which in turn lowers overpotentials and allows for great performance. When it comes to LMBs, stability and corrosion resistance are two of the most important aspects of design and operation. Materials for the battery's anode and cathode should have low solubility in the electrolyte to avoid mixing, and materials for the battery's container should be corrosion-resistant to guarantee stability and longevity. All things considered, LMBs are a viable technology for large-scale energy storage because of how efficiently and reliably they work [16,17].

LMBs differ from traditional solid-state batteries in many important ways due to the liquid metal interfaces that make them work. Because the electrodes are in a liquid state, any structural flaws or defects may be spontaneously repaired while they are in use, making this a major advantage. Dendrite growth is a typical concern in solid-state systems, but this feature substantially removes it. As a result of the lack of solid-solid interfaces and the strong ionic conductivity of the molten salt electrolyte, LMBs also display a high power density because of their low internal resistance. Enhanced mass transfer, decreased diffusion restrictions, and quick charge/discharge cycles are all results of fluid dynamics at liquid interfaces.

Longevity and cyclability are two more major benefits of LMBs. Due to the absence of mechanical stress, the liquid electrodes have low deterioration over time, resulting in a prolonged cycle life. Over thousands of cycles, experimental findings demonstrate consistent performance with almost no capacity loss. Additionally, due to their simple design, abundance of inexpensive materials, and excellent scalability, LMBs are cost-effective. Their scalability across different battery sizes is made possible by their simplicity, which makes them ideal for grid-scale energy storage. Another built-in advantage of LMBs is thermal management. This is because, depending on the operating temperature range, the heat produced during cycling may often keep the material in a molten state without the need for further heating from outside sources. By minimizing dependence on rare and costly components and making use of earth-abundant metals, LMBs provide a sustainable option from an environmental perspective. Sustainable energy storage systems would benefit greatly from their recyclability, which adds to their environmental appeal [16,18].

A simple, very effective, and environmentally friendly energy storage option, LMBs offer great promise for large-scale applications. Their dependable and efficient functioning is guaranteed by their three-layer construction, which is dictated by basic laws of thermodynamics and electrochemistry. For the advancement of renewable energy systems, LMBs are an essential technology due to their many benefits, including self-healing capabilities, high power density, and environmental friendliness. To meet the increasing needs of modern energy networks, researchers are expected to concentrate on improving thermal management, increasing manufacturing capacity, and optimizing materials.

3.3 ELECTRODE/ELECTROLYTE AND TEMPERATURE-DEPENDENT CHEMISTRY FOR LMBs

One could create a periodic table with potential liquid metal electrode possibilities, limited by three main criteria: The first requirement is that they be a liquid at normal pressure and temperatures; the second is that they be electrically conductive; the third is that they be nonradioactive and exist naturally as stable isotopes; and the fourth is that they be able to melt at temperatures below 1000 °C and boil at temperatures above 25 °C. The potential of electrode materials to deposit from water is the first criterion for categorizing them as positive or negative electrodes. Different deposition potentials apply to LMBs because they use solid, organic, or molten salt electrolytes. A metal is considered a negative electrode if its deposition potential is greater than –2.0 V, whereas it is considered a positive electrode if its potential is greater than

−1.0 V. Aluminum stands out from the others because it can do double duty. Another way to classify elements is by their electronegativity. Negative electrodes should be made of electropositive metals such as Li, K, Na, Ca, and Mg, and positive electrodes should be made of electronegative metals, semimetals, or chalcogens such as Bi, Te, Ga, Pb, Sn, Sb, or S. With melting points ranging from below 0 °C to over 800 °C, a wide variety of metals and alloys may be used to create LMBs with varying operating temperatures. Particularly appealing are alkali metal alloys due to their great elemental richness and low melting points [9]. The melting points of alloys based on Ga are relatively low. Despite their scarcity, Bi-based alloys are well-suited for the storage of alkali-ions at slightly increased temperatures because of their higher melting points and reduction potentials [16]. The operating temperature of LMBs is heavily dependent on research into the phase-change behavior of alloys, making this area of study vital. The charging and discharging operations of bimetallic cells are carried out by alloying and de-alloying reactions, respectively. Factors including valence electron quantity, ionization degree, and crystal structure dictate a metal's melting temperature, which its bond strength may approximately predict. Melting point depression describes the tendency for pure substances to have greater melting points than impure substances. This tendency is well shown by eutectic systems [19,20]. Consequently, as an alloy reaches temperatures higher than its melting point, its Gibbs free energy change curve will have either one minimum (representing a stable liquid phase) or two minima that share a tangent point (representing the boundaries of the liquidus phase on a phase diagram). The use of phase diagrams allows for the prediction of alloy phases produced by electrochemical processes. However, kinetic limits can cause voltage profiles to deviate when reaction temperatures are lower or current densities are greater. For instance, it has been demonstrated that fissures that are seen on solidified Ga metal surfaces during de-lithiation may mend themselves once the de-lithiation process is complete. This is popularly known as the "self-healing" ability [21,22]. While designing an LMB cell, it is essential to consider electrolyte's thermal characteristics, as they determine the operating temperature range. Since LiBs for portable energy storage became widely available in the 1990s, there has been a marked decrease in research on LMBs. On the other hand, the increasing need for stationary energy storage on a grid scale has sparked a renewed interest in LMBs during the 2010s. This fresh enthusiasm highlights the enormous promise of LMBs as a solution to the problems associated with contemporary energy storage.

3.3.1 High-Temperature LMBs

A revolutionary energy storage device suited for grid-scale applications, HT-LMBs are defined by their distinctive three-liquid-layer construction. While all three components of this system—the negative electrode, the molten salt electrolyte, and the positive electrode—are in a liquid state, their density differences and immiscibility cause them to spontaneously self-segregate into three separate layers. This self-segregation streamlines the design by removing the requirement for physical separators, all while ensuring stable operation and efficient ionic conductivity. The greatest melting point of the three components determines the minimum operating temperature of HT-LMBs. Battery performance and operation are greatly impacted by the molten salt electrolyte, which prevents short electrical circuits while enabling ionic transport and facilitating the transmission of metal ions between the negative and positive electrodes. Energy efficiency, operating temperature, rate capability, and Coulombic efficiency are all affected by the molten salt electrolyte in an HT-LMB system. HT-LMBs avoid the dendrite formation problems that plague solid-state batteries by keeping all components in a liquid state throughout operation, which allows for high ionic conductivity.

Prior studies on HT-LMBs have encountered many obstacles, as pointed out by Li et al. The factors contributing to self-discharge and decreased Coulombic efficiency include a low energy density, electrode materials with weak electronic conductivity, and active components that are highly soluble in molten salt solutions. As an example, LiBs still produced more power than Mg||Sb systems running at 700 °C (~25 Wh/kg), but Li-ion Bi- and sulfate-battery systems (~200 Wh/kg) were still behind. Furthermore, its impracticality for broad use was exacerbated by high working temperatures that went beyond 500 °C. The electronic conductivity of electrodes was greatly enhanced and Te solubility in molten salts was decreased by alloying Te with Sn, an approach suggested by Li et al. to tackle these problems. Thanks to this breakthrough, the Li||Te-Sn system was created, which outperformed previous systems such as Li||Bi (148.5 Wh/kg) and Mg||Sb (25 Wh/kg) with a discharge voltage of about 1.6 V and an energy density of 495 Wh/kg [23]. Stable cycling performance and nearly 100% capacity retention over 60 cycles were achieved by enhancing the adhesive characteristics of Sn using the Te-Sn alloy. This prevented floating solid discharge products in the molten electrolyte.

To further improve HT-LMB performance, Weppner et al. [24] presented alloying techniques for electrode materials that greatly improved electronic conductivity and decreased solubility in molten salts, substantially improving HT-LMB performance. These developments allowed for greater use of active materials, reduced self-discharge, and increased Coulombic efficiency. Improvements in ionic conductivity and thermal stability achieved by using new molten salt electrolytes further enhanced the system's performance, making HT-LMBs a more viable option for large-scale applications. To improve conductivity and stability, the electrodes were alloyed during synthesis, and impurities were removed from molten salt electrolytes by vacuum drying. The dependability and consistency of the system were guaranteed by performing the assembly in an inert environment to minimize oxidation. Thus, alloyed electrode systems were able to increase discharge voltages and energy densities while maintaining almost 100% Coulombic efficiency across longer cycles.

The Bi-Li system also showed a dramatic improvement over its predecessor, which had problems with thermodynamic stability, electrochemical efficiency, and material behavior. For Bi-Li alloys, especially at higher Li concentrations, there was a paucity of accurate thermodynamic data and trustworthy phase diagrams in previous research. Bi-Li systems have limited utility in energy storage technologies due to discrepancies in phase transition data caused by the thermal instability of intermediate compounds such as BiLi and BiLi3. These difficulties were tackled by Sangster et al. [25], who evaluated the Bi-Li phase diagram by means of electrochemical investigations, electrical resistivity tests, and advanced thermal analysis. These investigations successfully ascertained the Bi solubility limits and eutectic temperatures for Li. The study's comprehensive phase diagram of the Bi-Li system showed that the system's essential eutectic points were located at 243 and 175 °C, respectively. Both BiLi and $BiLi_3$, which were stable intermediate phases, were extensively studied. BiLi exhibited a 1:1 stoichiometry and a peritectic melting temperature of 415 °C, whereas $BiLi_3$ melted at 415 °C with a stoichiometry ranging from 73.8 to 75.2% Li.

A unique self-healing LMB was created by Ning et al. [26] to solve capacity deterioration and energy inefficiency. This battery uses a molten salt electrolyte of LiCl and LiF as its negative electrolyte, Bi as its positive electrode, and Li as its negative electrode. Using the Li-Bi alloy's special characteristics, such as rapid ionic diffusion and the reversible production of solid intermetallic compounds like Li_3Bi during discharge, the system ran at 550 °C. Full dissolution of the intermetallic occurred upon charging, returning the substance to its liquid form. The technology became long-lasting and economical for use on a grid scale when its self-healing mechanism fixed problems including energy inefficiency and capacity decline. With an impressive 96% capacity retention after 1000 cycles at a 3C rate, a Coulombic efficiency of 99.7%, and an energy efficiency of 70%, the results demonstrated extraordinary performance from the Li-Bi battery (Figure 3.2).

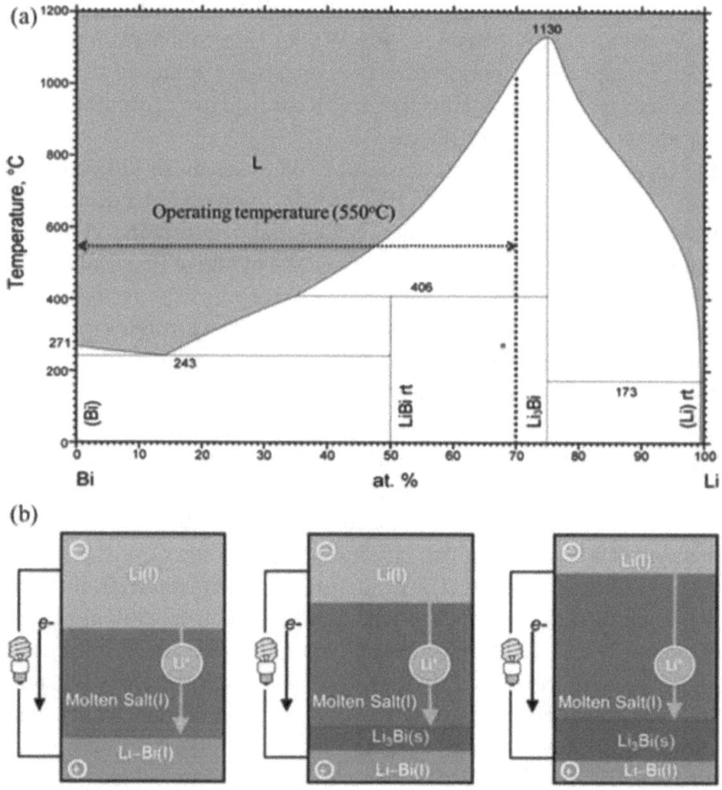

FIGURE 3.2 The mechanism of Li||Bi liquid metal battery cycling (a) Li-Bi phase diagram (b) Schematic of a cell during discharge showing the formation of the solid Li_3Bi layer.

Source: Adapted with permission [26]. Copyright 2015, Elsevier.

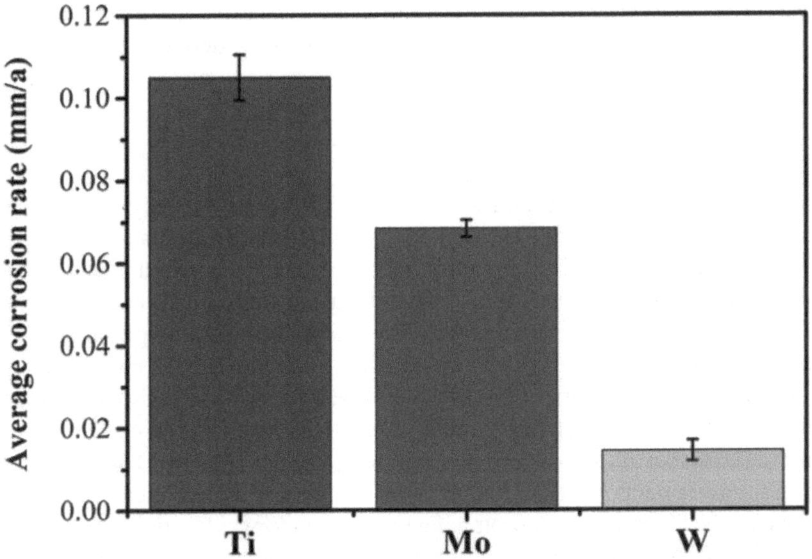

FIGURE 3.3 Ti, Mo, and W corrosion rates in the liquid Sb-Sn alloy.

Source: Adapted with permission [27]. Copyright 2021, American Chemical Society.

Corrosion in positive current collectors (PCCs) affects performance and long-term stability, which is another major difficulty with HT-LMBs. The strong electrical conductivity, corrosion resistance, and stability at high temperatures of Ti, Mo, and W were thoroughly examined as possible PCC materials by Cui et al. [27]. W showed better corrosion resistance than Ti and Mo, with an average corrosion rate of just 0.014 mm/a compared to 0.105 mm/a and 0.068 mm/a of Ti and Mo, respectively, according to the research (Figure 3.3). Over 800 cycles, the stability of Li||Sb–Sn LMBs with W PCCs was shown by long-term cycling testing. The LMBs maintained stable voltage plateaus and showed a low capacity loss.

Despite the substantial progress in HT-LMB technology, further research is necessary to optimize electrode materials, electrolyte compositions, and thermal management strategies. Innovations such as ternary Sb–Bi–Sn alloys, as introduced by Wang et al. [28], provide a promising avenue for enhancing battery performance by lowering the melting point while maintaining high energy densities. Additionally, the development of optimized electrolyte formulations, such as LiCl-LiBr-KBr compositions proposed by Xie et al. [29], demonstrates the potential to further reduce operational temperatures while maintaining superior ionic conductivity.

When it comes to grid-scale energy storage, HT-LMBs show a lot of potential as long-term, efficient, and scalable solutions, especially for applications that need reliable operation in harsh environments. They are an important technology in the shift to renewable energy sources since they keep working even when heated, have a long lifespan, and have strong safety measures. As technology improves in material selection, electrolyte optimization, and self-healing processes, HT-LMBs are expected to become more commercially viable and widely used for large-scale energy storage applications.

3.3.2 Medium-Temperature LMBs

The molten salt electrolytes used in most HT-LMBs have a very high melting point. Hence, the batteries must be operated at very high temperatures. The electrolyte often determines the operating temperature of the battery, even though metal or alloy electrodes such as Li, Na, Bi, Sn-Pb, and Si-Pb have low melting points. Researchers have investigated solid electrolytes as a potential solution to the problem of creating a molten salt electrolyte with a low melting point and good ionic conductivity. In these instances, the temperature at which LMBs operate is highly dependent on the melting point of the metal electrodes. There has been a lot of research over the past several decades into Na-based medium-temperature (MT)-LMBs that use solid electrolytes and operate below 350 °C. Two examples of these batteries are ZEBRA batteries and Na||S batteries. A promising new option for grid-scale energy storage is Li-based MT-LMBs. To solve the problems with traditional Na-S systems operating at high or low temperatures, Lu et al. [30] created an Na-S battery that operates at an intermediate temperature. Molten Na and S pose hazards to the safety of the user, degrade the material over time, and drive up the price of high-temperature Na-S batteries (300–350 °C). As a result of the breakdown of polysulfide, low-temperature Na-S batteries experience self-discharge and a quick decline in capacity. The research presented here used a thick β"-Al_2O_3 solid electrolyte separator and a catholyte based on tetraglyme to create an Na-S battery that operates at around 150 °C. A vapor-phase conversion was used to produce the β"-Al_2O_3 electrolyte, which successfully prevented inter-diffusion and side reactions while attaining a high ionic conductivity of 8.5×10^{-3} S cm^{-1} at 150 °C. The electrochemical experiments verified that the reduction of sodium polysulfide occurred in stages, with redox maxima at around 2.22 V and 1.75 V, respectively. The battery's Coulombic efficiency was 94% in the beginning, stabilizing at around 90% after 60 cycles, and its initial discharge capacity was 473 mAh g^{-1}. The enhanced stability and efficiency of the battery showcased the promise of Na-S batteries operating at intermediate temperatures for energy storage on a grand scale. A novel intermediate-temperature Na-S battery with a dual electrolyte system was created by Wang et al. [31] to address the drawbacks of conventional high- and low-temperature Na-S batteries. The battery uses an inorganic ionic liquid (Na[OTf]-Cs[TFSA]) and a β"-alumina solid electrolyte (BASE) to function at 150 °C. By enhancing polysulfide solubility and ionic conductivity, the ionic liquid (IL) improves the reactivity of polysulfide intermediates, whereas the BASE suppresses the polysulfide shuttle effect, which revealed an impressive initial discharge capacity of 795 mAh g^{-1} at 0.1 mA. This capacity was sustained over 1000 cycles, with an average of 381 mAh g^{-1}, and a Coulombic efficiency close to 100%. Overpotentials were reduced due to the IL's excellent salt transport, and polysulfide crossover was inhibited by the BASE structure. To improve energy density and cycle performance, future studies will optimize the composition of electrolytes, the solubility of polysulfides, and the design of cathodes.

Using an aerogel composed of nickel chloride and reduced graphene oxide ($NiCl_2$-rGO) as the cathode material, Li et al. [32] created a charged-state Na-nickel chloride (Na-$NiCl_2$) battery. Overuse of Ni and irreversible capacity loss from Ni exfoliation are problems with traditional Na-$NiCl_2$ batteries. A stable conductive network is provided by the hierarchical $NiCl_2$-rGO aerogel, which also boosts electrical conductivity. Tests using electrochemistry showed that the cathode made of $NiCl_2$-rGO aerogel had a high reversible capacity, reaching 116 mAh g^{-1} after 50 cycles and 128 mAh g^{-1} in a fixed capacity window with very low polarization. Reducing the Ni/NaCl molar ratio from 1.8 to 0.5, the improved cathode structure lowered material costs while retaining performance. In an effort to find a more economical replacement for Na-$NiCl_2$ systems, Li et al. [33] investigated ZEBRA batteries that use Na-iron chloride (Na-$FeCl_2$) and operate at intermediate temperatures. The chemistry of Na-$FeCl_2$ can improve cycle life by reducing degradation processes and reduce costs by 61%. To facilitate assembly in the discharged condition without extensive pre-treatment, the study incorporated sulfur-based additives to the Fe cathode. Results showed that electrochemical activation and capacity retention were both improved with 4- to 6-mol% sulfur. Over 100 cycles, the Na-$FeCl_2$ battery demonstrated an impressively high energy efficiency of over 92% and a discharge energy density of over 135 Wh kg^{-1}. This development makes ZEBRA batteries a more practical and affordable option for stationary energy storage. The researchers Baclig et al. [34] used a solid electrolyte of K-β"-Al_2O_3 and an Na-K alloy anode to create a high-voltage, RT liquid metal flow battery (Figure 3.4). The sealing and thermal management of Na-S and Na-$NiCl_2$ batteries provide difficulties at temperatures ranging from 270 to 350 °C. Mechanical stability and selective K^+ conduction were

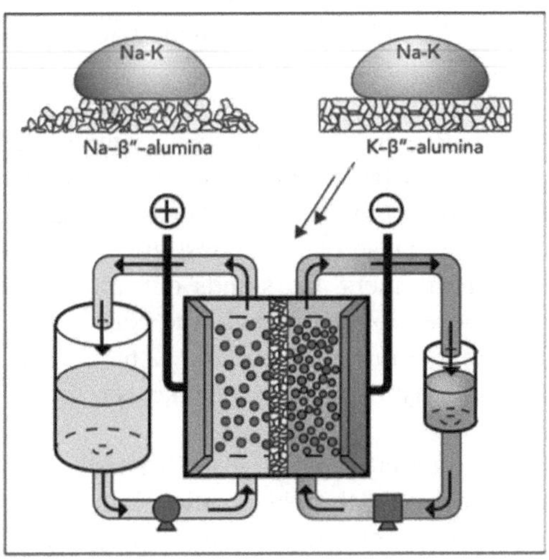

FIGURE 3.4 Schematic diagram of liquid metal flow battery using an Na-K alloy anode and K-β"-Al_2O_3 solid electrolyte.

Source: Adapted with permission [34]. Copyright 2018, Elsevier.

both shown to be maintained by K-β"-Al_2O_3 membranes in this work. At 22 °C, the battery produced 65 mW cm^{-2} of electricity and OCVs of 3.1–3.4V. Thinner membranes might decrease the ohmic resistance of the K-β"-Al_2O_3 membrane, which was the main factor restricting the process.

By refining the use of molten salt electrolyte and Na-Bi alloy anode, Yan et al. [35] created an Na-Bi battery that operates at 150–200 °C. A high ionic conductivity (1.4 S cm^{-1} at 150 °C) and a low melting point (127 °C) were characteristics of the electrolyte, which was composed of NaF, NaCl, and NaI in a ratio of 40:40:20 mol%. After 500 cycles, the electrochemical tests showed a consistent discharge voltage of 0.8V and a Coulombic efficiency of 94%. Improved efficiency and longer cycle life were achieved by reducing electrode corrosion caused by the uniform Na-Bi intermetallic phase. An NaCl-buffered ionic liquid and β"-Al_2O_3 separator were utilized by Yang et al. [36] to create an intermediate-temperature Na-copper chloride (Na/$CuCl_2$) battery that operates at 150 °C. Sintering at 700 °C was used to create the Cu-based cathode, which has an energy density of 750 mWh g^{-1} and a specific capacity of 285 mAh g^{-1}. The battery showed steady cycling, but it needed more tweaking to keep the electrolyte from deteriorating. The material stability, cycle life, ionic conductivity, and cost-effectiveness of MT-LMBs are all addressed in this research. To improve the feasibility of MT-LMBs for energy storage applications on a grid scale, future studies should concentrate on improving electrolyte compositions, cathode designs, and solid electrolyte structures.

3.3.3 Room-Temperature LMBs

Exploring the possibility of operating at ambient temperatures (0–40 °C) was an early goal of research into LMBs, with the hope of expanding its use beyond fixed energy storage devices [37]. Using solid-state battery technologies and tried-and-true solutions from LiBs, RT-LMBs may be created at these temperatures. Solid electrolytes and organic and inorganic liquid electrolytes are among the several electrolyte alternatives accessible with RT-LMBs, which is a major benefit. Ga and its alloys, Hg, Cs, Fr, Rb, and Na-K alloys are among the metals and alloys with melting points close to or below RT. Commercial feasibility in energy storage applications is hindered by the toxicity of Hg as well as the radiation dangers linked to Rb, Cs, and Fr. Because of its modest electrochemical potential compared to Li and its inherent safety, Ga stands out among these materials [38]. Anode candidates that show promise include the Na-K alloy, a eutectic combination that can melt at temperatures as low as −12.6 °C. The availability of raw ingredients and the simplicity of synthesis make Na-K alloys a cost-effective alternative to Li [39].

With a low reduction potential of −2.71 V for Na and −2.92 V for K compared to the usual hydrogen electrode, the Na–K alloy is one of the most attractive alternatives to Li anodes and efficiently reduces dendrite development problems [37]. As an Na anode, the Na–K alloy can attain a specific capacity of 629 mAh g^{-1}, and as a K anode, it can obtain 579 mAh g^{-1}, all because of the liquidus line in its phase diagram. Nevertheless, a significant obstacle in building an RT-LMB based on Na–K alloys is dealing with the liquid metal's wettability on diverse substrates, such as current collectors or solid electrolytes, and making sure it works with varied electrolytes [40]. Yu et al. developed a liquid metal anode by incorporating fibrous carbon substrates, which were immersed in the alloy. This process was facilitated by capillary forces and the chemical inducement of the K–C compound, effectively confining the liquid metal and enabling stable cycling with various cathodes [41]. Despite its extensive usage for molten Na at high temperatures in Na-BASE solid electrolyte owing to its strong Na-ion conductivity, it becomes pulverized when it comes into contact with Na-K alloy as a result of ion exchange. At high temperatures, the incompatibility of Na–Rb alloy with Na-BASE is due to the instability caused by the somewhat higher atomic sizes of K and Rb relative to Na. Luckily, the Na–K system works better with K-BASE because it has a lower free energy and stays stable when mixed with molten K and Na–K alloy [9].

Liquid alkali metal alloys are complicated for reasons beyond their physical qualities, namely their electrochemical properties. The primary role of the Na–K alloy as an anode for Na or K compounds is an important subject to consider since it comprises two alkali metals. The researchers Goodenough et al. discovered that the ionic preference of the cathode determines the amount of metal removed from the anode in a battery system that uses a Na–K alloy anode and different cathodes [42]. When the cathode favorably accepts Na^+ ions, the anode acts as an Na rechargeable cell; conversely, when the cathode preferentially accepts K^+ ions, the anode acts as a K rechargeable cell. Because of its dual-anode characteristic, Na–K alloy may be used as an anode in both Na-metal and K-metal LMBs without forming dendrites. Moreover, the electrolyte can affect charge selection by creating a solid-electrolyte interphase (SEI) that is rich in either Na or K at the electrode-electrolyte border. This interface allows the passage of one species while inhibiting the other [43]. Guo et al. created a ternary hybrid-cation battery based on this charge selection process. It combines an Na–K alloy anode, which is often utilized in LiBs. The selective charge behavior was governed by the ion distribution inside the SEI layer, since only Li ions could intercalate into the cathode and only K ions could strip and deposit on the anode in the hybrid electrolyte system. Ding et al. showed how a liquid Na–K anode may be formed in situ and looked at how anion effects in the SEI layer affect the process in another work [44]. By optimizing the electrolyte system, they successfully induced in-situ formation of the liquid metal through a galvanic replacement reaction at the electrode surface, providing a straightforward and effective method to protect alkali metal electrodes.

In addition to alkali metals, Ga and its alloys, including the Ga–In alloy, have melt points that are quite low and show potential electrochemically for energy storage uses. Ga has great promise in battery systems as a starting material for alloying processes with different types of alkali ions. The behavior of Ga electrodes in LiBs was initially studied, especially at around 40 °C, as pure Ga has a melting point of 29.8 °C, which is somewhat higher than RT [45]. According to the phase diagram, lithiation during the discharge process causes the rapid solidification of Ga, forming specific intermetallic phases at different compositions at RT [46]. The phase diagram shows that at 40 °C charge-discharge cycling, many voltage plateaus appear, which correspond to the development of Li_2Ga_7, LiGa, and Li_2Ga intermetallic phases. A self-healing process like that of HT-LMBs was demonstrated by Deshpande et al., who discovered that the Ga electrode's fissures could be mended after full de-lithiation [47]. At RT, bulk solid Ga electrodes have limited cyclability because the reaction and transport kinetics of solidified Ga are slower than those of its liquid phase. Due to their lower melting point, Ga–In alloys have been investigated as a potential solution to this constraint. For example, the eutectic Ga–In alloy is more suited for real-world battery uses since it melts at a somewhat lower temperature than pure Ga, at 15.3 °C. Furthermore, its use in battery systems is affected by the fact that the ternary Ga–In–Li alloy has a melting point higher than RT across its entire composition range. LMBs made using Ga-based alloys may have better self-healing capabilities, less dendrite formation, and longer life in designs for batteries operated at intermediate temperatures. To make sure these alloys are stable and efficient in real energy storage applications, more study is needed to enhance their composition and electrochemical performance [9].

The reversibility of the reaction makes alloy electrodes "self-healing" materials, which bode well for their potential use in LMBs. Conceptual prototypes have already utilized eutectic Ga–In as a liquid metal-air battery [48]. Furthermore, a self-healing anode for LiBs was presented by Wu et al., which utilized a Ga–Sn alloy and benefited from a lower melting point of 10.7 °C. While this method did reduce capacity somewhat, it greatly improved cyclability, allowing for up to 4000 stable cycles [49]. Efficient electronic and ionic transport channels were enabled by the study's use of a three-dimensional carbon framework with high porosity and surface area. Using this as a foundation, Yu et al. showed that a Ga–In LMB may attain high cyclability without the need for intricate structural frameworks. The researchers used ultrasonication to disseminate liquid metal into liquid metal nano particles, which improved stability over extended cycling by preventing pulverization and delamination from the substrate. For non-Li batteries, such as Zn-ion batteries and Al-ion batteries, liquid metals and alloys based on Ga also play an important role in solving interfacial problems [50]. Although most of the previous research used solid electrolytes, the liquid electrodes were either Na-K or Ga-based alloys. Nevertheless, a groundbreaking RT-LMB employing both Na-K and Ga-based liquid alloys was recently developed by Ding et al. Consisting of a negative electrode, electrolyte, and positive electrode, all of which naturally self-segregate into three separate layers owing to their immiscibility and density differences, this arrangement is comparable to the three-liquid-layer HT-LMB. When it came to maximizing volumetric energy density, the wetting behavior of liquid metals was essential. The wettability of gold was superior to that on Al, stainless steel, and Ga-In alloy surfaces. The adsorption energy between the Ga–In alloy and gold was almost six times greater than for the other two substrates, according to density functional theory estimates. In addition, electrostatic potential maps showed that the strongest contact and anchoring effect occurred on Au-coated surfaces because there were more binding sites for liquid metal atoms. More realistic RT-LMBs and MT-LMBs that include solid-liquid electrode-electrolyte interfaces may be designed with the help of these surface and interfacial chemistry findings [9].

3.4 CHALLENGES

The many advantageous characteristics of LMBs, such as their self-healing capabilities, low cost, high power density, and extended cycle life, make them an attractive candidate for grid-scale energy storage. Their metal electrodes in a liquid state improve performance and safety by preventing dendrite development, a problem with solid-state batteries [51]. Electrodes and electrolytes have seen improvements in their physicochemical features, including electronegativity, phase transitions, ionic conductivity, and wettability, thanks to developments in LMB chemistry during the last several decades. Both optimization of battery design and performance prediction have benefited greatly from these advancements and computational modeling. Improving battery efficiency and paving the way for widespread use of LMBs is a multi-pronged effort that involves chemical refinement, electrode microstructure optimization, and macrostructural parameter adjustment, among other things. The study of LMBs is an intriguing field of research due to their varied chemistry. By reducing the melting temperatures of specific metals, alloying techniques increase the variety of materials that might be used as electrodes. Eutectic alloys have the potential to lower the melting temperatures of both positive and negative electrodes, whereas Na-based and Ca-based alloys (Ca-Bi, Ca-Sb, Ca-Mg, Ca-Ge) have been studied as negative electrodes. To improve LMB performance, researchers are looking at a variety of cathode materials, including Sb, Si, Sn, Pb, Ga, In, and Te [52]. The improved electrochemical characteristics and phase stability made possible by multi-element fusible alloys allow for more leeway in the design of affordable, high-performance LMBs. Incorporating inexpensive metal components into efficient alloy electrodes guarantees great electrical conductivity and robust wettability on solid electrolytes and current collectors while simultaneously reducing prices. When compared to organic liquid electrolytes, molten salt electrolytes in HT-LMBs have better ionic conductivity (up to 3 S cm^{-3}), flame retardancy, and cost efficiency. These electrolytes enhance battery efficiency by promoting quick charge-transfer kinetics. Subvalent ions can develop, self-discharge can occur, and energy efficiency can be lowered when metals are soluble in hot, molten salts [53]. Scientists are working to find the sweet spot between melting temperatures, ionic conductivity, and material stability in electrolyte compositions to solve these problems. Electrode kinetics may be sluggish due to slower ion transport, even when lower melting points lower operating temperatures. This highlights the continued need for researching and developing electrolytes with intermediate melting points at a reasonable cost. Further research is needed to understand the links between solid ceramic electrolytes, material processing, defect chemistry, and transport parameters; yet they show promise as an option for RT and MT LMBs. By utilizing simulations to establish a connection between atomistic defect chemistry and macroscopic performance, Poletayev et al. investigated diffusion in fast-ion conductors such as β- and β-alumina [54]. Polymer-based electrolytes also offer advantages, including high ionic conductivity and transference numbers, making them suitable for Na^-, K^-, and Mg-based LMBs [55]. At lower temperatures, liquid metals have less wettability on solid electrolyte surfaces, which is a big problem for MT-LMBs and RT-LMBs that use these electrolytes. Interfacial adhesion is a problem for many alloys, but it is more severe for Na-K, Ga, and Ga-based alloys. As an example, the spherical droplets formed by Ga liquid metal, owing to its high surface tension, decrease its wettability and make electrode construction more complicated [49]. Choosing aqueous binders for Ga-based liquid metals and other materials that improve interfacial adhesion should be the focus of future work. There are comparable issues with wettability for Na-K alloys, which calls for creative engineering solutions [56]. Sealing and corrosion are additional concerns for HT-LMBs and MT-LMBs, as liquid metals react with oxygen and moisture at elevated temperatures, leading to degradation and safety risks. Effective sealing materials must be selected based on operating temperatures. Below 200 °C, polymers are viable, but above this threshold, degradation occurs, making them unsuitable for long-term applications. Traditional Na||S and ZEBRA batteries use Al-Al_2O_3-Al sandwich seals, lasting over 15 years [57]. However, these seals are incompatible with Li-based LMBs due to Li's reactivity with Al_2O_3. Glass and glass-ceramic seals offer an alternative, though they pose challenges related to thermal expansion differences and low stability [58]. Corrosion also affects battery components, including electrical collectors, containers, insulators, and seals [59]. Careful selection of positive electrode materials is required to avoid dissolution, as reactive metals such as Li and Na react easily with negative electrode current collectors. To make LMBs more marketable, researchers need to develop an all-encompassing evaluation framework that takes price, system complexity, safety, cycle life, efficiency, and energy density into account. The enormous promise of LMBs for large-scale energy storage is being piqued by ongoing developments in interface engineering, multiphysics modeling, and battery chemistry.

3.5 FUTURE PERSPECTIVES AND CONCLUSION

Presenting a comprehensive overview of rechargeable LMBs, this chapter delves into recent strides in battery chemistry and multiphysics modeling, shedding light on important obstacles that have been overcome in their creation. Dendrite-free functioning, easy cell manufacturing, effective material recovery, and dependence on earth-abundant materials are some of the key benefits of HT-LMBs, which use molten salt electrolytes and liquid metal electrodes. But to make them more efficient, their operating temperatures need to be reduced without compromising their three-liquid-layer design. Alternative methods and new battery chemistries can do this. To speed up the commercial deployment of HT-LMBs, it is essential that multiphysics modeling continues to progress so that large-scale design and operating strategies may be fine-tuned. Problems at the interface of liquid electrodes and solid electrolytes have hindered the broad use of MT-LMBs and RT-LMBs, which provide hope for attaining large energy densities. One of the key challenges in creating high-performance LMBs is creating solid electrolytes with high transference numbers and ionic conductivity. The total efficiency and lifetime of batteries are greatly affected by factors including cell layout and testing processes, in addition to battery chemistry. The intricate relationship between electrochemical and thermophysical processes can only be further understood and optimized for practical use with future advancements in multiphysics modeling. A more thorough comprehension of fluid dynamics, electrochemistry, and heat is necessary for the future advancement of LMBs. To optimize electrode designs and electrolyte compositions, advanced computational models that include electric potential distribution, electro-vortex flow, thermal convection, and mass transfer are essential. A more stable and scalable LMB technology will be possible with the help of improved modeling skills, which will shed light on how to improve charge-discharge efficiencies and reduce the impact of self-discharge phenomena. New chemistries with improved electrochemical stability, high energy density, and lower operating temperatures should be investigated in future studies to increase performance and practicality. An important step forward in the quest for realistic deployment beyond grid storage applications is the development of RT-LMBs and MT-LMBs using fusible alloys like Ga-based or Na-K systems. The development of alloy engineering techniques is also crucial for lowering melting points, increasing wettability, and guaranteeing material compatibility with both solid and liquid electrolytes. Further improvements to battery performance, lifespan, and safety can be achieved through the exploration of different liquid electrodes and innovative molten salt compositions. Due to their long cycle life, cheap materials, and eco-friendliness, LMBs have the capacity to transform grid-scale energy storage. Limiting the use of rare and costly resources, LMBs rely on earth-abundant and recyclable materials, in contrast to LiBs. With the use of optimized electrode materials and low-cost molten salt electrolytes, LMBs may be mass-produced at a reasonable cost, opening the door to large-scale renewable energy integration. They are an attractive choice for energy storage applications that need long-term storage due to their self-healing characteristics and resistance to dendrite development, which improve safety and dependability. The last ten years have seen remarkable strides in the creation of various types of LMBs, including HT-LMBs that increase energy density by optimizing molten salt electrolytes and using alloying strategies, MT-LMBs that lower operating temperatures by integrating solid electrolytes and new electrode materials, and RT-LMBs that enable RT operation by using liquid metal systems based on Na-K and Ga. The potential for LMBs to address the increasing need for sustainable, affordable, and efficient energy storage systems is highlighted by these achievements. Future developments in material choice, interface engineering, and scalable manufacturing techniques will be necessary for the commercialization of LMBs. To achieve broad acceptance, it is crucial to solve problems such as electrolyte stability, corrosion resistance, and optimizing costs. There may be new possibilities for electric mobility, smart grid integration, and stationary energy storage because of developments in fast-ion conductors and hybrid electrolyte systems. As studies progress, LMBs have the potential to significantly impact the shift toward a future energy system that is both more sustainable and robust.

REFERENCES

[1] M. Park, J. Ryu, W. Wang, J. Cho, Material design and engineering of next-generation flow-battery technologies, *Nat. Rev. Mater.* 2 (2016) 16080.

[2] W. Chen, G. Li, A. Pei, Y. Li, L. Liao, H. Wang, J. Wan, Z. Liang, G. Chen, H. Zhang, J. Wang, Y. Cui, A manganese–hydrogen battery with potential for grid-scale energy storage, *Nat. Energy.* 3 (2018) 428–435.

[3] M.E. Arroyo-de Dompablo, A. Ponrouch, P. Johansson, M.R. Palacín, Achievements, challenges, and prospects of calcium batteries, *Chem. Rev.* 120 (2020) 6331–6357.
[4] H. Gao, B.M. Gallant, Advances in the chemistry and applications of alkali-metal–gas batteries, *Nat. Rev. Chem.* 4 (2020) 566–583.
[5] J. Kim, D. Shin, Y. Jung, S.M. Hwang, T. Song, Y. Kim, U. Paik, LiCl-LiI molten salt electrolyte with bismuth-lead positive electrode for liquid metal battery, *J. Power Sources.* 377 (2018) 87–92.
[6] D. Chao, W. Zhou, F. Xie, C. Ye, H. Li, M. Jaroniec, S.-Z. Qiao, Roadmap for advanced aqueous batteries: From design of materials to applications, *Sci. Adv.* 6 (2025) eaba4098.
[7] H. Kim, D.A. Boysen, J.M. Newhouse, B.L. Spatocco, B. Chung, P.J. Burke, D.J. Bradwell, K. Jiang, A.A. Tomaszowska, K. Wang, W. Wei, L.A. Ortiz, S.A. Barriga, S.M. Poizeau, D.R. Sadoway, Liquid metal batteries: Past, present, and future, *Chem. Rev.* 113 (2013) 2075–2099.
[8] J. Muldoon, C.B. Bucur, T. Gregory, Quest for nonaqueous multivalent secondary batteries: Magnesium and beyond, *Chem. Rev.* 114 (2014) 11683–11720.
[9] S. Wu, X. Zhang, R. Wang, T. Li, Progress and perspectives of liquid metal batteries, *Energy Storage Mater.* 57 (2023) 205–227.
[10] S. Zhang, Y. Liu, Q. Fan, C. Zhang, T. Zhou, K. Kalantar-Zadeh, Z. Guo, Liquid metal batteries for future energy storage, *Energy Environ. Sci.* 14 (2021) 4177–4202.
[11] H. Yin, B. Chung, F. Chen, T. Ouchi, J. Zhao, N. Tanaka, D.R. Sadoway, Faradaically selective membrane for liquid metal displacement batteries, *Nat. Energy.* 3 (2018) 127–131.
[12] J. Zhou, J. Cheng, B. Wang, H. Peng, J. Lu, Flexible metal–gas batteries: A potential option for next-generation power accessories for wearable electronics, *Energy Environ. Sci.* 13 (2020) 1933–1970.
[13] J. Hao, B. Li, X. Li, X. Zeng, S. Zhang, F. Yang, S. Liu, D. Li, C. Wu, Z. Guo, An in-depth study of Zn metal surface chemistry for advanced aqueous Zn-ion batteries, *Adv. Mater.* 32 (2020) 2003021.
[14] D. Agarwal, R. Potnuru, C. Kaushik, V.R. Darla, K. Kulkarni, A. Garg, R.K. Gupta, N. Tiwari, K.S. Nalwa, Recent advances in the modeling of fundamental processes in liquid metal batteries, *Renew. Sustain. Energy Rev.* 158 (2022) 112167.
[15] D.J. Bradwell, Liquid metal batteries: Ambipolar electrolysis and alkaline earth electroalloying cells by, 2011. https://dspace.mit.edu/handle/1721.1/62741
[16] Y. Ding, X. Guo, G. Yu, Next-generation liquid metal batteries based on the chemistry of fusible alloys, *ACS Cent. Sci.* 6 (2020) 1355–1366.
[17] D.H. Kelley, T. Weier, Fluid mechanics of liquid metal batteries, *Appl. Mech. Rev.* 70 (2018) 020801.
[18] D.J. Bradwell, H. Kim, A.H.C. Sirk, D.R. Sadoway, Magnesium–antimony liquid metal battery for stationary energy storage, *J. Am. Chem. Soc.* 134 (2012) 1895–1897.
[19] Y. Ding, C. Zhang, L. Zhang, H. Wei, Y. Li, G. Yu, Insights into hydrotropic solubilization for hybrid ion redox flow batteries, *ACS Energy Lett.* 3 (2018) 2641–2648.
[20] D. Yu, Z. Xue, T. Mu, Eutectics: Formation, properties, and applications, *Chem. Soc. Rev.* 50 (2021) 8596–8638.
[21] R.D. Deshpande, J. Li, Y.-T. Cheng, M.W. Verbrugge, Liquid metal alloys as self-healing negative electrodes for lithium ion batteries, *J. Electrochem. Soc.* 158 (2011) A845.
[22] Y. Qi, C. Shen, Q. Hou, Z. Ren, T. Jin, K. Xie, A self-healing liquid metal anode for lithium-ion batteries, *J. Energy Chem.* 72 (2022) 522–531.
[23] H. Li, K. Wang, H. Zhou, X. Guo, S. Cheng, K. Jiang, Tellurium-tin based electrodes enabling liquid metal batteries for high specific energy storage applications, *Energy Storage Mater.* 14 (2018) 267–271.

[24] W. Weppner, R.A. Huggins, Determination of the kinetic parameters of mixed-conducting electrodes and application to the system Li3Sb, *J. Electrochem. Soc.* 124 (1977) 1569.
[25] J. Sangster, A.D. Pelton, The Bi-Li (Bismuth-Lithium) system, *J. Phase Equilibria.* 12 (1991) 447–450.
[26] X. Ning, S. Phadke, B. Chung, H. Yin, P. Burke, D.R. Sadoway, Self-healing Li–Bi liquid metal battery for grid-scale energy storage, *J. Power Sources.* 275 (2015) 370–376.
[27] K. Cui, W. Zhao, D. Zhou, S. Li, C. Liu, P. Li, X. Qu, Stable positive current collectors for Li||Sb–Sn liquid metal batteries, *ACS Appl. Energy Mater.* 4 (2021) 9013–9021.
[28] W. Zhao, P. Li, Z. Liu, D. He, K. Han, H. Zhao, X. Qu, High-performance antimony–bismuth–tin positive electrode for liquid metal battery, *Chem. Mater.* 30 (2018) 8739–8746.
[29] H. Xie, Z. Chen, P. Chu, J. Wang, Z. Li, H. Zhao, An elaborate low-temperature electrolyte design towards high-performance liquid metal battery, *J. Power Sources.* 536 (2022) 231527.
[30] X. Lu, B.W. Kirby, W. Xu, G. Li, J.Y. Kim, J.P. Lemmon, V.L. Sprenkle, Z. Yang, Advanced intermediate-temperature Na–S battery, *Energy Environ. Sci.* 6 (2013) 299–306.
[31] D. Wang, J. Hwang, C. Chen, K. Kubota, K. Matsumoto, R. Hagiwara, A β"-Alumina/inorganic ionic liquid dual electrolyte for intermediate-temperature sodium–sulfur batteries, *Adv. Funct. Mater.* 31 (2021) 2105524.
[32] Y. Li, L. Shi, X. Gao, J. Wang, Y. Hu, X. Wu, Z. Wen, Constructing a charged-state Na-NiCl2 battery with NiCl2/graphene aerogel composite as cathode, *Chem. Eng. J.* 421 (2021) 127853.
[33] G. Li, X. Lu, J.Y. Kim, V.V. Viswanathan, K.D. Meinhardt, M.H. Engelhard, V.L. Sprenkle, An advanced Na–FeCl2 ZEBRA battery for stationary energy storage application, *Adv. Energy Mater.* 5 (2015) 1500357.
[34] A.C. Baclig, G. McConohy, A. Poletayev, A. Michelson, N. Kong, J.-H. Lee, W.C. Chueh, J. Rugolo, High-voltage, room-temperature liquid metal flow battery enabled by Na-K/K-β″-Alumina stability, *Joule.* 2 (2018) 1287–1296.
[35] S. Yan, X. Zhou, H. Li, Y. Shen, Y. He, H. Zhou, K. Wang, K. Jiang, Utilizing in situ alloying reaction to achieve the self-healing, high energy density and cost-effective Li||Sb liquid metal battery, *J. Power Sources.* 514 (2021) 230578.
[36] L.-P. Yang, X.-M. Liu, Y.-W. Zhang, H. Yang, X.-D. Shen, Advanced intermediate temperature sodium copper chloride battery, *J. Power Sources.* 272 (2014) 987–990.
[37] L. Xue, H. Gao, W. Zhou, S. Xin, K. Park, Y. Li, J.B. Goodenough, Liquid K-Na alloy anode enables dendrite-free potassium batteries, *Adv. Mater.* 28 (2016) 9608–9612.
[38] X. Guo, Y. Ding, L. Xue, L. Zhang, C. Zhang, J.B. Goodenough, G. Yu, A self-healing room-temperature liquid-metal anode for alkali-ion batteries, *Adv. Funct. Mater.* 28 (2018) 1804649.
[39] H. Ye, Z.-J. Zheng, H.-R. Yao, S.-C. Liu, T.-T. Zuo, X.-W. Wu, Y.-X. Yin, N.-W. Li, J.-J. Gu, F.-F. Cao, Y.-G. Guo, Guiding uniform Li plating/stripping through lithium–aluminum alloying medium for long-life Li metal batteries, *Angew. Chemie Int. Ed.* 58 (2019) 1094–1099.
[40] L. Xue, W. Zhou, S. Xin, H. Gao, Y. Li, A. Zhou, J.B. Goodenough, Room-temperature liquid Na–K anode membranes, *Angew. Chemie.* 130 (2018) 14380–14383.
[41] L. Zhang, S. Peng, Y. Ding, X. Guo, Y. Qian, H. Celio, G. He, G. Yu, A graphite intercalation compound associated with liquid Na–K towards ultra-stable and high-capacity alkali metal anodes, *Energy Environ. Sci.* 12 (2019) 1989–1998.

[42] L. Xue, H. Gao, Y. Li, J.B. Goodenough, Cathode dependence of liquid-alloy Na–K anodes, *J. Am. Chem. Soc.* 140 (2018) 3292–3298.

[43] M. Lee, J. Hong, J. Lopez, Y. Sun, D. Feng, K. Lim, W.C. Chueh, M.F. Toney, Y. Cui, Z. Bao, High-performance sodium–organic battery by realizing four-sodium storage in disodium rhodizonate, *Nat. Energy.* 2 (2017) 861–868.

[44] Y. Ding, X. Guo, Y. Qian, H. Gao, D.H. Weber, L. Zhang, J.B. Goodenough, G. Yu, In situ formation of liquid metals via galvanic replacement reaction to build dendrite-free alkali-metal-ion batteries, *Angew. Chemie Int. Ed.* 59 (2020) 12170–12177.

[45] B.-W. Zhang, L. Ren, Y.-X. Wang, X. Xu, Y. Du, S.-X. Dou, Gallium-based liquid metals for lithium-ion batteries, *Interdiscip. Mater.* 1 (2022) 354–372.

[46] H. Okamoto, Ga-Li (Gallium-Lithium), *J. Phase Equilibria Diffus.* 27 (2006) 200–200.

[47] M. Zabrocki, W. Gąsior, A. Dębski, Thermodynamic properties of Ga-In-Li alloys—a potential material for liquid metal batteries, *J. Mol. Liq.* 332 (2021) 115765.

[48] G. Liu, J.Y. Kim, M. Wang, J.-Y. Woo, L. Wang, D. Zou, J.K. Lee, Soft, highly elastic, and discharge-current-controllable eutectic gallium–indium liquid metal–air battery operated at room temperature, *Adv. Energy Mater.* 8 (2018) 1703652.

[49] Y. Wu, L. Huang, X. Huang, X. Guo, D. Liu, D. Zheng, X. Zhang, R. Ren, D. Qu, J. Chen, A room-temperature liquid metal-based self-healing anode for lithium-ion batteries with an ultra-long cycle life, *Energy Environ. Sci.* 10 (2017) 1854–1861.

[50] L.E. Blanc, D. Kundu, L.F. Nazar, Scientific challenges for the implementation of Zn-ion batteries, *Joule.* 4 (2020) 771–799.

[51] Y. Zhu, H.Y. Hoh, S. Qian, C. Sun, Z. Wu, Z. Huang, L. Wang, M. Batmunkh, C. Lai, S. Zhang, Y.L. Zhong, Ultrastable zinc anode enabled by CO2-Induced interface layer, *ACS Nano.* 16 (2022) 14600–14610.

[52] T. Daeneke, K. Khoshmanesh, N. Mahmood, I.A. de Castro, D. Esrafilzadeh, S.J. Barrow, M.D. Dickey, K. Kalantar-zadeh, Liquid metals: Fundamentals and applications in chemistry, *Chem. Soc. Rev.* 47 (2018) 4073–4111.

[53] O. Bohnke, S. Ronchetti, D. Mazza, Conductivity measurements on nasicon and nasicon-modified materials, *Solid State Ionics.* 122 (1999) 127–136.

[54] A.D. Poletayev, J.A. Dawson, M.S. Islam, A.M. Lindenberg, Defect-driven anomalous transport in fast-ion conducting solid electrolytes, *Nat. Mater.* 21 (2022) 1066–1073.

[55] F. Chen, X. Wang, M. Armand, M. Forsyth, Cationic polymer-in-salt electrolytes for fast metal ion conduction and solid-state battery applications, *Nat. Mater.* 21 (2022) 1175–1182.

[56] Y. Ding, X. Guo, Y. Qian, L. Zhang, L. Xue, J.B. Goodenough, G. Yu, A liquid-metal-enabled versatile organic alkali-ion battery, *Adv. Mater.* 31 (2019) 1806956.

[57] M.G. Nicholas, R.M. Crispin, Diffusion bonding stainless steel to alumina using aluminium interlayers, *J. Mater. Sci.* 17 (1982) 3347–3360.

[58] S. Song, Z. Wen, Y. Liu, J. Lin, X. Xu, Q. Zhang, New glass-ceramic sealants for Na/S battery, *J. Solid State Electrochem.* 14 (2010) 1735–1740.

[59] G.J. Janz, R.P.T. Tomkins, Corrosion in molten salts: An annotated bibliography, *Corrosion.* 35 (2013) 485–504.

4 Fluid Mechanics of Liquid Metal Batteries

Phuoc-Anh Le

4.1 INTRODUCTION

The idea of a liquid metal battery (LMB), which consists of both a liquid electrolyte and electrodes, provides great potential for cost-effective storage [1]. This type of battery is believed to provide an inventive solution to grid-scale power storage issues. Because of its capacities, the electric grid can better integrate renewable resources. Moreover, LMBs can eliminate the need to construct new energy production, transmission, and distribution systems and are anticipated to increase the overall dependability of an aging power grid [2]. The estimated total costs per unit of stored energy are cheap and competitive if large cells can be built using readily available and affordable active materials. The increasing contribution of solar and wind power to the global energy scene makes large-scale storage an essential component of a working power system [3]. The key requirements for a successful technology in this context are minimal capacity fading and a high number of charge and discharge cycles at a low cost per unit of stored energy.

A battery featuring a totally liquid active structure has several benefits, such as stable stratification, which allows the battery to self-assemble when densities are selected appropriately [4]. Fast kinetics at liquid-liquid interfaces enable quick charging and discharging as well as high-rate capacity. Liquid phase makes structureless electrodes have almost infinite cyclability since they are immune to aging [5]. Since all of the battery's electrodes and electrolytes are liquid, fluid mechanics is essential to the development and operation of liquid metal batteries, a novel technology for grid-scale energy storage [5, 6]. Here, we summarize previous and ongoing work on the fluid mechanics of LMBs and identify areas that warrant further investigation. Connecting to a wealth of previous research on phenomena seen or predicted in LMBs, such as the Tayler instability, electro-vortex flow, Marangoni flow, and interface instabilities, could help the readers gain certain knowledge about this generation of large-scale energy storage. In relation to the LMB designs that are presently being researched, this chapter concentrates on phenomena, materials, length scales, and current densities. Moreover, this chapter attempts to highlight innovations that may result in new mechanisms or design enhancements.

DOI: 10.1201/9781003489726-4

4.2 BRIEF HISTORY OF THE ORIGIN AND WORKING PRINCIPLE OF LMBs

LMBs have an interesting history dating back to World War II, when some German scientists developed a battery that used a salt mixture as an electrolyte in bombs and rockets, which laid the foundation for the later generation of molten-salt batteries. Following the development of the previous scientists, the new generation of molten-salt batteries was restarted in the mid-1960s with the use of sodium for the negative electrodes, which had the outstanding advantages of a low boiling point and low cost. Then, a series of electrolytes were introduced with much in-depth research with practical applications for molten-salt batteries, such as sodium-nickel chloride ($Na-NiCl_2$), sodium-bismuth (Na-Bi), lithium-bismuth (Li-Bi), lithium-tin (Li-Sn), and lithium-antimony (Li-Sb) [7]. After a break of about 30 years, research efforts were redirected at the next generation of molten-salt batteries called liquid-metal batteries, with the remarkable announcement of the Sadoway group from MIT about the usage of magnesium-antimony (Mg-Sb) and lead-antimony (Pb-Sb) for liquid metal batteries [8, 9]. It is understandable that the LMBs have received attention and interest in the past decade, when there is increasingly high demand for energy use and storage to serve the robust development of heavy industry and large-scale renewable energy storage, such as solar and wind energy.

Figure 4.1a shows an overview of a standard LMB cell, consisting of two liquid metal electrodes separated by a layer of molten salt electrolyte, and this structure

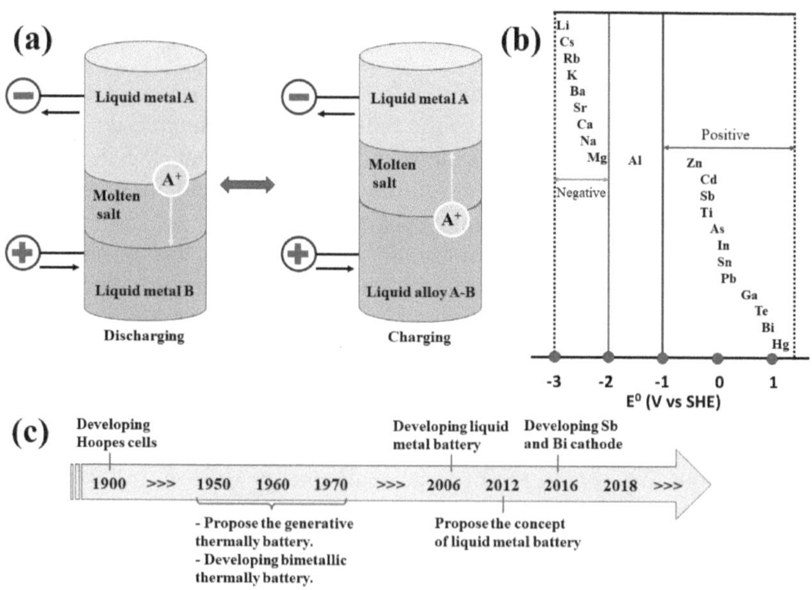

FIGURE 4.1 (a) A diagram of one LMB cylindrical cell following the charge and discharge process; (b) the candidate list of negative and positive electrodes for an LMB with an aqueous molten salt electrolyte; and (c) the timeline development of LMBs.

has a clear separation during both charging and discharging due to the difference in density and immiscibility [10–12]. One standard LMB cell has a liquid metal A as the negative electrode, a liquid metal B as the positive electrode, and one intermediate molten salt electrolyte layer. One LMB cell produces a voltage when its two electrodes have potential energy between them while it reaches operating temperature [13, 14]. The cell voltage pulls electrons from the liquid electrode during battery discharge in order to power the external load, leading to the electrons returning to the electrode. A salt-metal alloy is created internally as a result of liquid metal ions passing through the molten salt and attaching to the electrode ions. Electrons are pushed in the opposite direction during recharging, which removes molten salt ions from the alloy and redeposits them onto the upper layer. According to their deposition potential in an aqueous electrolyte, prospective electrode materials are first classified as positive or negative electrodes (Figure 4.1b); however, because LMBs employ molten salt electrolytes, these deposition potentials are difficult to consider [15]. Figure 4.1c depicts the fascinating and strange development of the LMBs, the type of batteries that seemed to have passed into history. After a long hiatus in development since the emergence of the three-liquid-layer Hoopes cell in the 1920s, the LMBs have made a surprising and powerful comeback with new hope and new responsibilities.

In detail, when the liquid metal A discharges, it contributes n electrons (e^-) to the electrical circuit, where n_e represents the charge number (oxidation reaction) [16, 17]:

$$A \rightarrow A^{n^+} + ne^- . \tag{4.1}$$

After that, ions A^{n^+} diffuse through the layer of molten salt electrolyte to the alloy, where an electrochemical process known as a reduction reaction takes place at the positive electrode-electrolyte interface [16, 17]:

$$A^{n^+} + ne^- \rightarrow A. \tag{4.2}$$

The metals that can be used as negative and positive electrodes are shown in Figure 4.1b, and aluminum (Al) has the unique advantage of being used as both negative and positive electrodes. Positive electrodes are defined as metals with a deposition potential more than –1.0 V, and negative electrodes are defined as those having a deposition potential less than –2.0 V [14].

For LMBs, it is crucial to consider the choice, characteristics, and functionality of the appropriate molten-salt electrolytes. The ideal qualities for the candidate electrolytes used in the LMBs are as follows: (i) excessive ionic conductivity; (ii) low metal solubility with a low melting temperature; (iii) appropriate density to separate the three liquid layers; and (iv) no side reactions in the positive and negative electrodes within the voltage and temperature window of operation [16]. Typically, liquid metal electrodes meet the following specifications: (i) having stable operating temperatures ranging from room temperature (27 °C) to 1000 °C; (ii) having a greater electrical conductivity than a normal molten salt electrolyte ($\sigma > 1$ S cm^{-1}); and (iii) being isotopically naturally stable and nonradioactive [14, 17]. The development of electrolytes for liquid metal batteries is a long research process with many breakthroughs. Up to now, two types of electrolytes have been researched the most: molten salt electrolytes and ionic liquid electrolytes.

In fact, LMBs are actually highly complex hybrid physical systems. In summary, LMB is an electrochemical cell—a device that conducts electrochemical processes. Because the components are moving while the cell operates, it is exposed to mass transport. Heat transport is induced by the cell's heating. Phase shifts can result from localized fluctuations in concentration and temperature. A magnetic field is produced by the electrical current and charge transfer that takes place during operation. The resultant Lorentz force can cause flows and magnetohydrodynamic instabilities in large cells. The composition of the alloy affects the electrical potential of the cell, which might change locally. Furthermore, the interfaces may deform, leading the system to be multiphase [18, 19].

4.3 FLUID MECHANISM

The design and operation leading to the performance of LMBs strongly depend on the fluid mechanism because the electrode and also the electrolyte exist entirely in the liquid phase. Because of this, the fluid dynamics of LMBs is highly complex and involves a wide range of fluid flow phenomena, such as solutal convection in the cathode; Marangoni convection, caused by surface-tension differences due to concentration and temperature gradients; and Rayleigh-Bénard convection triggered by the electrolyte layer's Joule heating. The remainder of this chapter will deal mainly with fluid mechanics through various important hydromagnetic aspects including Tayler instability, interface instabilities, electro-vortex flow, and Marangoni flow.

4.3.1 Tayler Instability

A current-driven kink-type instability known as the Tayler instability occurs when an electrical current flows through an incompressible, viscous, and resistive liquid conductor, producing an azimuthal magnetic field that becomes unstable [20, 21]. This limits the largest possible size of large-scale liquid metal batteries [22].

Here, a cylindrical liquid metal battery cell was considered, with two metal electrodes and an electrolyte layer of molten salt. The entire cylindrical cell can be considered a homogeneous electric wire with a current density oriented uniformly about the J_0 axis and creating a magnetic field with an azimuthal component along B_C. At this time, the Lorentz force arises due to J_0 and B_0 both pointing inward, creating an unstable longitudinal regime and also non-concentric vortices in the transverse direction. The formula for this kind of instability is the square of the Hartmann number, which represents the ratio of magnetic to viscous forces in the fluid mechanics of liquid metals [18, 20]:

$$H_a = B_0 R \sqrt{\frac{\sigma}{\rho \nu}}, \tag{4.3}$$

where $B_0 = B_\varphi(R)$ is the azimuthal magnetic field at the cylinder radius R, while σ, ρ, and ν are the conductivity, density, and kinematic viscosity of the fluid for a cylinder of radius R.

The beginning of current-driven instabilities for a solely azimuthal magnetic field is mostly determined. The instability of the kink type is contingent upon the radial (r) dependency of B_φ. The relevant requirement for the kink-type instability (azimuthal wavenumber m = 1) was deduced by Vandakurov and Tayler [23, 22]:

$$\frac{\partial\left(rB_\varphi^2(r)\right)}{\partial r} > 0. \tag{4.4}$$

The technology of LMBs, which are suggested for grid-scale energy storage, is vulnerable to Tayler instability. If the ensuing flow were powerful enough, it might rupture the electrolyte layer and kill the battery [24]. Tayler instability will first appear in the higher metal layer; if the flow is too strong, it could penetrate the electrolyte layer, leading to a short circuit. At present, numerous strategies can be used to prevent or lessen the Tayler instability. For example, by deftly managing the battery current to avoid the requirement given by Eq. (4), the instability may be avoided. A cylindrical annulus battery with an empty center bore can be constructed as an alternative to a cylindrical LMB that carries an axial current. A greater current is needed prior to beginning, since the bore, which is now unloaded, does not contribute to the induced magnetic field B_φ [24]. A better approach would be to route the battery current back into the bore in the opposite direction to prevent the Tayler instability; however, ohmic losses in this wire will reduce the available voltage. Another idea is to shake the liquid azimuthally to reduce the Tayler instability effect, but this is difficult for a large-scale liquid metal cell [25, 26]. A recent simulation study calculated that for a model LMB using Mg electrodes, Tayler instability occurs when the diameter exceeds 0.43 m at a standard current density of $J = 300\ \text{mA cm}^{-2}$ and collapses if the cell radius exceeds 3 m [27]. In practice, with such a large radius, more research will be needed before commercialization can be predicted. When a significant axial current flows through liquid metal. The Tayler instability might not yet have an impact on technology because of the current magnitude involved and the huge aspect ratio of modern LMB designs. However, it needs to be properly taken into account for bigger, next-generation batteries.

4.3.2 Electro-Vortex Flow

Considering a standard cylindrical LMB cell as a conductor of uniform cross-section, the current flows through the positive electrode in the transverse direction and creates a magnetic field in the liquid, and flow interacts with the current to produce a Lorentz force acting on the liquid metal. This Lorentz force drives the liquid flow, also known as electro-vortex flow (Figure 4.2). The electro-vortex flow in a three-layered LMB occurs close to the current collectors and pushes the eddy flow toward the battery wall. This phenomenon creates eddy flows in the battery, which enhance the mixing of the liquid metal layers, thereby minimizing the polarization effect and improving the mass transfer process in the liquid electrodes. However, if this eddy

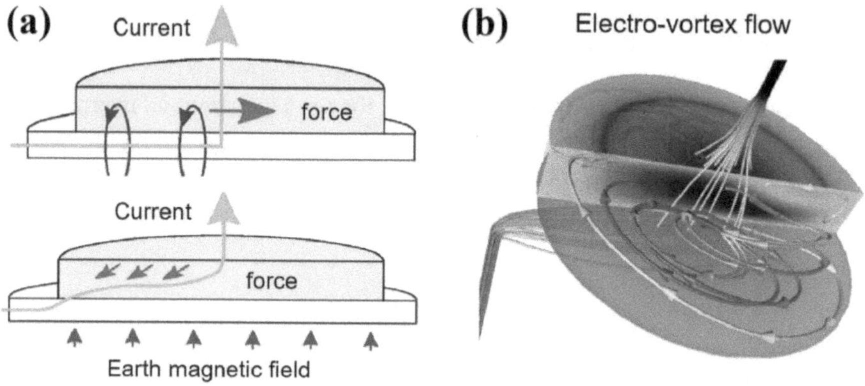

FIGURE 4.2 (a) A magnetic field is created in the fluid by the current flowing through the positive electrode in a lateral direction. This magnetic field combines with the current flow to produce the Lorentz force acting on the liquid metal, which results in electro-vortex flow. (b) Electro-vortex flow.

Source: Adapted with permission [3], Copyright 2023, Elsevier.

flow is too strong, it will deform the liquid electrolyte layer and affect the operation of the battery.

In order to demonstrate that flow is driven by a Lorentz force with nonzero curl, the momentum conservation of an incompressible Newtonian fluid subjected to a force per unit mass is provided: F [19]:

$$\frac{\partial u}{\partial t}+(u.\nabla)u=-\frac{1}{\rho}\nabla p+v\nabla^2 u+F\ , \tag{4.5}$$

Where u is the fluid velocity. Thus, a stationary fluid ($u = 0$) will only conserve momentum if

$$\frac{1}{\rho}\nabla p=F\ . \tag{4.6}$$

Herein, Kelley et al. [19] indicated that $F=-g\hat{z}$, where $\hat{z}$ is assumed to be vertical, is the gravitational force. The well-known result that $p = \rho gh$ is the hydrostatic pressure distance h below the fluid surface may therefore be obtained by directly integrating Eq. (6). However, since the curl of a gradient is always zero, Eq. (6) is unsolvable if $\nabla \times F \neq 0$. If a fluid is operated upon by a force F for which $\nabla \times F \neq 0$, we infer that our initial assumption that $u = 0$ was incorrect. Therefore, the fluid must flow.

Because the current diverges from the negative current collector to the casing, which acts as the positive current collector, electro-vortex flow most often happens in ordinary LMBs. In simulations, electro-vortex flow may be facilitated or hindered by altering the dimensions and aspect ratio of the current collectors and electrodes. Finite electro-vortex flow can affect even tiny liquid metal batteries that handle mild currents, which may make it a more crucial engineering factor than Tayler instability. The thin fluid layers and small current collectors in the LMB cause the current distributions to diverge more, which strengthens the electro-vortex flow [21]. Because intermetallic particles usually have poorer conductivity than the surrounding melt, electro-vortex flow might possibly result from their melting at the interface between the positive electrode and electrolyte. In addition, another observation that should be noted during the study of flow in metal liquid batteries is the correlation between electro-vortex fluid and thermal convection driven. Joule-heated thermal convection and electro-vortex flow are rivals as well. At the top of the battery, where current diverges from the negative current collector, electro-vortex flow tends to produce a poloidal roll that is oriented inward at the top and descends along the central axis. In contrast, a poloidal roll that rises along the center axis—where current and heat are concentrated—is typically the result of Joule heating. As current rises, both opposing phenomena become more intense. Since the work done by electro-vortex flow grows proportionately with current and the rate of Joule heating grows as the square of the current, we would anticipate thermal effects to predominate in the limiting situation of very high current. However, since this is a very complicated phenomenon that depends on a wide range of variables, including the volume of the complete battery cell, the melting temperature, and the spatial distribution of current, it must be carefully studied through a combination of modeling and experiments.

4.3.3 Interface Instabilities

In fact, the largest resistance of a battery cell comes from the electrolyte layer, which has a much lower conductivity than the electrode layers. Therefore, the resistance of the electrode/electrolyte interface is always large, and any displacement of the electrolyte layer will affect the current distribution throughout the cell.

Interface instability is the term used to describe "metal pad roll instability" or "sloshing instability," which is the result of conducting fluids present when electrical currents are flowing through them [28–32]. Small interface deformations between the aluminum and cryolite cause this well-known interfacial instability in aluminum reduction cells [28, 29]. This causes a horizontal disturbance current in liquid aluminum. An undesirable consequence of the bus bars supplying the cell with current is a vertical magnetic background field. This field interacts with the horizontal current to produce Lorentz force. This force propels a revolving interface in a horizontal direction that might become unstable in some circumstances.

For example, one could create a vertical magnetic field B_z depicted in one LMB cylinder cell. The formerly uniform current density will disperse in order to reduce the overall resistance if one or both of the electrolyte/electrode interfaces are tilted from the horizontal position. Current flows across the electrolyte almost vertically because of its poor conductivity. Currents that are oriented horizontally (I_h) will mostly show up in the metal layers. These horizontal currents produce azimuthal Lorentz forces in conjunction with the vertical magnetic field B_z. With the aid of continuous cell wall redirections, the forces result in movements that eventually become spinning waves. A pure electrode layer makes up the bottom layer of a conventional LMB, where the spinning wave is produced. Lorentz forces are generated in the anode at the top and cathode at the bottom layers of LMBs, with opposite directions. However, due to the significantly smaller density differential and hence weaker restoring gravity force than the electrolyte/cathode interface, the anode/electrolyte contact will primarily create waves for common material combinations. In a typical wave form, the surface of the wave is colored blue for low vertical current density and red for high vertical current density. The location with the lowest distance h_{min} between the electrodes has the highest vertical current density. Here, the commencement of the wave and the break of the electrolyte layer may be roughly defined by the Seles criteria, which is represented by the value β [18]:

$$\beta = \frac{I.B_z}{h_0 h_A \left(\rho_E - \rho_A\right) g}, \tag{4.7}$$

where h_0 represents the electrolyte's starting height, h_A is the anode metal layer's height, and ρ_E and ρ_A the anode metal's and electrolyte's densities, respectively.

The minimal electrolyte height, h_{min}, normalized by its starting value, h_0, would be shown on the ordinate. Various curves correspond to modifications of various single parameters. A wave forms for β above but close to β_{cr}, but it eventually reaches a saturated state where h_{min}/h_0 reaches a fixed finite value. For $\beta_{cr} \approx 2$, electrolyte rupture only occurs later. Although β is still a good measure to represent electrolyte layer rupture, it is far from ideal.

As long as the magnetic background field and current are high enough, this interface instability can manifest in any cell and eventually lead to an LMB short-circuit. It is important to consider the occurrence of sloshing in actual cells with restricted current density if the diameter is more than a few centimeters, not to mention the ambition to build battery cells with diameters up to meters. Thus, one of the most significant instabilities in the functioning of LMBs is interface instability.

4.3.4 The Other Fluid Mechanics of LMBs

Since the fluid mechanics phenomena in liquid metal batteries are very diverse and complex, depending on many factors, in this chapter we briefly discuss other phenomena such as Marangoni flow, thermal convection, and magnetoconvection with the belief that it will interest the reader and encourage further study of these topics.

Marangoni flow is where a temperature gradient generates a surface tension gradient that can significantly propel fluid movement, and Tayler instability is where a vertical current's high magnetic pressure may disrupt the fluid and lead its flow to kink [33, 34]. In general, the surface tension variations brought on by concentration and temperature gradients cause Marangoni convection. Higher surface tension areas draw fluid along the interface from lower surface tension areas if the surface tension changes spatially. The "Marangoni flow" is the result of viscosity couples that move to the inner fluid. Due to its dependence on temperature, chemical composition, and other factors, surface tension can change across space.

Thermal convection is also thought to be one of the factors causing instability in an LMB cell. Spatial temperature gradients lead to density gradients because nearly all known fluids expand when heated. If the gradients are big enough in the presence of gravity, denser fluid sinks and lighter fluid floats, leading to thermal convection. The flow form is typically characterized by large-scale convection rolls [35–37]. In addition to thermal convection, there is also magnetoconvection, which is when convection is affected by both electric current and magnetic field. However, this is a large and complex topic, so it should be presented in a separate and more in-depth manner. Liquid metal batteries consisting of three different layers are actually much more complicated to study in terms of thermal convection and magnetic convection than those with a single layer of liquid metal. Here, when the battery is activated, it is necessary to heat the battery cell to the melting temperature of the liquid inside by means of heaters installed outside the battery case [38]. Thermal convection affects strongly the electrolyte, which undergoes significant heating during the discharge/charge process, particularly in its top half [34]. Moreover, thermal convection may introduce intermetallic elements from the electrolyte to contaminate the negative electrode, according to experiments that have caused concern [39].

4.4 CONCLUSION

In the future, liquid metal batteries show great promise as a cost-effective, large-scale energy storage solution for energy systems. One of the attractive features of the idea is scalability at the cell level, even if current battery cell sizes are still small. Larger cells will exhibit a multiplicity of flow phenomena, all of which need to be carefully comprehended in order to take advantage of advantageous characteristics such as mixing and to avoid potentially hazardous ones such as electrolyte layer ruptures.

Due to space constraints, this chapter can only present some of the most general knowledge within the limits with the hope that readers have a brief, concise, and easy-to-understand view of the issue raised. With the spirit of seeking knowledge, we hope that readers can refer to more useful documents, from the major to the in-depth, on liquid metal batteries and their fluid mechanisms.

4.5 ACKNOWLEDGMENTS

The author is grateful to Professor Ram K. Gupta, Pittsburg State University, for the invitation. The author would like to express his appreciation to the Institute of Chemistry, Vietnam Academy of Science and Technology, for providing time and support during the research process.

REFERENCES

[1] D. Agarwal, R. Potnuru, C. Kaushik, V. R. Darla, K. Kulkarni, A. Garg, R. K. Gupta, N. Tiwari, K. S. Nalwa, Recent advances in the modeling of fundamental processes in liquid metal batteries, *Renewable and Sustainable Energy Reviews* 158 (2022) 112167.

[2] Y. Tian, S. Zhang, The renaissance of liquid metal batteries, *Matter* 3(6) (2020) 1824–1826.

[3] S. Wu, X. Zhang, R. Wang, T. Li, Progress and perspectives of liquid metal batteries, *Energy Storage Materials* 57 (2023) 205–227.

[4] F. Li, L. Q. Chi, W. X. Bo, L. Xuan, J. W. Tao, Overview on the liquid metal battery for grid-level large-scale energy storage, *Key Engineering Materials* 723 (2016) 572–578.
[5] O. E. G. Brizuela, C. Duczek, N. Weber, W. Nash, M. Sarma, K. E. Einarsrud, A continuous multiphase model for liquid metal batteries, *Journal of Energy Storage* 73 (2023) 109147.
[6] Y. Ding, X. Guo, G. Yu, Next-generation liquid metal batteries based on the chemistry of fusible alloys, *ACS Central Science* 6(8) (2020) 1355–1366.
[7] H. Li, H. Yin, K. Wang, S. Cheng, K. Jiang, D. R. Sadoway, Liquid metal electrodes for energy storage batteries, *Advanced Energy Materials* 6(14) (2016) 1600483.
[8] D. J. Bradwell, H. Kim, A. H. C. Sirk, D. R. Sadoway, Magnesium–antimony liquid metal battery for stationary energy storage, *Journal of the American Chemical Society* 134(4) (2012) 1895–1897.
[9] K. Wang, K. Jiang, B. Chung, T. Ouchi, P. J. Burke, D. A. Boysen, D. J. Bradwell, H. Kim, U. Muecke, D. R. Sadoway, Lithium–antimony–lead liquid metal battery for grid-level energy storage, *Nature* 514 (2014) 348–350.
[10] R. F. Ashour, D. H. Kelley, A. Salas, M. Starace, N. Weber, T. Weier, Competing forces in liquid metal electrodes and batteries, *Journal of Power Sources* 378 (2018) 301–310.
[11] J. Xu, A. M. Martinez, K. S. Osen, O. S. Kjos, O. E. Kongstein, G. M. Haarberg, Electrode behaviors of Na-Zn liquid metal battery, *Journal of The Electrochemical Society* 164(12) (2017) A2335–A2340.
[12] Y. Ding, X. Guo, Y. Qian, L. Xue, A. Dolocan, G. Yu, Room-temperature all-liquid-metal batteries based on fusible alloys with regulated interfacial chemistry and wetting, *Advanced Materials* 32 (2020) 2002577.
[13] H. Li, K. Wang, S. Cheng, K. Jiang, High performance liquid metal battery with environmentally friendly antimony-tin positive electrode, *ACS Applied Materials Interfaces* 8(20) (2016) 12830–12835.
[14] H. Kim, D. A. Boysen, J. M. Newhouse, B. L. Spatocco, B. Chung, P. J. Burke, D. J. Bradwell, K. Jiang, A. A. Tomaszowska, K. Wang, W. Wei, L. A. Ortiz, S. A. Barriga, S. M. Poizeau, D. R. Sadoway, Liquid metal batteries: Past, present, and future, *Chemical Reviews* 113(3) (2013) 2075–2099.
[15] A. J. Bard, R. Parsons, J. Jordan, *Standard Potentials in Aqueous Solution*, Marcel Dekker: New York, 1985.
[16] H. Liu, X. Zhang, S. He, D. He, Y. Shang, H. Yu, Molten salts for rechargeable batteries, *Materials Today* 60 (2022) 128–157.
[17] S. Bénard. *Flows and mixing in liquid metal batteries.* Fluid mechanics [physics.class-ph], Université Paris-Saclay, 2023. English. NNT: 2023UPAST139. tel-04409555v2.
[18] T. Weier, A. Bund, W. E. Mofid, G. M. Horstmann, C. C. Lalau, S. Landgraf, M. Nimtz, M. Starace, F. Stefani, N. Weber, Liquid metal batteries—materials selection and fluid dynamics, *IOP Conference Series: Materials Science and Engineering* 228 (2017) 012013.
[19] D. H. Kelley, T. Weier, Fluid mechanics of liquid metal batteries, *Applied Mechanics Reviews* 70(2) (2018) 020801.
[20] N. Weber, V. Galindo, J. Priede, F. Stefani, T. Weier, The influence of current collectors on Tayler instability and electro-vortex flows in liquid metal batteries, *Physics of Fluids* 27 (2015) 014103.
[21] F. Stefani, V. Galindo, C. Kasprzyk, S. Landgraf, M. Seilmayer, M. Starace, N. Weber, T. Weier, Magnetohydrodynamic effects in liquid metal batteries, *IOP Conference Series: Materials Science and Engineering* 143 (2016) 012024.
[22] M. Seilmayer, F. Stefani, T. Gundrum, T. Weier, G. Gerbeth, M. Gellert, G. Rüdiger, Experimental evidence for a transient Tayler instability in a cylindrical liquid-metal column, *Physical Review Letters* 108 (2012) 244501.

[23] Y. V. Vandakurov, Theory for the stability of a star with a toroidal magnetic field, *Soviet Astronomy* 16 (1972) 265.

[24] F. Stefani, T. Weier, T. Gundrum, G. Gerbeth, How to circumvent the size limitation of liquid metal batteries due to the Tayler instability, *Energy Conversion and Management* 52(8–9) (2011) 2982–2986.

[25] R. J. Tayler, The adiabatic stability of stars containing magnetic fields-I. Toroidal fields, *Monthly Notices of the Royal Astronomical Society* 161 (1973) 365.

[26] U. Shumlak, C. W. Hartman, Sheared flow stabilization of the $m = 1$ kink mode in Z pinches, *Physical Review Letters* 75(18) (1995) 3285–3288.

[27] W. Herreman, C. Nore, L. Cappanera, J. L. Guermond, Tayler instability in liquid metal columns and liquid metal batteries, *Journal of Fluid Mechanics* 771 (2015) 79–114.

[28] T. Sele, Instabilities of the metal surface in electrolytic alumina reduction cells, *Metallurgical Transactions B* 8 (1977) 613–618.

[29] P. A. Davidson, Overview overcoming instabilities in aluminium reduction cells: A route to cheaper aluminium, *Materials Science and Technology* 16(5) (2000) 475–479.

[30] S. Molokov, G. El, A. Lukyanov, Classification of instability modes in a model of aluminium reduction cells with a uniform magnetic field, *Theoretical and Computational Fluid Dynamics* 25 (2011) 261–279.

[31] S. Molokov, The nature of interfacial instabilities in liquid metal batteries in a vertical magnetic field, *Europhysics Letters* 121 (2018) 44001.

[32] N. Weber, P. Beckstein, W. Herreman, G. M. Horstmann, C. Nore, F. Stefani, T. Weier, Sloshing instability and electrolyte layer rupture in liquid metal batteries, *Physics of Fluids* 29 (2017) 054101.

[33] D. Camel, P. Tison, J. J. Favier, Marangoni flow regimes in liquid metals, *Acta Astronautica* 13 (1986) 723–726.

[34] T. Köllner, T. Boeck, J. Schumacher, Thermal Rayleigh-Marangoni convection in a three-layer liquid-metal-battery model, *Physical Review E* 95 (2017) 053114.

[35] F. Chilla, J. Schumacher, New perspectives in turbulent Rayleigh-Bénard convection, *The European Physical Journal E* 35 (2018) 58.

[36] G. Ahlers, Turbulent convection, *Physics* 2 (2009) 74.

[37] E. Bodenschatz, W. Pesch, G. Ahlers, Recent developments in Rayleigh-Bénard convection, *Annual Review of Fluid Mechanics* 32(1) (2000) 709–778.

[38] J. S. Cheng, I. Mohammad, B. Wang, J. M. Forer, D. H. Kelley, Laboratory model of electrovortex flow with thermal gradients for liquid metal batteries. *Experiments in Fluids* 63(178) (2022).

[39] M. S. Foster, Laboratory studies of intermetallic cells, *Regenerative EMF Cells* (Advances in Chemistry), Vol. 64, C. E. Crouthamel and H. L. Recht, eds., American Chemical Society, Washington, DC (1967) 136–148.

5 Li-Based Liquid Metal Batteries

Md. Hafizul Islam, Saiful Islam, and Mohammad Harun-Ur-Rashid

5.1 INTRODUCTION

Energy storage technologies play a pivotal role in addressing the growing global demand for electricity, particularly in the context of renewable energy integration. With the increasing shift toward solar, wind, and other intermittent renewable sources, reliable and efficient energy storage has become essential for stabilizing the grid, ensuring a consistent energy supply, and minimizing energy waste. Traditional energy storage technologies, such as lead-acid and flow batteries, have made significant advancements; however, challenges such as high costs, limited cycle life, and performance degradation persist, especially when scaled for grid-level applications [1]. New energy storage solutions are needed to meet the demands of long-duration storage, rapid response to grid fluctuations, and cost-efficiency in large-scale deployment. The concept of liquid metal batteries (LMBs) traces back to the 1960s, initially explored for their potential in grid-scale energy storage and renewable energy support. Early research focused on the use of molten metals such as magnesium and antimony, but it was in the early twenty-first century that the concept evolved significantly, particularly with the development of Li-based systems. The introduction of lithium as a liquid electrode material, combined with molten salts, drastically improved energy density and electrochemical performance. The history of the LMB is one of a cutting-edge technology that was first imagined in a different political and economic environment with the goal of enabling flexibility in addressing the limitations of a civilization that was about to enter the nuclear era and electrifying daily life [2]. Researchers at institutions like MIT have been at the forefront of these advancements, pushing the boundaries of LMB design to achieve higher efficiencies, greater scalability, and better adaptability to modern energy infrastructure. However, the success of Li-based LMBs (Li-LMBs) lies in their simplicity and robustness, with features such as fast ionic transport, self-healing characteristics, high energy densities, high Coulombic efficiencies, low self-discharge, and ability to operate at high temperatures with minimal degradation, marking a significant milestone in the pursuit of efficient large-scale energy storage [3, 4]. These batteries operate at elevated temperatures, maintaining the electrode materials in a liquid state, which avoids issues such as electrode cracking and dendrite formation—common failure modes in solid-state batteries. LMBs, and particularly Li-based variants, promise a revolutionary pathway to overcoming many of the

DOI: 10.1201/9781003489726-5

limitations of traditional battery technologies, offering enhanced performance in large-scale applications [5].

This chapter aims to explore the technological foundations, operational mechanisms, and potential applications of Li-LMBs. It will provide a comprehensive overview of the fundamental principles underlying their design, the materials and processes that enable their unique performance characteristics, and the challenges and opportunities associated with their commercialization. Emphasis will be placed on the role of Li-LMBs in grid-scale energy storage and industrial applications, and their broader impact on the global transition toward renewable energy. Through this exploration, the chapter seeks to offer insights into how Li-LMBs can address the challenges facing current energy storage technologies, positioning them as a key player in the future of sustainable energy systems.

5.2 FUNDAMENTALS OF LI-LMBs

5.2.1 Basic Principles of Operation

Li-LMBs operate on a fundamental electrochemical principle wherein liquid metal electrodes are separated by a molten salt electrolyte. The battery consists of three key layers: a liquid metal anode, a molten salt electrolyte, and a liquid metal cathode. These components are kept at high temperatures, ensuring that both the anode and the cathode remain in a liquid state, and ions can move freely through the molten salt. Upon discharge, lithium atoms from the anode are ionized, releasing electrons to an external circuit, while the lithium ions travel through the electrolyte to the cathode, where they recombine with electrons. The reverse process occurs during charging. The key advantage of Li-LMBs lies in the liquid-liquid interface, which allows for highly efficient and self-healing operations over extended cycles, without the mechanical stresses that solid electrodes endure [6].

5.2.2 Principle of Screening Liquid Metal Electrodes

The periodic chart shown in Figure 5.1a highlights the potential liquid metal electrode choices, which are constrained by the three specifications listed below:

- liquid at suitable temperatures, that is, the boiling point should be greater than 25 °C and the melting point should be less than 1000 °C at normal pressure;
- electrically conductive, having an ionic conductivity (σ >1 S cm^{-1}) that is at least higher than that of a common molten salt electrolyte;
- stable and naturally occurring isotope, meaning it is not radioactive.

The deposition potential of the electrode material from an aqueous solution is used to first classify candidate electrode materials into positive or negative electrodes. (Figure 5.1c). Negative electrodes (A metals) are electrode materials with a deposition potential more than −2.0 V, while positive electrodes (B) are electrode materials with a potential greater than −1.0 V. An alternative method of classifying

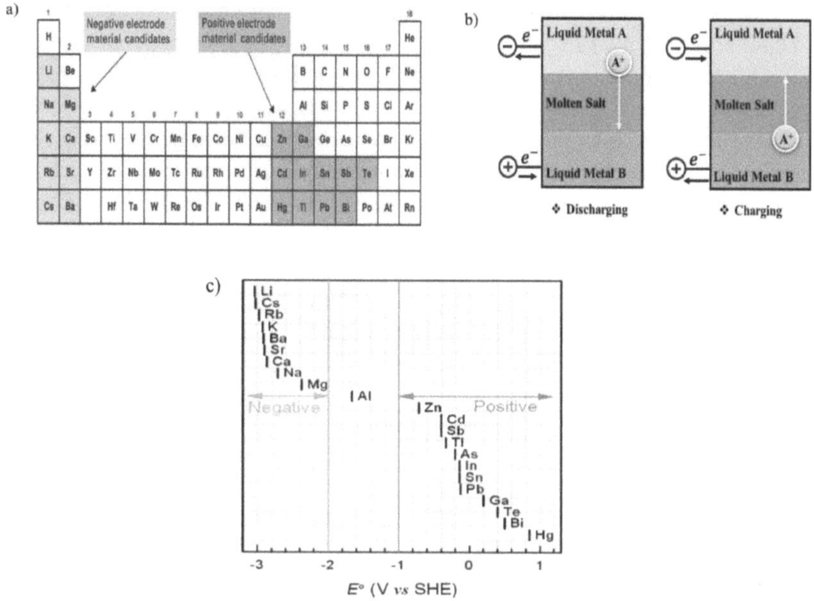

FIGURE 5.1 (a) Potential electrode materials for LMBs: negative (orange) and positive (green), (b) diagram showing the charge and discharge states of a LMB, (c) the proposed negative (orange) and positive (green) LMB electrode materials are chosen based on their deposition potentials in comparison to the typical hydrogen electrode in aqueous electrolytes.

elements is to consider their electronegativity: more electronegative metals, semimetals, or chalcogen (e.g., Bi, Te, Ga, Pb, Sn, Sb, S) would be candidates for the positive electrode, and more electropositive metals (e.g., Li, K, Na, Ca, Mg) would be among the candidates for the negative electrode. The fundamentals of a typical bimetallic three-liquid-layer Li-LMB are as follows: when the electrodes are discharged, the cations are conducted across the molten salt electrolyte (interlayer) to the positive electrode (bottom layer) as electrons are released to an external circuit, causing the negative electrode layer to decrease in thickness as metal A (top layer) is electrochemically oxidized ($A \rightarrow A^{z+} + ze^-$). The liquid A-B alloy ($A^{z+} + ze^- \rightarrow$ A (in B)) is formed by the electrochemical reduction of the cations, which thickens the positive electrode layer concurrently. The LMB cell's thermodynamic driving force, or cell voltage, is derived from the strong interaction between the two electrodes. When charging, the same procedure is reversed. Within the broad meaning of LMB, A and B can be both metal and alloy in a liquid state, or A can be a metal or alloy in a liquid state while B can be a nonmetal substance in a solid states [5].

5.2.3 Key Components and Their Roles

The design and performance of Li-LMBs rely on the careful selection of materials for each component. These materials enable efficient electrochemical reactions while ensuring stability, conductivity, and durability at elevated temperatures. It consists of

TABLE 5.1
Component Materials Used in Various Li-ion Based Batteries [10]

Anode	Cathode	Electrolyte
Lithium metal	Li-Mn-O	$LiPF_6$
Carbon (Hard)	FeS_2	$LiAsF_6$
Carbon (graphite)	$LiCoO_2$	$LiClO_4$
Synthetic Graphite	$LiNiMnCoO_2$	Polyethylene Oxide
Titanate (Lithium)	V_2O_5	$LiCF_3SO_3$
Silicon-based materials	Electronic Conducting Polymers	–
Tin-based alloys	$LiFePO_4$	–

positive and negative current collectors, electrolyte, separator, anode, and cathode. The components listed in Table 5.1 are used in different Li-ion batteries. The lithium ions travel from the cathode to the anode. Through a separator, positively charged lithium ions migrate from the anode to the cathode and vice versa. Free electrons are produced by the movement of the lithium ions. At the positive current collector, charge is created by the movement of free electrons. The direction of the flow of electric current is from the positive current collector to the negative current collector. Separators impede the passage of electrons within batteries.

5.2.3.1 Anode: Lithium as a Liquid Metal

The anode in Li-LMBs is composed of liquid lithium. Lithium's low atomic weight and high electrochemical potential make it an ideal candidate for the anode material. In its liquid state, lithium offers high mobility and an efficient charge transfer mechanism, contributing to the overall high energy density of the battery. The liquid nature of the anode also helps prevent issues such as dendrite formation, a common problem in solid lithium-ion batteries, which can lead to short circuits and battery failure. Moreover, the liquid state of lithium at the operating temperature (~400–500 °C) allows for superior charge transfer and ion mobility, contributing to the overall high efficiency of the battery [7]. In some configurations, lithium alloys such as lithium-tin (Li-Sn) or lithium-lead (Li-Pb) have been explored as alternative anode materials. These alloys can provide enhanced stability and reduce lithium loss through volatilization during prolonged cycling at high temperatures. The use of alloys also helps to tailor the melting point and mechanical properties of the anode, offering flexibility in battery design for specific applications.

5.2.3.2 Cathode: Choice of Metal

The cathode material in Li-LMBs typically consists of a denser, liquid metal such as antimony (Sb), lead (Pb), or bismuth (Bi). The choice of cathode material is influenced by factors such as its electrochemical potential, density, and melting point. For instance, antimony offers high capacity and favorable thermodynamics when paired with lithium. The cathode acts as the accepting material for lithium ions during discharge, forming a liquid alloy with the incoming lithium ions, and releasing them during charging. The liquid nature of the cathode ensures good surface contact and high ionic mobility, contributing to the battery's efficiency and longevity [5].

Other metals, such as tin (Sn) and zinc (Zn), have been investigated to optimize specific properties such as energy density, cycle life, and voltage stability. The performance of the cathode depends not only on the material's ability to form reversible alloys with lithium but also on its compatibility with the electrolyte and its durability under prolonged cycling.

5.2.3.3 Electrolyte: Ionic Conductivity and Stability

The electrolyte in Li-LMBs is a molten salt that separates the liquid anode and cathode, facilitating the transport of lithium ions between them while preventing electronic conductivity. The electrolyte must possess high ionic conductivity, thermal stability, and chemical inertness. Commonly used molten salts include mixtures of lithium chloride (LiCl) and potassium chloride (KCl) or lithium fluoride (LiF) and sodium fluoride (NaF) [8]. These salts have low melting points and high ionic conductivity, allowing for efficient lithium-ion transport at the operating temperature of the battery. Additionally, molten salts are chemically stable and do not degrade under the harsh conditions inside the battery. One of the main advantages of using molten salts as electrolytes is their ability to create a self-regulating interface between the liquid anode and cathode. This natural separation prevents short circuits while still allowing for the free movement of ions. The properties of the molten salts, such as viscosity and density, can be adjusted by varying the composition of the electrolyte to optimize performance for specific applications.

Solid electrolytes represent an alternative approach to enhancing the safety and stability of Li-LMBs, particularly in environments where liquid electrolytes may be less desirable. Solid electrolytes, such as ceramic or glass-based materials, provide high ionic conductivity while acting as a physical barrier between the liquid anode and cathode. This prevents direct contact and potential short-circuiting, even if there are disturbances within the battery. While solid electrolytes can offer improved safety and thermal stability, they introduce additional challenges. Maintaining adequate ionic conductivity through a solid electrolyte at elevated temperatures can be difficult, and the manufacturing process for solid-state LMBs is more complex.

5.2.3.4 Separator: Role in Preventing Short Circuits

While traditional solid-state batteries utilize porous polymer separators, Li-LMBs rely on the intrinsic separation provided by the molten salt electrolyte. However, the electrolyte must act as a barrier that prevents the physical mixing of the liquid metal anode and cathode while still allowing for the free flow of ions. This separation is critical to maintaining the integrity of the battery and preventing catastrophic failure through short circuits. The stability and immiscibility of the molten salt with the liquid metals are crucial to achieving this separation without requiring additional mechanical barriers [9].

5.3 ELECTROCHEMICAL REACTIONS

The operation of Li-LMBs is governed by electrochemical reactions that occur during charge and discharge cycles. These reactions involve the transfer of lithium ions between the anode and the cathode through the molten salt electrolyte, while electrons travel through an external circuit to generate electrical energy.

5.3.1 Charge and Discharge Mechanisms

During discharge, lithium at the anode undergoes oxidation, releasing lithium ions (Li^+) and electrons (e^-). The lithium ions travel through the electrolyte and react with the cathode material (e.g., Sb), forming an intermetallic compound or alloy. Simultaneously, the electrons travel through the external circuit, providing electrical energy to the load. The reaction at the cathode is typically a reduction process where lithium ions are incorporated into the liquid cathode material.

At the anode: $Li \rightarrow Li^+ + e^-$;
At the cathode: $Li^+ + e^- + Sb \rightarrow Li_xSb$.

The reverse process occurs during charging. Electrical energy is supplied to the battery, forcing the lithium ions to migrate back to the anode and undergo reduction. The cathode undergoes oxidation, releasing lithium ions back into the electrolyte. This cycling process allows the battery to store and release electrical energy repeatedly with high efficiency and minimal degradation.

At the cathode: $Li_xSb \rightarrow Li^+ + Sb + e^-$;
At the anode: $Li^+ + e^- \rightarrow Li$.

5.3.2 Thermodynamics and Kinetics

The thermodynamics of Li-LMBs are governed by the free energy changes associated with the oxidation and reduction reactions at the anode and cathode. The efficiency of these reactions is influenced by the relative chemical potentials of the anode and cathode materials, as well as the stability of the electrolyte. The liquid state of both electrodes enhances the kinetics of the electrochemical reactions, allowing for faster ion transport and reducing the energy barriers associated with phase transitions.

The basic thermodynamics of the materials used for the positive and negative electrodes determine the theoretical voltage of every electrochemical cell. Around 100 different binary alloy electrode combinations are feasible for LMB systems, and each one has a different voltage discharge profile. By analyzing the thermodynamic characteristics of binary alloy systems, it is possible to determine which chemistries have higher cell voltages and, thus, higher cell efficiencies at faster charge-discharge rates. The electrochemical cell of a LMB may be expressed generally as,

$$A(l) \mid AX_z(l) \mid A\ (\text{in } B), \tag{5.1}$$

where A (e.g., Li, Na, K, Mg, Ca) is the negative electrode metal, B (e.g., Zn, Cd, Hg, Al, Ga) is the positive electrode metal, and AX_z is an alkali or alkaline-earth molten salt electrolyte. For this cell, the generic negative and positive half-cell reactions are

$$\begin{aligned} &\text{Negative } A(l) = A^{z+} + ze^-, \\ &\text{Positive } A^{z+} + ze^- = A \text{ (in B)}, \end{aligned} \tag{5.2}$$

and the overall cell reaction is

$$\text{cell } A(l) = A \text{ (in B)}. \tag{5.3}$$

The thermodynamic driving force is the change in partial molar Gibbs free energy,

$$\Delta\bar{G}_{cell} = \Delta\bar{G}_{A(in\ B)} - \Delta\bar{G}_{A(l)}, \tag{5.4}$$

where the partial molar Gibbs energy $\bar{G}_i$ for each component i is given by

$$\begin{aligned} \Delta\bar{G}_{A(in\ B)} &= G^{\circ}{}_{A(l)} + RT \ln a_{A(in\ B)}, \\ \Delta\bar{G}_{A(l)} &= G^{\circ}{}_{A(l)} + RT \ln a_{A(l)}\ (a_{A(l)} = 1), \end{aligned} \tag{5.5}$$

where a_i is the activity, $G^{\circ}{}_{A(l)}$ the standard chemical potential, R the gas constant, and T temperature. From the Nernst equation,

$$\Delta\bar{G}_{cell} = -\ zFE_{\ cell,eq}, \tag{5.6}$$

and equations 5.4 and 5.5, the cell equilibrium voltage is related to the change in partial molar Gibbs free energy as

$$E_{\ cell,eq} = -\Delta\bar{G}_{cell}/(zF) = -(Rt/(zF)) \ln a_{A(in\ B)}, \tag{5.7}$$

where F is the Faraday constant and z the number of electrons. Conceptually, the thermodynamic driving force for cell discharge can be interpreted as emanating from a strong interaction of metal A with metal B, in which the activity of A can be extremely low. This is manifest in the form of a high equilibrium cell voltage. Experimental measurements of enthalpies of reaction, electromotive force, vapor pressure, and chemical equilibrium have been made to determine the thermodynamic activities of most binary alloys as a function of mole fraction and temperature $[a_A = f(x_A,T)]$ [2].

5.4 DESIGN AND FABRICATION OF LI-LMBs

The design and fabrication of Li-LMBs are integral to their performance, longevity, and scalability. These batteries are composed of simple yet highly specialized components, engineered to maintain their liquid state during operation while minimizing degradation and energy losses.

5.4.1 Battery Architecture

The architecture of Li-LMBs revolves around a three-layer structure, consisting of two liquid metals (anode and cathode) and a molten salt electrolyte. This structure simplifies the construction of both single-cell and multi-cell designs [11].

5.4.1.1 Single-Cell Design

In a single-cell Li-LMB, the design consists of three key layers: a low-density liquid metal (often lithium) serving as the anode, a molten salt electrolyte in the middle, and a high-density liquid metal cathode (such as lead or bismuth) at the bottom. These layers naturally separate due to their differing densities, creating a self-segregating system. The anode is typically contained at the top of the cell, while the electrolyte occupies the middle, allowing for the selective transport of lithium ions during charge and discharge cycles. The dense cathode material forms the bottom layer. Single cells are housed in cylindrical or prismatic containers made from materials that can withstand the operating temperatures, typically between 400 °C and 700 °C. Efficient containment and temperature management are key to preserving the integrity of the layers, as thermal gradients could lead to mixing of the layers or degradation of materials. The single-cell design offers high reversibility, rapid response times, and simplicity of construction, which are advantageous for research and testing purposes [5].

5.4.1.2 Multi-Cell Design

The multi-cell configuration is designed to scale up the performance of LMBs for grid-scale applications. In multi-cell designs, individual cells are connected in series or parallel to meet desired voltage and capacity requirements. These designs typically involve multiple single cells stacked in a modular array within a shared housing. Each cell retains the three-layer architecture but is connected to a central bus bar that manages the flow of current between cells. To maintain the integrity of each layer and avoid cross-contamination, special sealing and insulation materials are used to isolate the cells thermally and electrically. The multi-cell configuration must also address the potential for uneven heating across the system, which could lead to performance issues such as hot spots or incomplete phase separation in the electrodes. Advanced thermal management systems and uniform heating methods, such as external heaters or the use of heat-retaining materials, are employed to mitigate these risks [12].

5.4.2 Manufacturing Processes

The manufacturing of Li-LMBs involves several steps, including material synthesis, purification, and cell assembly. These processes must meet stringent requirements to ensure that the materials maintain their liquid state under operating conditions without significant degradation.

5.4.2.1 Material Synthesis and Purification

High-purity materials are critical to the performance of LMBs. The liquid metals used in the anode and cathode are synthesized and purified to eliminate impurities that could interfere with the electrochemical reactions. For the lithium anode, purification methods such as electrorefining or distillation are employed to remove trace elements and ensure a high degree of purity. Similarly, the cathode material, whether it is lead, bismuth, or another metal, must be refined to reduce contaminants that could decrease its conductivity or cause undesirable side reactions [13, 14].

The molten salt electrolyte, often composed of lithium salts such as lithium chloride or lithium fluoride, is synthesized with precise control over composition. Impurities in the electrolyte can lead to corrosion of the container or reduce ionic conductivity. Synthesis processes typically involve the careful selection of precursor materials, followed by controlled melting, mixing, and casting under inert atmospheres to prevent oxidation or contamination [15].

5.4.2.2 Cell Assembly Techniques

The assembly of liquid metal cells begins with the creation of individual components such as the anode, cathode, and electrolyte, followed by the integration of these components into a single container. The container material must be thermally stable and chemically inert to the molten materials. Common choices include stainless steel or other high-temperature alloys lined with ceramic or glass to prevent corrosion.

During assembly, the electrolyte is typically inserted as a solid, followed by the introduction of the anode and cathode metals. The battery is then heated to its operating temperature, causing the materials to melt and self-segregate into their respective layers. Proper alignment of the layers is critical to ensure efficient electrochemical activity. Advanced cell assembly methods use automated machinery for precise material placement, sealing, and thermal treatment. Quality control measures during assembly are essential to ensure that no contaminants or structural flaws are introduced that could compromise battery performance [16].

5.4.3 Design Considerations

The design of Li-LMBs must account for several factors, including thermal management, structural integrity, and safety measures. These considerations are critical for maintaining stable operation, extending the battery's lifespan, and ensuring safe handling and usage.

5.4.3.1 Thermal Management

Since Li-LMBs operate at elevated temperatures (typically between 400 °C and 700 °C), effective thermal management is essential. The battery must be designed to maintain these temperatures without excessive energy input, as maintaining the liquid state of the materials is crucial for performance [17]. Heat retention is managed through insulation materials, such as ceramic or refractory linings, that surround the battery cells and prevent heat loss. External heaters may be used during startup to bring the battery to its operating temperature. Once the battery is in operation, heat generated from the internal electrochemical reactions often suffices to maintain the required temperature, reducing the need for external heating. However, cooling systems may also be necessary to prevent overheating, especially in multi-cell arrays where heat buildup can occur. Advanced thermal management designs include heat pipes, liquid cooling loops, or phase-change materials that can absorb excess heat and distribute it evenly throughout the battery [18].

5.4.3.2 Structural Integrity

The structural integrity of Li-LMBs is paramount. The container and internal components must resist corrosion, thermal cycling, and mechanical stress. Materials selection is therefore crucial; materials such as stainless steel, nickel alloys, or refractory metals are often used for the external housing, while ceramics or glass are used for internal linings to prevent corrosion from molten salts. Mechanical stability is another critical factor. The materials must withstand thermal expansion and contraction during startup, shutdown, and cycling without cracking. For large-scale batteries, the design must account for the weight and distribution of the molten materials to avoid mechanical failure [19].

5.4.3.3 Safety Measures

Safety is a key concern in the design of Li-LMBs, particularly because of the high operating temperatures and the use of reactive materials. Measures must be taken to prevent failures that could lead to leaks, fires, or explosions. Key safety features include robust sealing mechanisms to prevent the escape of molten materials and gases, pressure relief valves to avoid dangerous pressure buildup, and automated shutdown systems to detect and respond to overheating or other failures. In multi-cell systems, fail-safes are designed to isolate individual cells in the event of a malfunction, preventing a cascading failure that could compromise the entire battery. These designs may include thermal barriers between cells, redundant cooling systems, and active monitoring of temperature, pressure, and cell performance [20].

5.5 PERFORMANCE AND EFFICIENCY

Li-LMBs have garnered significant attention due to their exceptional performance and efficiency metrics, which are critical for a wide range of energy storage applications [21]. These batteries are characterized by their high energy density, power density, Coulombic efficiency, and long cycle life, making them highly suitable for both grid-scale storage and high-demand industrial applications. This section delves into the key performance metrics, factors affecting battery performance, and degradation mechanisms, providing a comprehensive overview of the efficiency and durability of Li-LMBs.

5.5.1 Key Performance Metrics

5.5.1.1 Energy Density

One of the most critical metrics for any battery technology is energy density, which refers to the amount of energy stored per unit volume or mass [22]. Li-LMBs exhibit a high energy density, often surpassing traditional battery technologies such as lead-acid and nickel-cadmium batteries. The high energy density of LMBs is primarily due to the inherent properties of lithium, which is one of the lightest and most electropositive elements, allowing for the storage of a significant amount of energy in a relatively small volume. The energy density of these batteries can vary based on the specific materials and design used, but generally, they offer the following ranges:

Gravimetric Energy Density: This refers to the amount of energy stored per unit mass, typically measured in watt-hours per kilogram (Wh/kg). Li-LMBs can achieve gravimetric energy densities ranging from 200 to 400 Wh/kg.

Volumetric Energy Density: This refers to the amount of energy stored per unit volume, typically measured in watt-hours per liter (Wh/L). These batteries can have volumetric energy densities in the range of 500 to 900 Wh/L.

5.5.1.2 Power Density

Power density is another crucial performance metric, indicating the rate at which energy can be delivered by the battery. Li-LMBs exhibit remarkable power densities, reaching up to 200 W/cm^2 [23], significantly surpassing traditional lithium-ion batteries, which typically range from 10 to 20 W/cm^2 [24]. This exceptional power density is due to the battery's unique design, featuring liquid electrodes that operate at high temperatures (400–700 °C), facilitating rapid ion transport and minimal internal resistance. The fast charge and discharge cycles, enabled by the liquid state of the electrodes, further enhance the power density, making these batteries highly efficient for applications requiring rapid energy delivery, such as grid storage and electric vehicles.

5.5.1.3 Coulombic Efficiency

Coulombic efficiency, or charge/discharge efficiency, measures the ratio of the charge extracted from a battery to the charge put into it during a cycle [25]. High Coulombic efficiency is indicative of minimal energy loss during operation, leading to greater overall efficiency of the battery system. Li-LMBs typically exhibit high Coulombic efficiency, often exceeding 90%. In some studies, these batteries have maintained over 97.1% Coulombic efficiency even after 1,000h of operation [26]. This high efficiency is crucial for applications where energy losses need to be minimized, such as in long-duration energy storage systems and renewable energy integration, where every unit of stored energy is valuable. Table 5.2 lists the performance metrics of Li-LMBs compared to other energy storage systems

5.5.1.4 Cycle Life

Cycle life refers to the number of complete charge-discharge cycles a battery can undergo before its capacity degrades to a specified level, usually 80% of its original capacity. Li-LMBs are designed for long cycle life, which is a significant advantage over other battery technologies that may degrade more rapidly with repeated cycling [27]. The long cycle life of LMBs makes them particularly well-suited for grid-scale energy storage, where batteries are expected to operate reliably over many years. This durability also reduces the need for frequent replacements, lowering the total cost of ownership and enhancing the sustainability of energy storage systems.

TABLE 5.2
Performance Metrics of Li-LMBs Compared to Other Energy Storage Systems

Performance metrics	Li-LMBs	Lithium-ion Batteries (Li-ion)	Lead-Acid Batteries	Sodium-Sulfur (NaS) Batteries
Energy Density (Wh/kg)	100–200	150–250	30–50	150–240
Power Density (W/kg)	500–1,000	200–2,000	100–300	150–230
Efficiency (%)	75–85	85–95	70–80	75–90
Cycle Life (cycles)	>10,000	500–10,000	300–1,500	2,500–4,500
Self-discharge Rate (%/month)	<1%	1–2%	3–20%	<1%
Operating Temperature (°C)	300–500	20–60	-20 to 50	300–350
Response Time	Fast (milliseconds)	Fast (milliseconds)	Moderate (seconds)	Fast (milliseconds)
Scalability	High (suitable for grid-scale)	Moderate	Low	High
Cost ($/kWh)	150–250 (projected for large scale)	200–600	100–200	200–400
Safety	High (inherent stability)	Moderate	Low	Moderate

5.5.2 Factors Affecting Battery Performance

5.5.2.1 Temperature

Temperature plays a crucial role in the performance of Li-LMBs. These batteries typically operate at elevated temperatures, where the liquid metal components remain in a molten state, allowing for efficient ion transport. However, maintaining an optimal operating temperature is critical; too low a temperature can lead to solidification of the metal components, while too high a temperature can cause thermal degradation of the battery materials [28]. Thermal management systems are therefore essential in ensuring that the battery operates within its optimal temperature range, maximizing both performance and lifespan.

5.5.2.2 Current Density

Current density, defined as the amount of current per unit area of the electrode, significantly impacts the performance of LMBs. High current densities can lead to faster charge and discharge cycles, which is beneficial for applications requiring rapid energy delivery. However, operating at excessively high current densities can also accelerate degradation processes, such as dendrite formation in lithium-based systems, which can lead to short circuits and reduced cycle life. Optimizing current density is therefore a key consideration in the design and operation of Li-LMBs.

5.5.2.3 Cycle Life

Several factors influence the cycle life of Li-LMBs, including the choice of materials, the quality of the electrolyte, and the design of the battery architecture. Material selection is critical, as the anode, cathode, and electrolyte must all be able to withstand repeated cycling without significant degradation. The electrolyte, in particular, must maintain high ionic conductivity while resisting decomposition over time. Additionally, the battery's architecture must be designed to minimize mechanical stress and prevent the formation of defects that could lead to capacity loss or failure.

5.5.2.4 Thermal Management

Thermal management is a significant factor affecting the performance and efficiency of Li-LMBs. Due to their operating principles, these batteries require a stable thermal environment to maintain the liquid state of the metal components. Effective thermal management systems are needed to dissipate heat generated during operation and to prevent the battery from overheating, which could lead to thermal runaway and potential safety hazards [29]. Advanced cooling systems, thermal insulation, and the use of heat exchangers are common strategies employed to manage the thermal environment within these batteries.

5.5.3 Degradation Mechanisms and Mitigation Strategies

5.5.3.1 Dendrite Formation

One of the primary degradation mechanisms in Li-LMBs is dendrite formation. Dendrites are needle-like structures that can form on the lithium anode during repeated charge-discharge cycles [30]. If these dendrites grow too long, they can pierce the separator and cause a short circuit, leading to battery failure. Various strategies are being explored to mitigate dendrite formation, including the use of protective coatings on the anode, modifying the electrolyte composition, and optimizing the current density during operation to minimize the conditions that promote dendrite growth.

5.5.3.2 Electrolyte Decomposition

Electrolyte decomposition in Li-LMBs is a significant concern, as it can lead to capacity fade and reduced battery life. This process occurs when the electrolyte, typically a salt dissolved in an organic solvent, reacts with the highly reactive lithium metal. Such reactions can form a solid electrolyte interphase (SEI) layer, which, while protective, may continuously grow, consuming electrolyte and active lithium [31]. In liquid metal systems, the elevated operating temperatures exacerbate these reactions, accelerating decomposition. Mitigating this issue requires electrolyte formulations that are thermally stable and chemically inert to prolong battery performance and enhance safety.

5.5.3.3 Mitigation Strategies

To extend the lifespan of Li-LMBs and maintain their performance, several mitigation strategies are being employed. These include the development of advanced materials with improved stability and durability, the use of coatings and additives to protect the anode and electrolyte, and the optimization of battery operating conditions to reduce the impact of degradation mechanisms [32]. Additionally, ongoing research into battery design and architecture aims to create systems that are more resilient to the stresses of repeated cycling and high-power operation.

5.6 APPLICATIONS OF LI-LMBs

Li-LMBs offer versatile solutions for diverse applications, including grid-scale energy storage and backup power systems, enhancing efficiency and reliability across renewable energy integration and high-demand industries.

5.6.1 Grid-Scale Energy Storage

Li-LMBs are vital for grid-scale energy storage, seamlessly integrating renewable energy sources. They stabilize power supply by storing excess energy during low demand, then releasing it during peak times, effectively balancing loads and shaving peaks. This enhances grid reliability, supports renewable integration, and optimizes energy distribution, contributing to a more sustainable energy infrastructure.

5.6.1.1 Integration with Renewable Energy Sources

Li-LMBs play a crucial role in integrating renewable energy sources into the power grid. Their high energy density and fast charge/discharge capabilities allow them to efficiently store and dispatch energy from intermittent sources such as solar and wind [33]. By capturing excess energy during peak production times and releasing it when generation is low, these batteries stabilize the grid and ensure a consistent energy supply. This capability supports a seamless transition to renewable energy, reducing dependency on fossil fuels and enhancing grid reliability. Their long cycle life further contributes to a sustainable energy infrastructure, optimizing the use of renewable resources.

5.6.1.2 Load Balancing and Peak Shaving

Li-LMBs are pivotal for load balancing and peak shaving in electrical grids [34]. By storing excess energy during off-peak periods, they alleviate grid stress during high-demand times. This capability enables utilities to manage fluctuating energy loads effectively, reducing the need for costly peaking power plants and stabilizing energy prices. Their rapid charge and discharge rates allow for quick response to load changes, ensuring a balanced and reliable power supply. This functionality helps optimize grid performance, supports efficient energy distribution, and enhances overall grid stability, making them a valuable asset in modern energy management strategies.

5.6.2 Industrial and Commercial Applications

Li-LMBs are crucial for industrial and commercial applications, offering reliable backup power for critical infrastructure and excelling in high-demand environments such as manufacturing and mining. Their durability and rapid response ensure uninterrupted operations during outages, while their ability to handle frequent deep cycles makes them ideal for sustaining energy-intensive processes, enhancing both efficiency and resilience.

5.6.2.1 Backup Power Systems

Li-LMBs are transforming backup power systems with their high reliability, rapid discharge capability, and long operational life. These batteries ensure seamless power delivery during outages, making them ideal for critical infrastructure, data centers, and industrial operations [35]. Their robust design allows for frequent cycling without significant degradation, providing consistent performance over extended periods. Additionally, their scalability and compact form factor make them suitable for various backup applications, from small commercial setups to large-scale industrial facilities. As energy demands grow, these batteries offer a dependable solution for maintaining uninterrupted power in critical situations.

5.6.2.2 High-Demand Industrial Applications

Li-LMBs are pivotal in high-demand industrial applications, offering unparalleled energy density, durability, and rapid response [2]. These batteries excel in industries requiring consistent, high-power output, such as manufacturing, mining, and transportation. Their ability to handle frequent deep discharge cycles without significant loss in capacity ensures reliable performance under strenuous conditions. Additionally, they support large-scale energy storage for industrial grids, stabilizing power supply and enhancing operational efficiency [36]. The scalability and long lifespan of these batteries make them ideal for integrating renewable energy sources into industrial operations, driving sustainability while meeting the rigorous demands of modern industries [37]. Table 5.3 shows the comparison of Li-LMBs in different applications.

TABLE 5.3
Comparison of Li-LMBs in Different Applications

Aspect	Grid Energy Storage	Electric Vehicles (EVs)	Portable Electronics	Space Applications
Typical Liquid Metals	Lithium (Li), Antimony (Sb)	Lithium (Li), Sodium (Na)	Lithium (Li), Sodium (Na)	Lithium (Li), Mercury (Hg)
Energy Density	High energy density, suited for large-scale storage	Moderate energy density, optimized for high-power output	Lower energy density, optimized for compactness	High energy density, critical for space missions
Power Density	Moderate to high, designed for stability over time	High, optimized for rapid acceleration and power delivery	High, with a focus on quick bursts of energy	High, with emphasis on reliability and power density

TABLE 5.3 *(Continued)*
Comparison of Li-LMBs in Different Applications

Aspect	Grid Energy Storage	Electric Vehicles (EVs)	Portable Electronics	Space Applications
Cycle Life	Long cycle life, suitable for frequent charging/discharging	Moderate cycle life, tailored for automotive use	Shorter cycle life, designed for portability	Long cycle life, crucial for space longevity
Temperature Tolerance	Broad temperature range, can handle fluctuations	Moderate tolerance, typically managed by vehicle systems	Narrower temperature range, influenced by device design	Extreme temperature tolerance, adapted for space conditions
Cost	Relatively high initial cost, but cost-effective over long term	Moderate cost, balancing performance and affordability	Higher cost per unit, impacted by miniaturization	Very high cost, due to specialized materials and design
Safety	Generally safe, but risks with high temperatures and reactivity	Moderate safety, with risks managed through engineering	Lower safety risks, due to smaller scale and design safeguards	High safety standards, with rigorous testing and controls
Scalability	Highly scalable for large grid installations	Scalable for vehicle use, with ongoing advancements	Limited scalability, more suited for small devices	Limited scalability, tailored for specific missions
Environmental Impact	Can be environmentally friendly with proper recycling	Considerable, with ongoing improvements in battery recycling	Moderate, with increasing focus on sustainable materials	High, with a focus on minimizing impact in space environments
Key Applications	Utility-scale energy storage, renewable energy integration	Consumer and commercial electric vehicles	Mobile devices, laptops, wearable tech	Satellites, space probes, space stations

5.7 FUTURE PROSPECTS AND CHALLENGES

5.7.1 Emerging Trends in Li-LMBs

Li-LMBs are at the forefront of energy storage innovation, driven by the increasing demand for efficient, scalable, and cost-effective solutions [38]. Emerging trends in this field focus on enhancing battery performance, reducing costs, and expanding applications. Innovations in materials science, such as the development of new alloys and more stable electrolytes, are paving the way for batteries with higher energy densities and longer lifespans. Additionally, advancements in manufacturing techniques are enabling the production of batteries with more complex architectures, which can optimize performance and reliability [39].

Another significant trend is the integration of these batteries into smart grids and renewable energy systems. As the world shifts toward renewable energy sources such as solar and wind, the need for efficient energy storage solutions becomes more critical. Li-LMBs, with their rapid response times and ability to handle frequent charge-discharge cycles, are well-suited for this role. Their scalability also makes them an attractive option for grid-scale applications, where they can help to stabilize the grid and manage fluctuations in energy supply and demand.

5.7.2 Technical Challenges

Li-LMBs present several technical challenges that must be addressed to enable their broader adoption and commercial viability. These challenges stem primarily from material stability and scalability, which are crucial for ensuring long-term operation, safety, and cost-effectiveness.

5.7.2.1 Material Stability

Material stability is one of the most critical challenges in the development of Li-LMBs. The highly reactive nature of lithium, particularly in its liquid state, makes it prone to unwanted side reactions with the electrolyte and other cell components. These reactions can lead to the formation of interfacial layers that impede ionic conductivity, ultimately reducing the battery's performance and lifespan. Furthermore, the choice of cathode materials also plays a significant role in stability [40]. Some materials may degrade over time, especially under high-temperature operating conditions, leading to a decrease in energy efficiency and overall battery performance. Addressing these stability issues requires the development of advanced materials that can withstand the harsh electrochemical environment within the battery.

5.7.2.2 Scalability and Cost Reduction

Scalability and cost reduction are pivotal for the commercialization of Li-LMBs [5]. The challenge lies in scaling up the production of these batteries from laboratory settings to industrial scales while maintaining high performance and reliability. One of the key issues is the manufacturing process, which must be optimized to reduce costs without compromising the quality of the battery components. This includes the synthesis of high-purity lithium and the selection of cost-effective cathode materials. Additionally, the design of the battery architecture must be such that it can be easily scaled while ensuring uniform temperature distribution and minimizing the risks of thermal runaway [41]. Innovations in manufacturing processes, material sourcing, and battery design are essential to overcoming these challenges and making Li-LMBs a feasible option for large-scale energy storage applications.

5.7.3 Environmental and Economic Considerations

The environmental and economic implications of Li-LMBs are significant and must be carefully considered as this technology advances. On the environmental front, the extraction and processing of lithium and other materials used in these batteries can have a substantial impact [42]. Sustainable mining practices and the development of recycling technologies are crucial to minimize the environmental footprint of these batteries. Additionally, the long lifespan of Li-LMBs means that they generate less waste compared to other energy storage technologies, which is a positive environmental aspect.

Economically, the adoption of Li-LMBs could have far-reaching implications [43]. As these batteries become more widespread, they have the potential to disrupt traditional energy markets by enabling more efficient and reliable energy storage solutions. This could lead to a shift in how energy is produced, stored, and consumed, with significant economic benefits for industries that adopt this technology. However, the initial costs of developing and deploying these batteries could be high, and careful economic analysis is required to ensure that the benefits outweigh the costs.

5.7.4 Research Directions and Opportunities

The future of Li-LMBs depends on continued research and development in several key areas. One of the most promising areas of research is the development of new materials that can improve the performance and stability of these batteries. This includes exploring alternative anode and electrolyte materials, as well as developing advanced coatings and additives that can enhance the durability and efficiency of the battery components. Another important area of research is in battery design and architecture. As mentioned earlier, advancements in manufacturing techniques are enabling the production of more complex battery architectures that can optimize performance. Continued research in this area could lead to the development of batteries that are more efficient, reliable, and scalable. Finally, there are significant opportunities for research into the integration of Li-LMBs into existing energy systems. This includes exploring how these batteries can be used to support renewable energy sources, stabilize the grid, and provide backup power in critical applications. Research into the economic and environmental implications of these batteries is also crucial, as it will help to guide the development of policies and regulations that can support the adoption of this technology.

5.8 CONCLUSION

Li-LMBs present a transformative approach to energy storage, offering distinct advantages such as high energy density, long lifespan, and intrinsic safety features. Their ability to operate at elevated temperatures with minimal degradation, coupled with the self-healing properties of liquid metal electrodes, positions them as a robust solution for large-scale applications, particularly in grid stabilization and renewable energy integration. The simplicity of their design, modular scalability, and rapid charge-discharge capabilities further enhance their potential for industrial and commercial deployment. However, to fully realize the potential of Li-LMBs, challenges such as improving material stability, reducing production costs, and enhancing scalability must be addressed. Continued advancements in electrolyte composition, thermal management systems, and manufacturing processes will be critical to overcoming these obstacles. Research aimed at optimizing material choices and refining cell architectures will play a pivotal role in enhancing performance, ensuring longevity, and minimizing environmental impact. Li-LMBs represent a promising step toward sustainable and efficient energy storage systems. Their successful commercialization and adoption will not only contribute to the global transition toward renewable energy but also provide a reliable solution to meet the growing demand for long-duration energy storage in both grid-scale and industrial applications.

REFERENCES

1. M. A. Rahman, J. H. Kim, and S. Hossain, Recent advances of energy storage technologies for grid: A comprehensive review. *Energy Storage*, 2022. **4**(6): p. e322.
2. H. Kim, D. A. Boysen, J. M. Newhouse, B. L. Spatocco, B. Chung, P. J. Burke, D. J. Bradwell, K. Jiang, A. A. Tomaszowska, and K. Wang, Liquid metal batteries: Past, present, and future. *Chemical Reviews*, 2013. **113**(3): pp. 2075–2099.
3. T. Kim, W. Song, D.-Y. Son, L. K. Ono, and Y. Qi, Lithium-ion batteries: Outlook on present, future, and hybridized technologies. *Journal of Materials Chemistry A*, 2019. **7**(7): pp. 2942–2964.
4. D. Agarwal, R. Potnuru, C. Kaushik, V. R. Darla, K. Kulkarni, A. Garg, R. K. Gupta, N. Tiwari, and K. S. Nalwa, Recent advances in the modeling of fundamental processes in liquid metal batteries. *Renewable and Sustainable Energy Reviews*, 2022. **158**: pp. 112167.
5. S. Wu, X. Zhang, R. Wang, and T. Li, Progress and perspectives of liquid metal batteries. *Energy Storage Materials*, 2023. **57**: pp. 205–227.
6. D.-H. Liu, Z. Bai, M. Li, A. Yu, D. Luo, W. Liu, L. Yang, J. Lu, K. Amine, and Z. Chen, Developing high safety Li-metal anodes for future high-energy Li-metal batteries: Strategies and perspectives. *Chemical Society Reviews*, 2020. **49**(15): pp. 5407–5445.
7. Y. Tang, L. Zhang, J. Chen, H. Sun, T. Yang, Q. Liu, Q. Huang, T. Zhu, and J. Huang, Electro-chemo-mechanics of lithium in solid state lithium metal batteries. *Energy & Environmental Science*, 2021. **14**(2): pp. 602–642.
8. R. Roper, M. Harkema, P. Sabharwall, C. Riddle, B. Chisholm, B. Day, and P. Marotta, Molten salt for advanced energy applications: A review. *Annals of Nuclear Energy*, 2022. **169**: p. 108924.
9. J. Xing, S. Bliznakov, L. Bonville, M. Oljaca, and R. Maric, A review of nonaqueous electrolytes, binders, and separators for lithium-ion batteries. *Electrochemical Energy Reviews*, 2022. **5**(4): p. 14.
10. A. Daya, and S. P. Sathiyan. Review on Li-ion based battery chemistry: Challenges and opportunities. in *IOP Conference Series: Materials Science and Engineering*. 2022. IOP Publishing.
11. S. Fazal, F. Ahmad, M. I. Khan, A. Shanableh, and S. Manzoor, A comprehensive review on the challenges associated with lithium-ion batteries and their possible solutions. *New Journal of Chemistry*, 2024. **48**: pp. 14104–14127.
12. A. R. Itagi, R. Kallimani, K. Pai, S. Iyer, and O. L. Lopez, Cell balancing for the transportation sector: Techniques, challenges, and future research directions. *arXiv preprint arXiv:2404.13890*, 2024. https://www.researchgate.net/publication/380003941_Cell_Balancing_for_the_Transportation_Sector_Techniques_Challenges_and_Future_Research_Directions?channel=doi&linkId=662740f8f7d3fc2874770c01&showFulltext=true
13. X. Zhang, A. Han, and Y. Yang, Review on the production of high-purity lithium metal. *Journal of Materials Chemistry A*, 2020. **8**(43): pp. 22455–22466.
14. F. Sadeghfar, and M. Ghaedi, Photocatalytic treatment of pollutants in aqueous media. in *Interface Science and Technology*. 2021, Elsevier. pp. 725–759.

15. K. Sridharan, and T. Allen, Corrosion in molten salts. in *Molten Salts Chemistry*. 2013, Elsevier. pp. 241–267.
16. B. Acebedo, M. C. Morant-Miñana, E. Gonzalo, I. Ruiz de Larramendi, A. Villaverde, J. Rikarte, and L. Fallarino, Current status and future perspective on lithium metal anode production methods. *Advanced Energy Materials*, 2023. **13**(13): pp. 2203744.
17. X. Zhou, S. Yan, X. He, H. Zhou, J. Ning, H. Li, K. Wang, and K. Jiang, Low-temperature, high cycling stability, and high Coulombic efficiency liquid metal batteries enabled by lithium halide-potassium halide molten salt electrolytes. *Energy Storage Materials*, 2023. **61**: p. 102889.
18. R. Singh, K. Shrivastava, T. Ichikawa, and A. Jain, Liquid-metal batteries for next generation, in *Handbook of Energy Materials*. 2022, Springer. pp. 1–22.
19. J. Wu, X. Chen, W. Fan, X. Li, Y.-W. Mai, and Y. Chen, Rationally designed alloy phases for highly reversible alkali metal batteries. *Energy Storage Materials*, 2022. **48**: pp. 223–243.
20. N. Khan, C. A. Ooi, A. Alturki, M. Amir, and T. Alharbi, A critical review of battery cell balancing techniques, optimal design, converter topologies, and performance evaluation for optimizing storage system in electric vehicles. *Energy Reports*, 2024. **11**: pp. 4999–5032.
21. Y. Fan, X. Chen, D. Legut, and Q. Zhang, Modeling and theoretical design of next-generation lithium metal batteries. *Energy Storage Materials*, 2019. **16**: pp. 169–193.
22. R. Borah, F. R. Hughson, J. Johnston, and T. Nann, On battery materials and methods. *Materials Today Advances*, 2020. 6: p. 100046.
23. Y. Bar-Cohen, *High Temperature Materials and Mechanisms*. Vol. 44. 2014, CRC Press Boca Raton, FL, USA.
24. J. Rabaey, *Low Power Design Essentials*. 2009, Springer Science & Business Media.
25. B. D. Adams, J. Zheng, X. Ren, W. Xu, and J. G. Zhang, Accurate determination of coulombic efficiency for lithium metal anodes and lithium metal batteries. *Advanced Energy Materials*, 2018. **8**(7): p. 1702097.
26. W. Zhang, Y. Guo, T. Yang, Y. Wang, X. Kong, X. Liao, and Y. Zhao, High voltage and robust lithium metal battery enabled by highly-fluorinated interphases. *Energy Storage Materials*, 2022. **51**: pp. 317–326.
27. N. Nitta, F. Wu, J. T. Lee, and G. Yushin, Li-ion battery materials: Present and future. *Materials Today*, 2015. **18**(5): pp. 252–264.
28. A. Gharehghani, M. Rabiei, S. Mehranfar, S. Saeedipour, A. Mahmoudzadeh Andwari, A. García, and C. M. Reche, Progress in battery thermal management systems technologies for electric vehicles. *Renewable and Sustainable Energy Reviews*, 2024. **202**: p. 114654.
29. Z. Rao, P. Lyu, P. Du, D. He, Y. Huo, and C. Liu, Thermal safety and thermal management of batteries. *Battery Energy*, 2022. **1**(3): p. 20210019.
30. J. Steiger, *Mechanisms of dendrite growth in lithium metal batteries*. 2015. https://d-nb.info/106673691X/34
31. H. Adenusi, G. A. Chass, S. Passerini, K. V. Tian, and G. Chen, Lithium batteries and the solid electrolyte interphase (SEI)—progress and outlook. *Advanced Energy Materials*, 2023. **13**(10): p. 2203307.
32. M. M. Kabir and D. E. Demirocak, Degradation mechanisms in Li-ion batteries: A state-of-the-art review. *International Journal of Energy Research*, 2017. **41**(14): pp. 1963–1986.
33. I. Scott and S.-H. Lee, Battery energy storage, in *Large Energy Storage Systems Handbook*. 2011, CRC Press and Taylor and Francis Group. pp. 153–178.

34. X. Fan, B. Liu, J. Liu, J. Ding, X. Han, Y. Deng, X. Lv, Y. Xie, B. Chen, and W. Hu, Battery technologies for grid-level large-scale electrical energy storage. *Transactions of Tianjin University*, 2020. **26**: pp. 92–103.
35. R. D. Trevizan, J. Obert, V. De Angelis, T. A. Nguyen, V. S. Rao, and B. R. Chalamala, Cyberphysical security of grid battery energy storage systems. *IEEE Access*, 2022. **10**: pp. 59675–59722.
36. J. Arteaga, H. Zareipour, and V. Thangadurai, Overview of lithium-ion grid-scale energy storage systems. *Current Sustainable/Renewable Energy Reports*, 2017. **4**: pp. 197–208.
37. J. T. Frith, M. J. Lacey, and U. Ulissi, A non-academic perspective on the future of lithium-based batteries. *Nature Communications*, 2023. 14(1): p. 420.
38. M. D. Ahmed and K. M. Maraz, Revolutionizing energy storage: Overcoming challenges and unleashing the potential of next generation Lithium-ion battery technology. *Materials Engineering Research*, 2023. **5**(1): pp. 265–278.
39. E. Ayerbe, M. Berecibar, S. Clark, A. A. Franco, and J. Ruhland, Digitalization of battery manufacturing: Current status, challenges, and opportunities. *Advanced Energy Materials*, 2022. **12**(17): p. 2102696.
40. J. Biemolt, P. Jungbacker, T. van Teijlingen, N. Yan, and G. Rothenberg, Beyond lithium-based batteries. *Materials*, 2020. 13(2): p. 425.
41. B. R. D. Goswami, Y. Abdisobbouhi, H. Du, F. Mashayek, T. A. Kingston, and V. Yurkiv, Advancing battery safety: Integrating multiphysics and machine learning for thermal runaway prediction in lithium-ion battery module. *Journal of Power Sources*, 2024. **614**: p. 235015.
42. V. Flexer, C. F. Baspineiro, and C. I. Galli, Lithium recovery from brines: A vital raw material for green energies with a potential environmental impact in its mining and processing. *Science of the Total Environment*, 2018. **639**: pp. 1188–1204.
43. A. Machín, M. C. Cotto, F. Díaz, J. Duconge, C. Morant, and F. Márquez, Environmental aspects and recycling of solid-state batteries: A comprehensive review. *Batteries*, 2024. **10**(7): p. 255.

6 Na-Based Liquid Metal Batteries

Montree Sawangphruk

6.1 INTRODUCTION

6.1.1 Overview of Energy Storage Technologies

Energy storage technologies play a pivotal role in modern energy systems, offering solutions to balance supply and demand, improve energy efficiency, and enhance the reliability and resilience of the power grid [1, 2]. As the world shifts toward renewable energy sources such as wind and solar power, which are intermittent by nature, the importance of effective energy storage has become paramount. These technologies enable the capture of energy produced at one time for use at a later time, bridging the gap between energy availability and consumption patterns. Broadly, energy storage technologies can be categorized into mechanical, chemical, electrical, and thermal storage, each with its own set of mechanisms, applications, and advantages.

Mechanical storage includes pumped hydroelectric storage (PHS), which is the most established and widely used form of large-scale energy storage [3]. It involves pumping water to a higher elevation during periods of low energy demand and then releasing it to generate electricity when demand is high. Other mechanical storage methods include compressed air energy storage (CAES), where air is compressed and stored under pressure, and flywheel energy storage systems, which store energy in the rotational motion of a spinning mass.

Electrochemical storage is represented by various types of batteries, such as lithium-ion, sodium-sulfur, and flow batteries [4, 5]. These devices store energy in the form of chemical potential energy and convert it back into electrical energy when needed. Batteries are crucial for portable electronics, electric vehicles, and grid storage applications, offering high energy density and efficiency [6–12]. On the other hand, supercapacitors, a form of electrochemical energy storage, store energy through electrostatic interactions at the interface between the electrode and electrolyte, without involving significant chemical reactions [13–15]. This is why they are sometimes referred to as electrical double-layer capacitors (EDLCs). However, due to the electrochemical nature of the charge storage process, they fall under the broader category of electrochemical energy storage devices. Besides, their hybrids nature metal-ion capacitors can provide both high energy and power densities [16–18].

Electrical storage encompasses technologies such as superconducting magnetic energy storage (SMES), which stores energy in the magnetic field created by the flow of direct current in a superconducting coil, and capacitors, which store potential energy in an electric field [3].

DOI: 10.1201/9781003489726-6

Thermal storage involves capturing heat or cold for later use, which can significantly enhance the efficiency of heating and cooling systems [19, 20]. This category includes sensible heat storage, latent heat storage using phase change materials, and thermochemical storage.

Each energy storage technology offers unique benefits and challenges, making them suitable for different applications, from smoothing out the variability of renewable energy sources and providing backup power to enabling energy arbitrage and supporting grid stability. As technological advancements continue, the efficiency, capacity, and cost-effectiveness of these storage solutions are expected to improve, playing a critical role in the transition to a more sustainable and resilient energy future.

6.1.2 The Importance of Grid-Scale Energy Storage

The significance of grid-scale energy storage cannot be overstated in the context of modern energy systems. As the transition toward renewable energy sources accelerates, the challenge of their intermittent nature becomes increasingly apparent. Solar and wind resources, while abundant and sustainable, do not always produce power in alignment with consumption patterns. Herein lies the critical role of grid-scale energy storage: it acts as a buffer, ensuring that the electricity generated from these renewable sources is available precisely when it is needed, thereby enhancing grid reliability and resilience. For example, at Vidyasirimedhi Institute of Science and Technology (VISTEC) in Wang Chan Valley, Rayong, Thailand, a grid-scale energy storage system has been developed to address the growing demands for reliable, sustainable energy in modern urban environments, as shown in Figure 6.1.

FIGURE 6.1 A grid-scale energy storage system at Vidyasirimedhi Institute of Science and Technology (VISTEC), Wang Chan Valley, Rayong, Thailand. The image showcases the integration of solar photovoltaic panels with energy storage solutions, playing a crucial role in promoting sustainable energy systems and supporting the development of resilient and sustainable urban environments.

This system plays a pivotal role in ensuring the resilience and efficiency of energy infrastructure, contributing to the reduction of carbon footprints, and enhancing energy security. By harnessing advanced technologies, the system supports the integration of renewable energy sources such as solar and wind, stabilizing power grids and reducing dependency on fossil fuels. As cities continue to expand and evolve, such innovations are essential in creating sustainable urban ecosystems that promote long-term environmental health and energy independence, aligning with global goals for a greener future.

6.1.3 Introduction to LMBs

Liquid metal batteries (LMBs) represent a groundbreaking advancement in the realm of energy storage technologies, distinguished by their unique design and operational principles [21–28]. These innovative batteries consist of three liquid layers: two metal electrodes separated by a molten salt electrolyte. The distinct densities and immiscibility of these materials allow them to form stable, stratified layers that function at elevated temperatures. This characteristic structure enables the seamless flow of ions within the battery, facilitating efficient energy storage and release. The core appeal of LMBs lies in their potential for grid-scale energy storage. They are especially suited for applications requiring large capacities and long durations, such as stabilizing renewable energy sources on the electrical grid. The liquid state of both electrodes ensures a high rate of ionic conductivity and allows for rapid charge and discharge cycles. Furthermore, the fluid nature of the electrodes can mitigate degradation over time, promising a longer lifespan compared to traditional solid-state batteries. A significant advantage of LMBs is their scalability. The technology can be easily scaled up to meet higher energy storage requirements simply by increasing the size of the battery cells. This scalability, combined with the relatively low cost and abundance of suitable materials, particularly for sodium-based LMBs, makes them a compelling option for large-scale energy storage solutions. Moreover, LMBs operate based on the principles of simplicity and efficiency. Their operation requires fewer moving parts and less maintenance than complex mechanical storage systems, translating to potentially lower operational costs. The high operating temperature of LMBs, while a challenge in terms of materials and safety, contributes to their high efficiency by minimizing ohmic losses within the cell. In the broader context of transitioning to a more sustainable energy landscape, LMBs offer a promising pathway. They can significantly enhance the reliability and flexibility of renewable energy systems by providing a stable, cost-effective storage solution. As research and development in this field continue, the future of LMBs looks bright, with the potential to play a pivotal role in achieving a balanced, resilient, and sustainable energy grid [23].

6.1.4 Significance of Sodium in LMBs

Sodium stands out as a pivotal element in the development of LMBs due to its unique combination of physical and chemical properties, which are ideally suited to energy storage applications [21, 24, 29]. Its abundance and relatively low cost make sodium an attractive option for large-scale battery systems, potentially offering a more economical solution compared to other materials such as lithium. One of the key advantages of sodium is its excellent electrochemical performance. Sodium has a high ionic conductivity, which enables efficient charge and discharge processes within LMBs. This characteristic is crucial for maintaining the high-power output and rapid response times required for grid-scale energy storage. Additionally, sodium's low melting point (97.8°C) compared to other metals used in LMBs reduces the operational temperature range of the batteries, which can help in minimizing energy losses and improving overall efficiency. The significance of sodium also extends to sustainability and environmental impact. As a more readily available resource, sodium-based batteries offer a scalable and potentially less environmentally disruptive alternative to rare or toxic materials used in other types of batteries. Furthermore, the use of sodium can contribute to the diversification of energy storage solutions, reducing reliance on any single resource and enhancing energy security. In summary, the role of sodium in LMBs is multifaceted, encompassing economic, operational, and environmental benefits. Its incorporation into LMBs underlines the potential for sustainable, efficient, and cost-effective energy storage solutions, essential for supporting the integration of renewable energy into the grid and advancing toward a more sustainable energy future.

For example, sodium-based liquid metal batteries (Na-LMBs) utilize a metallic sodium negative electrode and a multi-cationic ternary molten chloride salt (LiCl-NaCl-KCl) as the electrolyte, which reduces sodium's solubility, lowers the self-discharge rate, and enables operation at lower temperatures (~450°C). The incorporation of a dual-active Bi9Sb alloy positive electrode improves material utilization, allowing the battery to operate stably for over 700 cycles with a 97% coulombic efficiency and 80% active material utilization. The Na-LMB stands out for its cost-efficiency, with a levelized cost of storage (LCOS) of less than 0.029 $/kWh, making it a highly competitive option for large-scale energy storage. Continued optimization could further enhance performance and reduce costs, marking a significant advancement toward sustainable, cost-effective energy storage [29].

6.2 FUNDAMENTALS OF SODIUM-BASED LMBs

Sodium-based LMBs emerge as a cutting-edge solution for grid-scale energy storage, harnessing the unique properties of liquid sodium to facilitate efficient, large-capacity energy storage [21, 24, 29]. Characterized by their distinctive three-layer configuration, these batteries consist of a liquid sodium anode, a molten salt electrolyte, and a cathode made from another metal, all operating at elevated temperatures. This design enables rapid ionic conductivity and high charge/discharge rates, essential for managing the intermittency of renewable energy sources. The abundance and low cost of sodium, coupled with the scalability and longevity of LMBs, underscore their potential in transforming energy storage and ensuring a sustainable, reliable power grid.

6.2.1 Basic Operation Principles

The basic operation principles of sodium-based LMBs are founded on a unique and innovative design that sets them apart from traditional battery technologies [21]. At

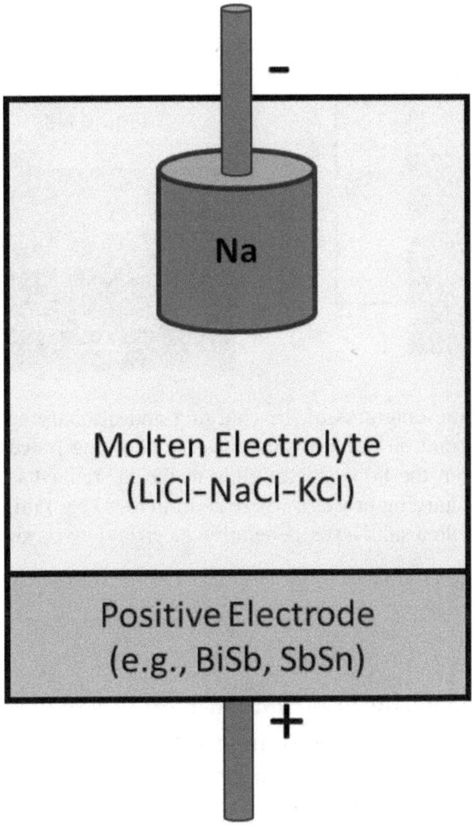

FIGURE 6.2 Schematic diagram of sodium-based LMBs showing the three main components: a positive electrode (e.g., BiSb, SbSn), a molten salt electrolyte (LiCl-NaCl-KCl), and a negative electrode made of sodium (Na). The molten electrolyte facilitates the movement of sodium ions between the electrodes, enabling the storage and discharge of electrical energy.

the heart of their operation is a three-layer structure, comprising a positive electrode (cathode), a molten salt electrolyte, and a negative electrode (anode), as shown in Figure 6.2, with each layer possessing distinct properties and roles in the battery's functionality [23, 24].

The three-layer structure of sodium-based liquid metal batteries relies on the differing densities and immiscibility of its components, allowing natural stratification. The anode is liquid sodium, known for its excellent conductivity and low melting point (97.8°C). This liquid state at the battery's operational temperature (200–500°C) enables efficient electron flow and high ionic conductivity. The electrolyte, a molten salt, facilitates ion conduction while preventing short circuits by separating the anode and cathode. The cathode, made from various materials, is selected based on the battery's application. Sodium's abundance and low cost make it ideal for scalable energy storage solutions.

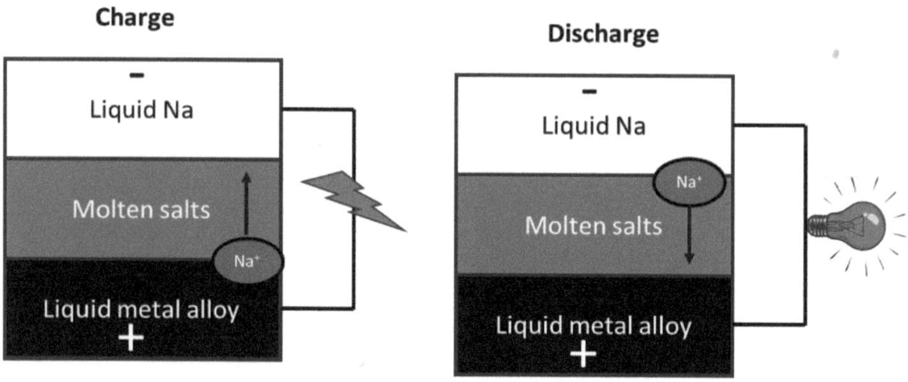

FIGURE 6.3 Schematic diagrams of the charging and discharging processes in sodium-based LMBs. The diagram on the left illustrates the charging process, where sodium ions (Na^+) move upward from the liquid metal alloy to the molten salts layer. The diagram on the right shows the discharging process, where sodium ions (Na^+) migrate back to the liquid metal alloy from the molten salts layer, generating electricity to power a device, as indicated by the lit bulb.

6.2.1.1 Charging Mechanism of Sodium-Based LMBs

During the charging process of sodium-based LMBs, sodium atoms at the negative electrode undergo oxidation, losing electrons to form sodium ions (Figure 6.3, left). These ions migrate through the molten salt electrolyte, which remains in a liquid state due to the battery's high operating temperature. The released electrons travel through the external circuit to the positive electrode, facilitating the storage of electrical energy as chemical potential energy. As sodium ions accumulate in the molten salt layer, the battery continues charging until the process is complete.

6.2.1.2 Discharging Mechanism of Sodium-Based LMBs

When sodium-based LMBs discharge, stored chemical energy is converted into electrical energy (Figure 6.3, right). Sodium ions migrate from the molten salt electrolyte to the liquid metal alloy electrode, undergoing reduction by accepting electrons from the external circuit. These electrons, stored during charging, reduce sodium ions back to neutral atoms, allowing electron flow through the circuit to power devices. This process efficiently converts chemical energy to electrical energy, continuing until the battery is depleted and needs recharging.

Sodium-based LMBs operate at high temperatures, where the liquid state of both electrodes and the molten salt electrolyte minimizes internal resistance and enhances ion mobility. The molten electrolyte eliminates the need for a solid separator, preventing issues like dendrite formation. These features make sodium-based LMBs highly efficient, with long cycle life and rapid charge-discharge capabilities, making them ideal for large-scale energy storage applications.

6.2.2 Advantages of Using Sodium

Sodium plays a key role in sodium-based LMBs, offering numerous advantages for revolutionizing energy storage [21]. Its abundance and cost-effectiveness make sodium an attractive material for large-scale applications [23, 24]. As one of the most abundant elements on Earth, primarily sourced from salt (NaCl), sodium's widespread availability provides significant cost advantages over materials like lithium. This economic feasibility lowers production costs and supports the scalability of sodium-based LMBs, making them ideal for renewable energy storage solutions.

In addition to its affordability, sodium's electrochemical properties further enhance its appeal. Sodium offers high ionic conductivity, ensuring efficient electron flow within the battery, which is critical for high-performance energy storage. Its relatively low melting point (97.8°C) reduces the energy required to maintain the battery's operating temperature, boosting efficiency. Sodium's electrochemical compatibility with various electrolytes and cathodes allows for flexible battery designs, optimizing performance across applications.

These combined benefits of abundance, cost-effectiveness, and favorable electrochemical properties make sodium-based LMBs well-suited for grid-scale energy storage. They present a promising solution for supporting the growing demand for renewable energy storage and advancing the shift toward a sustainable energy future.

6.2.3 Comparison with Other Battery Technologies

Sodium-based LMBs present several advantages, making them a compelling solution for grid-scale energy storage. Unlike lithium-ion batteries, which excel in portable electronics and electric vehicles due to high energy density, sodium-based LMBs are ideal for large-capacity, long-duration energy storage. This is critical for renewable energy integration, where the focus is on storing vast amounts of energy economically, rather than compact storage for devices [24]. While lead-acid batteries are reliable and low-cost, sodium-based LMBs offer superior cycle life and operational efficiency, making them a more sustainable option with lower lifetime costs. Compared to flow batteries, which are scalable, sodium-based LMBs feature simpler designs and lower operating temperatures, reducing maintenance and operational challenges. Additionally, sodium's abundance, non-toxicity, and non-flammability enhance the safety and environmental advantages of these batteries, positioning them as a sustainable choice for large-scale storage. The LCOS in Table 6.1 is a key metric in evaluating energy storage solutions [29, 30]. Sodium-based LMBs stand out with a remarkably low LCOS of \$0.029/kWh, with potential for further reduction to \$0.02/kWh. This makes them highly competitive for grid-scale applications, especially when compared to other technologies like lead-acid (\$3.12/kWh), vanadium redox flow (\$0.78/kWh), and sodium-sulfur (\$0.40/kWh). While lithium-ion batteries have an impressive LCOS of \$0.07/kWh, the lower cost and efficiency of sodium-based LMBs provide an economical and sustainable solution for grid-scale energy storage, supporting renewable energy integration into the grid.

TABLE 6.1
Cost Analysis of Energy Storage Technologies: LCOS for Some Energy Storage Technologies

System	LCOS ($/kWh)
Pb-Acid Battery	3.12
V Redox Flow Battery	0.78
Na-S Battery	0.4
Li-ion Battery	0.07
Ni-Cd Battery	1.1
Pumped Storage	1.76
Na-LMB	0.029
Na-LMB (Optimized)	0.02

6.3 COMPONENTS OF SODIUM-BASED LMBs

Sodium-based LMBs are distinguished by their innovative three-component structure, designed to harness the unique properties of sodium for efficient energy storage [24]. At their core, these batteries feature a liquid sodium anode that offers high conductivity and affordability, thanks to sodium's abundance and electrochemical efficiency. A molten salt electrolyte serves as the central layer, facilitating ion transfer while maintaining separation of the electrodes to prevent short circuits. The choice of cathode material varies, aiming to optimize the battery's performance characteristics. This combination of components ensures sodium-based LMBs provide a reliable, cost-effective solution for large-scale energy storage, aligning with the growing demand for renewable energy integration and grid stabilization.

6.3.1 The Cathode: Materials and Properties

The cathode in sodium-based LMBs plays a pivotal role in defining the battery's overall performance, including its efficiency, capacity, and longevity [24, 31]. The selection of cathode materials is governed by several critical criteria: thermal stability, density, and electrochemical compatibility with the sodium anode and molten salt electrolyte. These factors ensure that the battery operates effectively and safely over a wide range of temperatures and cycling conditions [32].

Thermal Stability. Given the high operational temperatures of sodium-based LMBs, typically ranging from 200°C to 500°C, the cathode material must remain chemically stable and maintain its structural integrity. This stability is crucial for preventing degradation that could lead to a decrease in battery efficiency and lifespan. Materials that can withstand these conditions while maintaining their electrochemical properties are preferred.

Density. The cathode material must have an appropriate density to ensure it forms a distinct layer above the molten sodium anode and beneath the electrolyte layer.

This stratification is vital for the battery's operation, allowing for the efficient transfer of ions between the anode and cathode during charge and discharge cycles.

Electrochemical Compatibility. The material should exhibit good electrochemical compatibility with sodium, facilitating efficient charge transfer processes without undesirable side reactions. This compatibility helps in maximizing the battery's energy density and efficiency.

Common Cathode Materials and Innovations. Various materials have been explored as cathodes in sodium-based LMBs, with antimony (Sb) and bismuth (Bi) being among the most common due to their favorable electrochemical properties and relative abundance. Recent innovations have focused on exploring new materials like alloys and compounds that can offer improved performance, such as higher energy density or reduced cost. Researchers are also investigating the use of mixed-metal and composite cathodes that can operate at lower temperatures, enhancing the battery's safety and reducing its energy consumption.

The continuous exploration and development of new cathode materials and configurations underscore the dynamic nature of research in sodium-based LMBs, with the goal of optimizing their performance for large-scale energy storage applications.

6.3.2 The Electrolyte: Key to Ion Transport

The electrolyte in sodium-based LMBs is a crucial component, acting as the medium through which ions are transported between the cathode and the anode during the battery's charge and discharge cycles [29, 33–36]. Typically, the electrolyte in these batteries is made from molten salts, chosen for their ability to facilitate efficient ion exchange at the high operating temperatures required for LMBs [23, 24].

Molten salt electrolytes are composed of salts that are liquid at the operational temperatures of LMBs, generally ranging between 200°C and 500°C. These electrolytes must possess high ionic conductivity to allow for the swift movement of sodium ions between electrodes, contributing to the battery's overall efficiency and power output. Moreover, they must remain chemically inert to both the anode and the cathode materials to prevent degradation or unwanted chemical reactions. Commonly used salts include those based on chloride, fluoride, or bromide, which are selected based on their melting points, electrochemical windows, and compatibility with other battery components.

Despite the advantages molten salt electrolytes bring to LMBs, they present several challenges. One major issue is their long-term stability, especially given the corrosive nature of some salts at elevated temperatures. This can lead to the deterioration of battery components and a decrease in battery life. Additionally, finding salts that are both highly conductive and chemically compatible with sodium and the selected cathode materials can be difficult. The electrolyte must not only conduct ions efficiently but also not react with the electrode materials in a way that could hinder battery performance or safety.

Advancements in electrolyte research are focusing on developing new salt mixtures that lower the operational temperature of LMBs, thereby reducing material compatibility issues and enhancing the stability of the electrolyte. Researchers are also exploring additives that can improve the wettability of the electrolyte on the electrodes, further enhancing ion transport and battery efficiency. These ongoing research efforts aim to overcome the current limitations and unlock the full potential of sodium-based LMBs for large-scale energy storage applications.

6.3.3 The Anode: Liquid Sodium

The anode in sodium-based LMBs, composed of liquid sodium, is fundamental to their operation and contributes significantly to their overall performance and efficiency [21, 29, 31, 34, 37]. The choice of sodium for the anode material is strategic, leveraging its advantageous properties under the operational conditions of LMBs, as well as important safety considerations inherent in handling and utilizing sodium in its liquid state [23, 24].

Sodium becomes liquid at temperatures above its melting point of 97.8°C, which is considerably lower than the operational temperature range of LMBs (typically between 200°C to 500°C). In this liquid state, sodium offers excellent ionic conductivity, enabling efficient electron flow during the battery's charge and discharge cycles. This high conductivity is crucial for the rapid and efficient transport of ions across the electrolyte, contributing to the battery's high-power output and fast response times. Additionally, sodium's low density and compatibility with various molten salt electrolytes and cathode materials make it an ideal candidate for the anode material in LMBs [23].

While sodium's properties offer many benefits, they also present unique safety challenges. Sodium reacts vigorously with water and can ignite when exposed to air, especially at elevated temperatures. These reactivity concerns necessitate stringent handling and operational protocols to ensure safety. LMBs are designed with robust containment systems to prevent sodium exposure to the environment. Additionally, the operational design includes features to minimize the risk of leakage or accidental exposure, such as double containment vessels and inert gas atmospheres to displace oxygen and moisture.

Safety protocols for handling and maintenance also emphasize the importance of dry, oxygen-free environments and the use of personal protective equipment (PPE) to safeguard against sodium's reactive nature. Advancements in battery design and safety mechanisms continue to mitigate these risks, enabling the safe and effective use of sodium-based LMBs in various applications, particularly for grid-scale energy storage, where their benefits can be fully realized.

6.4 DESIGN AND CONSTRUCTION

The design and construction of sodium-based LMBs are tailored to leverage the unique properties of liquid sodium and molten salt electrolytes for efficient, large-scale energy storage [23, 31, 36]. Central to their architecture is a three-layer stratification consisting of a liquid sodium anode, a molten salt electrolyte, and a cathode, designed to operate at elevated temperatures. This design ensures high ionic conductivity and rapid charge-discharge cycles, crucial for grid applications. Attention to thermal management, materials selection for longevity, and scalability underpins their construction. Moreover, safety features and containment strategies are meticulously integrated to manage the reactive nature of sodium, demonstrating the sophisticated engineering behind LMBs aimed at transforming energy storage solutions.

6.4.1 Cell Design Considerations

The cell design of sodium-based LMBs is critical to their performance, safety, and longevity, particularly due to their operation at elevated temperatures and the reactive nature of their components [23, 31]. Design considerations such as thermal management and material selection are therefore paramount.

6.4.1.1 Thermal Management for High-Temperature Operations

Effective thermal management is crucial for maintaining the operational integrity and efficiency of LMBs. These batteries operate at elevated temperatures (between 200°C and 500°C), which are necessary to keep the sodium anode and the molten salt electrolyte in their liquid states [23]. Efficient thermal management ensures uniform temperature distribution within the cell, preventing hotspots that could lead to material degradation or failure. Strategies include the use of insulating materials to minimize heat loss and the integration of cooling systems in larger battery installations to dissipate excess heat. Additionally, the design may incorporate phase change materials or heat pipes to regulate temperature effectively, ensuring that the battery operates within its optimal temperature range for maximum efficiency and lifespan.

6.4.1.2 Material Selection for Longevity and Performance

The selection of materials for the components of LMBs is driven by the need to withstand the corrosive and high-temperature environment of the battery [23]. Materials must be chemically compatible with sodium and the chosen electrolyte, resisting corrosion and degradation over many charge-discharge cycles. For the containment and structural components, high-temperature alloys, ceramics, or composites are often chosen for their strength, thermal stability, and resistance to sodium and electrolyte corrosion. The electrolyte and electrode materials are selected not only for their electrochemical performance but also for their stability and reactivity (or lack thereof) with other cell components. This careful selection of materials ensures the battery's functionality over its intended lifespan, reducing maintenance needs and enhancing overall performance.

Through meticulous thermal management and material selection, designers of sodium-based LMBs can optimize their performance, safety, and durability, addressing some of the primary challenges associated with high-temperature battery operations and paving the way for their wider adoption in energy storage applications.

6.4.2 Scalability and Modular Construction

The scalability and modular construction of sodium-based LMBs are fundamental attributes that enhance their applicability for large-scale energy storage solutions [23]. These characteristics not only allow for flexible system sizing and easier integration into existing energy infrastructure but also facilitate the implementation of robust safety and containment measures.

6.4.2.1 Scalability and Modular Construction

Sodium-based LMBs inherently support scalability due to their liquid-state components and straightforward design. The ability to increase energy storage capacity is as simple as scaling up the size of the battery cells or combining multiple cells into larger modules. This modular approach enables the customization of energy storage systems to meet specific requirements, whether for grid stabilization, renewable energy integration, or peak-shaving applications. Modular construction also simplifies manufacturing, maintenance, and potential upgrades, as individual modules can be easily replaced or added without disrupting the entire system.

6.4.2.2 Integration into Energy Storage Systems

The modular nature of sodium-based LMBs facilitates their integration into a variety of energy storage systems. These batteries can be designed to fit specific spatial constraints and configured to deliver the desired power and energy capacities. Their elevated temperature operation and efficiency make them particularly suitable for pairing with renewable energy sources, providing a buffer to absorb excess production and release energy during demand peaks or low production periods.

6.4.2.3 Safety and Containment Strategies

Safety and containment are paramount in the design of sodium-based LMBs, given the reactive nature of sodium and the high operational temperatures. Modular construction aids in implementing effective safety strategies by isolating potential failure points and minimizing risks. Containment vessels are designed with robust materials resistant to elevated temperatures and sodium corrosion. Additionally, safety measures include inert gas atmospheres to prevent oxidation, emergency cooling systems to manage thermal runaway, and leak detection systems to promptly address any breaches. These strategies ensure that even as LMB systems scale up, they maintain the highest safety standards, protecting both the system and its surroundings.

6.5 CHALLENGES AND FUTURE DIRECTIONS

Sodium-based LMBs present a promising avenue for advancing energy storage solutions, yet they face specific challenges that must be addressed to realize their full potential [24]. Key issues include enhancing material compatibility to prevent corrosion, reducing operational temperatures to improve safety and efficiency, and advancing electrolyte formulations for better performance. Future directions focus on overcoming these hurdles through material science innovations, developing more robust and cost-effective construction methods, and exploring novel chemistries to lower operational risks and costs. As research progresses, these efforts aim to optimize LMBs for broader commercial use, ensuring they play a pivotal role in the transition toward sustainable and resilient energy systems [31].

6.5.1 Material and Operational Challenges

Sodium-based LMBs, while promising for large-scale energy storage applications, encounter specific material and operational challenges that necessitate innovative solutions to unlock their full potential. Among these challenges, corrosion and material compatibility, alongside the need to manage high operating temperatures, stand out as critical areas requiring focused research and development.

6.5.1.1 Corrosion and Material Compatibility

One of the primary challenges in the development and deployment of sodium-based LMBs is the corrosion of materials used in their construction. The highly reactive nature of liquid sodium, combined with the corrosive properties of molten salt electrolytes, can lead to the degradation of container materials, electrodes, and other battery components. This corrosion not only reduces the efficiency and lifespan of the battery but also raises safety concerns due to the potential for leaks and failures. Addressing this issue involves the identification and development of new materials and alloys that are resistant to corrosion while maintaining high conductivity and compatibility with the battery's electrochemical environment. Advances in surface coatings and barrier technologies also offer promising strategies for enhancing material longevity and compatibility.

6.5.1.2 High Operating Temperatures

Sodium-based LMBs operate at elevated temperatures (typically between 200°C and 500°C) to maintain the sodium and electrolyte in a liquid state, which is essential for their operation. However, these elevated temperatures pose challenges for system design, including the need for insulation to minimize heat loss, materials that can withstand prolonged exposure to such temperatures, and safety measures to manage the risks of overheating and thermal runaway. Furthermore, the energy required to maintain these temperatures can impact the overall energy efficiency of the system. Research is ongoing to develop LMBs that operate at lower temperatures without compromising performance, through the exploration of alternative electrolytes and electrode materials that have lower melting points and are thermally stable.

Overcoming these material and operational challenges is crucial for advancing sodium-based LMBs toward wider commercial adoption, promising to significantly impact the future of energy storage and management.

6.5.2 Advances in Materials Science and Engineering

The evolution of sodium-based LMBs is closely tied to advances in materials science and engineering, particularly in the development of next-generation electrolytes and electrode materials, as well as innovations aimed at enabling lower temperature operations. These advancements are pivotal in addressing the inherent challenges of LMBs, pushing the boundaries of their performance, safety, and applicability using many approaches including machine learning [38] and a hybrid data-driven approach [39].

6.5.2.1 Next-Generation Electrolytes and Electrode Materials

Innovative research in materials science is paving way for the discovery and optimization of novel electrolytes and electrode materials that promise enhanced efficiency, longevity, and safety for sodium-based LMBs. For electrolytes, the focus has been on developing new compositions that offer lower melting points, higher ionic conductivities, and improved stability and compatibility with sodium and other electrode materials. These efforts include exploring different salt mixtures, incorporating additives to reduce corrosion, and enhancing the electrochemical window to prevent undesirable side reactions.

Parallel advancements are being made in electrode materials, with research aimed at identifying alternatives that can reduce costs, improve energy density, and enhance the battery's charge and discharge rates. Innovations include the use of alloy and composite materials that offer better performance and compatibility with the electrolyte, as well as modifications to the surface chemistry of the electrodes to improve their reactivity and stability.

6.5.2.2 Lower Temperature Sodium-Based LMBs

Reducing the operational temperature of sodium-based LMBs without compromising their performance is a significant area of focus. Operating at lower temperatures can mitigate many of the material compatibility and safety issues associated with high-temperature operation. Advances in this area involve the development of new electrolyte and electrode materials that remain liquid or highly conductive at reduced temperatures. Such innovations not only enhance the safety profile of LMBs but also potentially reduce the energy required for heating, improving the overall efficiency and viability of the technology.

These advances in materials science and engineering are crucial for the next generation of sodium-based LMBs, offering paths to overcome current limitations and unlock new possibilities for energy storage solutions. Through continued research and development, sodium-based LMBs are poised to play a more prominent role in meeting the world's growing energy storage needs.

6.6 CONCLUSION

Sodium-based LMBs offer a transformative solution for large-scale energy storage, particularly as renewable energy sources become more widespread. These batteries, featuring a liquid sodium anode and molten salt electrolyte, provide efficient, durable, and scalable energy storage. Their rapid charge/discharge capabilities with minimal degradation make them suitable for applications requiring large energy capacities and long-term reliability. Additionally, sodium's abundance, low cost, and environmental friendliness enhance the economic viability of LMBs, positioning them as a key solution in the transition to sustainable energy systems. As renewable sources like wind and solar grow, the need for reliable, scalable energy storage to ensure grid stability will increase. Sodium-based LMBs can store large amounts of energy and release it as needed, smoothing supply fluctuations and supporting renewable energy availability.

Their modular design and scalability also allow them to be adapted for various applications, from grid support to backup power. However, challenges remain, such as material corrosion, high operating temperatures, and the need for better electrolyte and electrode materials. Continued research and development, along with supportive policies and incentives, are essential for overcoming these issues. Raising public and investor awareness of LMB technology's benefits will also be crucial in driving market adoption. In conclusion, sodium-based LMBs stand at the forefront of energy storage technology, capable of significantly enhancing the integration of renewable energy and contributing to a sustainable, carbon-neutral energy future.

REFERENCES

[1] X. Luo, J. Wang, M. Dooner, J. Clarke, Overview of current development in electrical energy storage technologies and the application potential in power system operation, *Applied Energy* 137 (2015) 511–536.

[2] T.M. Gür, Review of electrical energy storage technologies, materials and systems: Challenges and prospects for large-scale grid storage, *Energy and Environmental Science* 11(10) (2018) 2696–2767.

[3] R. Amirante, E. Cassone, E. Distaso, P. Tamburrano, Overview on recent developments in energy storage: Mechanical, electrochemical and hydrogen technologies, *Energy Conversion and Management* 132 (2017) 372–387.

[4] K.C. Divya, J. Østergaard, Battery energy storage technology for power systems: An overview, *Electric Power Systems Research* 79(4) (2009) 511–520.

[5] G.L. Soloveichik, Battery technologies for large-scale stationary energy storage, *Annual Review of Chemical and Biomolecular Engineering* 2 (2011) 503–527.

[6] N. Joraleechanchai, T. Sangsanit, K. Homlamai, P. Krapong, M. Sawangphruk, Insight into the effect of thick graphite electrodes towards high-performance cylindrical Ni-rich NCA90 Li-ion batteries, *Journal of Energy Chemistry* 87 (2023) 322–333.

[7] N. Anansuksawat, T. Sangsanit, S. Prempluem, K. Homlamai, W. Tejangkura, M. Sawangphruk, How uniform particle size of NMC90 boosts lithium ion mobility for faster charging and discharging in a cylindrical lithium ion battery cell, *Chemical Science* 15(6) (2024) 2026–2036.

[8] K. Homlamai, N. Anansuksawat, T. Sangsanit, S. Prempluem, K. Santisuk, W. Tejangkura, M. Sawangphruk, Gas evolution analyses of Ni-rich Li-ion and Li-metal batteries at cylindrical jelly-roll configurations, *Journal of Power Sources* 617 (2024) 235150.

[9] S. Prempluem, T. Sangsanit, K. Santiyuk, K. Homlamai, W. Tejangkura, R. Songthan, N. Anansuksawat, M. Sawangphruk, Complex reaction mechanisms of electrolyte decomposition at large-scale Ni-rich Li-ion battery cells including electrode crosstalk effect, *Journal of Power Sources* 606 (2024) 234538.

[10] T. Sangsanit, K. Homlamai, N. Joraleechanchai, S. Prempluem, W. Tejangkura, M. Sawangphruk, Non-flammable electrolyte for large-scale Ni-rich Li-ion batteries: Reducing thermal runaway risks, *Journal of Power Sources* 594 (2024) 234021.

[11] T. Sangsanit, K. Santiyuk, R. Songthana, K. Homlamai, S. Prempluem, W. Tejangkura, M. Sawangphruk, Initial formaldehyde generation as a predictive marker for long-term stability of Ni-rich Li-ion batteries under abusive conditions, *Journal of Power Sources* 611 (2024) 234770.

[12] R. Songthan, T. Sangsanit, K. Santiyuk, K. Homlamai, W. Tejangkura, M. Sawangphruk, Revisiting the effect of natural and artificial graphite on the performance of Ni-rich Li-ion batteries at coin and cylindrical cells, *Journal of the Electrochemical Society* 171(5) (2024) 050524.

[13] M. Sawangphruk, P. Srimuk, P. Chiochan, A. Krittayavathananon, S. Luanwuthi, J. Limtrakul, High-performance supercapacitor of manganese oxide/reduced graphene oxide nanocomposite coated on flexible carbon fiber paper, *Carbon* 60 (2013) 109–116.

[14] P. Wuamprakhon, J. Phojaroen, T. Sangsanit, K. Santiyuk, K. Homlamai, W. Tejangkura, M. Sawangphruk, Unveiling a novel decomposition pathway in propylene carbonate-based supercapacitors: Insights from a Jelly Roll configuration study, *ChemSusChem* (2024) https://doi.org/10.1002/cssc.202400053.

[15] S. Kaenket, P. Suktha, K. Kongsawatvoragul, T. Sangsanit, P. Wuamprakhon, R. Songthan, W. Tejangkura, K. Santiyuk, K. Homlamai, M. Sawangphruk, Large-scale production of 18650 cylindrical supercapacitors: Effects of separators, electrode thickness, electrolyte additives, and testing protocols, *Journal of Power Sources* 581 (2023) 233512.

[16] P. Wuamprakhon, R. Songthan, T. Sangsanit, K. Santiyuk, J. Phojaroen, K. Homlamai, W. Tejangkura, M. Sawangphruk, Designing electrolytes for enhancing stability and performance of lithium-ion capacitors at large-scale cylindrical cells, *Journal of Power Sources* 622 (2024) 235331.

[17] A. Phukhrongthung, M. Sawangphruk, P. Iamprasertkun, C. Santhaweesuk, C. Puchongkawarin, S. Luanwuthi, Rocking chair-type aqueous sodium-ion capacitors with biomass-derived activated carbon and Na3V2(PO4)2F3 nanoflower in a water-in-salt electrolyte, *Journal of Energy Storage* 80 (2024) 110369.

[18] C. Aphirakaramwong, P. Wuamprakhon, T. Sangsanit, W. Tejangkura, R. Songthan, M. Sawangphruk, Optimizing electrode efficiency in lithium titanate: Investigating the impact of electrolyte additives on cylindrical Li-ion capacitor cells, *Energy Technology* 12(9) (2024) 2301661.

[19] S.M. Hasnain, Review on sustainable thermal energy storage technologies, part I: Heat storage materials and techniques, *Energy Conversion and Management* 39(11) (1998) 1127–1138.

[20] S. Kuravi, J. Trahan, D.Y. Goswami, M.M. Rahman, E.K. Stefanakos, Thermal energy storage technologies and systems for concentrating solar power plants, *Progress in Energy and Combustion Science* 39(4) (2013) 285–319.

[21] X. Lu, G. Li, J.Y. Kim, D. Mei, J.P. Lemmon, V.L. Sprenkle, J. Liu, Liquid-metal electrode to enable ultra-low temperature sodium-beta alumina batteries for renewable energy storage, *Nature Communications* 5 (2014) 4578.

[22] G.M. Haarberg, J. Thonstad, Electrochemical properties of metal-molten salt mixtures, *Journal of Applied Electrochemistry* 19(6) (1989) 789–801.

[23] H. Kim, D.A. Boysen, J.M. Newhouse, B.L. Spatocco, B. Chung, P.J. Burke, D.J. Bradwell, K. Jiang, A.A. Tomaszowska, K. Wang, W. Wei, L.A. Ortiz, S.A. Barriga, S.M. Poizeau, D.R. Sadoway, Liquid metal batteries: Past, present, and future, *Chemical Reviews* 113(3) (2013) 2075–2099.

[24] H. Li, H. Yin, K. Wang, S. Cheng, K. Jiang, D.R. Sadoway, Liquid metal electrodes for energy storage batteries, *Advanced Energy Materials* 6(14) (2016) 1600483.

[25] K. Wang, K. Jiang, B. Chung, T. Ouchi, P.J. Burke, D.A. Boysen, D.J. Bradwell, H. Kim, U. Muecke, D.R. Sadoway, Lithium–antimony–lead liquid metal battery for grid-level energy storage, *Nature* 514(7522) (2014) 348–350.

[26] X. Zhou, S. Yan, X. He, H. Zhou, J. Ning, H. Li, K. Wang, K. Jiang, Low-temperature, high cycling stability, and high Coulombic efficiency liquid metal batteries enabled by lithium halide-potassium halide molten salt electrolytes, *Energy Storage Materials* 61 (2023) 102889.

[27] Y. Zhou, G. Li, B. Li, X. Ning, Operando formation of multi-channel positive electrode achieved via tellurium alloying in liquid metal battery, *Energy Storage Materials* 53 (2022) 927–936.

[28] Y. Zhou, X. Ning, Improving wettability at positive electrodes to enhance the cycling stability of bi-based liquid metal batteries, *Small* 20(3) (2024) 2304528.

[29] H. Zhou, H. Li, Q. Gong, S. Yan, X. Zhou, S. Liang, W. Ding, Y. He, K. Jiang, K. Wang, A sodium liquid metal battery based on the multi-cationic electrolyte for grid energy storage, *Energy Storage Materials* 50 (2022) 572–579.

[30] M.M. Rahman, A.O. Oni, E. Gemechu, A. Kumar, Assessment of energy storage technologies: A review, *Energy Conversion and Management* 223 (2020) 113295.

[31] R. Fetzer, T. Zhang, F. Lindner, A. Weisenburger, G. Müller, How cell design affects the performance of sodium-antimony-bismuth liquid metal batteries, *Journal of Power Sources* 591 (2024).

[32] T. Dai, Y. Zhao, X.-H. Ning, R. Lakshmi Narayan, J. Li, Z.-W. Shan, Capacity extended bismuth-antimony cathode for high-performance liquid metal battery, *Journal of Power Sources* 381 (2018) 38–45.

[33] W. Ding, Q. Gong, S. Liang, R. Hoffmann, H. Zhou, H. Li, K. Wang, T. Zhang, A. Weisenburger, G. Müller, A. Bonk, Multi-cationic molten salt electrolyte of high-performance sodium liquid metal battery for grid storage, *Journal of Power Sources* 553 (2023) 233823.

[34] Q. Gong, W. Ding, A. Bonk, H. Li, K. Wang, A. Jianu, A. Weisenburger, A. Bund, T. Bauer, Molten iodide salt electrolyte for low-temperature low-cost sodium-based liquid metal battery, *Journal of Power Sources* 475 (2020) 228674.

[35] D. Ruiz-Martínez, T. Lana-Villarreal, R. Gómez, Liquid ammoniates as efficient electrolytes for room-temperature rechargeable sodium-metal batteries based on an organic cathode, *ACS Applied Energy Materials* 4(7) (2021) 6806–6814.

[36] H. Zhou, H. Li, S. Yan, X. Zhou, W. Zhang, S. Feng, Y. He, K. Jiang, K. Wang, In situ displacement reactions of molten sodium anode and multi-cationic halide electrolytes enabling high-performance liquid metal batteries, *Chemical Engineering Journal* 485 (2024) 149786.

[37] S.H. Shin, J.H. Lee, J.K. Lee, J.H. Kim, Correlation between carbon activity and carbon content in α-iron in liquid sodium negative electrode of liquid metal battery, *Journal of Power Sources* 297 (2015) 105–112.

[38] H. Zhou, B. Li, M. Yu, S. Li, G. Fan, X. Ning, Accelerated design of electrodes for liquid metal battery by machine learning, *Energy Storage Materials* 56 (2023) 205–217.

[39] Q. Shi, L. Zhao, E. Zhang, J. Xia, H. Li, K. Wang, K. Jiang, The future capacity prediction using a hybrid data-driven approach and aging analysis of liquid metal batteries, *Journal of Energy Storage* 67 (2023) 107637.

7 Ga-Based Liquid Metal Batteries

Kassie Nigus Shitaw, Wei-Nien Su, and Bing Joe Hwang

7.1 INTRODUCTION

Structural instability of electrode materials is one of the challenges in solid electrode-based traditional rechargeable batteries. This instability can be resolved with the use of liquid metal batteries, consequently affording an ultra-long cycle life. However, conventional LMBs require a high operating temperature (>350 °C) to liquefy the metal electrodes and maintain the high conductivity of molten salt electrolytes [1]. Figure 7.1 reveals the components and the charging/discharging process, operating temperatures, and advancement of conventional LMB batteries over time. Figure 7.1a demonstrates the schematic design of conventional LMBs, in which the molten salt electrolyte is employed, and the two different liquid metals (LMs) act as negative and positive electrodes [2]. The high operating temperature in conventional LMBs primarily leads to uncontrollable chemical reactions, electrode corrosion, leakage, safety concerns, and a deficiency in sealing technology—all critical issues [3]. Besides, the solubility of electrodes at high operating temperatures causes self-discharge, which is also a challenge in molten salt-based high-temperature LMBs. In addition, an extra energy input is required to melt the metals and molten salt electrolytes in the conventional LMBs. All the above challenges hinder the practical applications of conventional LMBs. Figure 7.1b shows progress, operating temperatures, and discharge voltage platform of various LMB systems over time [4].

Recently, room-temperature LMs or alloys have attracted prominent attention for their potential use in flexible and stretchable batteries due to their fluidity and conductivity near room temperature. In addition, room-temperature LMs are used as a self-healing material in various rechargeable battery technologies. The metal gallium (Ga) is relatively nontoxic. It has a low melting point (28.8 °C) with negligible vapor pressure at its melting temperature [5]. However, the conductivity of pure Ga metal is lower than that of liquid-state Ga-based alloys. Therefore, because of their intrinsic properties, including low toxicity, fluidity, chemical stability, flexibility, accessibility, low melting point, and high electronic and thermal conductivity, Ga-based LMBs received attention in developing novel room-temperature rechargeable battery technologies [6]. For example, NaK||GaIn and NaK||Galinstan systems can operate near room temperature and are the most stable and safe. This chapter briefly discusses the synthesis and properties of Ga-based LMs, cell designing, fundamental

DOI: 10.1201/9781003489726-7

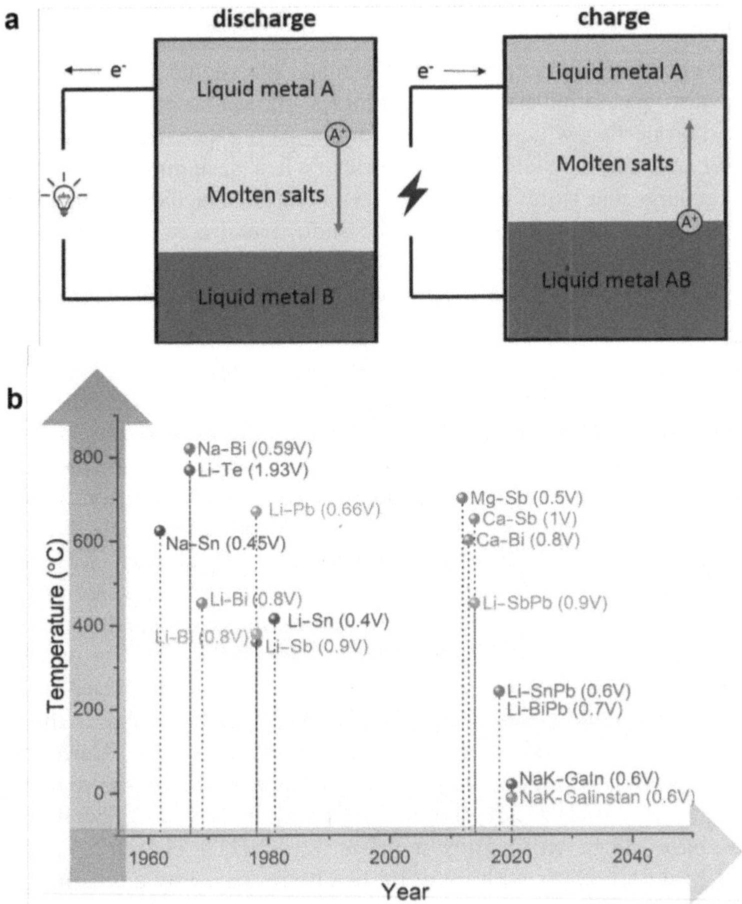

FIGURE 7.1 (a) Schematic of liquid metal batteries in the discharging and charging process. (b) Operating temperature advancements of various LMBs over time.

Sources: (a) Adapted with permission [2], Copyright (2016), Wiley-VCH Verlag GmbH & Co. KGaA, Weinheim. (b) Adapted with permission [4], Copyright (2020), American Chemical Society.

advantages, and limitations of Ga-based LMBs. It also gives future perspectives for developing highly promising room-temperature Ga-based LMBs.

7.2 FUNDAMENTAL PROPERTIES OF GA-BASED EUTECTIC ALLOYS

Pure Ga metal has a melting point of 29.8 °C and is comparatively less toxic under normal circumstances [5, 7]. The formation of Ga alloys is indispensable to creating new Ga-based metallic composites with new physical properties.

7.2.1 Physical Properties of Ga Metal and Ga-Based Eutectic Alloys

The physicochemical characteristics of the materials should be considered when developing next-generation battery technologies [8]. Ga-based LMs have imperative physical properties, which are crucial for enhancing the energy density, surface conductivity, and cycling stability of batteries. When designing suitable Ga-based alloys for rechargeable batteries, key factors to consider include the melting point, deformability, metal affinity, and electronic and thermal conductivity of Ga-based LMs. Table 7.1 summarizes the physical properties of various Ga-based LMs, such as melting point, density, viscosity, electronic and thermal conductivity, and surface tension.

7.2.1.1 Low Melting Point

Ga-based eutectic alloys have a lower melting point than room temperature, suggesting that they have fluidic properties [5]. Due to these features, Ga-based LMs have garnered significant attention for use as surface modification materials in battery technologies. Moreover, Ga-based LMs exhibit a low vapor pressure and are generally considered safe for human health during battery fabrication and disposal [9]. Ga-based LMs, such as GaSn, can enhance the kinetics of redox reactions at the interface and the electrode reaction kinetics. In addition, the Li||GaSn system achieves high discharge voltage (0.77 V at 1C) with a lower voltage polarization at different current densities, resulting in better rate performance with high coulomb efficiency [10]. Tables 7.1 and 7.2 display the composition ratios of mixed metals and the melting points of various Ga-based LMs. A small change in the composition ratio significantly affects the melting point of a specific alloy. For instance, the melting point of GaSn (92:8 wt%) is 20 °C, while the GaSn (88:12 wt%) alloy melts at 17 °C [11]. Therefore, a desirable Ga-based LM with a lower melting point can be prepared by adjusting the alloy composition ratios.

TABLE 7.1
Physical Properties of Ga Metal and Ga-Based Eutectic Alloys and Their Mixing Elemental Compositions

Ga metal/ Ga-based eutectic alloys	M.p. (°C)	Density (at 25 °C)	Viscosity (10^6 m² s⁻¹)	Electrical conductivity (10^6 S m⁻¹)	Thermal conductivity (W m⁻¹ K⁻¹)	Surface tension (10^{-3} N m⁻¹)
Ga	29.8	6.09	0.324	6.73	29.3	707
GaSn	22.0	6.17	—	—	—	670
GaIn	15.5	6.28	0.27	—	26.6	624
GaInSn	−19.0	6.44	0.298	3.46	16.5	718
GaInSnZn	7.6	6.5	0.711	2.8	—	500

Source: Adapted with permission [5], Copyright (2024), Wiley-VCH GmbH.

7.2.1.2 Metal Affinity

Different types of metallic elements may not be homogeneously mixed due to the difference in their chemical and physical characteristics [5]. Therefore, the mixing enthalpy of metals, which indicates their affinity, is crucial in forming the desired Ga-based LM alloy. Interestingly, the mixing enthalpy between Ga and other metallic elements is relatively negative, which reduces the alloying Gibbs free energy. This inspires the strong atomic affinity of mixed metals and makes them more miscible, which benefits in designing a suitable Ga-based LM [12]. Because of the good affinity of Ga metal to alkali metals, alloy layers, such as GaLi [13] and GaNa [14], can be in-situ formed at the anode surface during electrochemical cycling of Ga-containing Li or Na metal batteries.

7.2.1.3 Deformability

Ga-based LMs are easily deformable due to their fluidic nature. This intrinsic deformability enables their use in various flexible and stretchable batteries. Ga-based LMs inherently self-heal the distorted surfaces, a vigorous property of Ga-based LMs for rechargeable battery technologies [15, 16]. The damaged surface of substrates can be recovered due to solid phase transformation upon electrochemical reactions [13].

7.2.1.4 Electrical and Thermal Conductivity

Ga-based LMs are good electronic (106/Ω m) and thermal conductive (×60 times water) [17]. Their thermal stability and high conductivity make them exhibit a wide operating temperature range, with a melting point reduced to 10 °C and a boiling point of more than 2000 °C. As a result, the thermal stability, low melting point, and high boiling point properties of Ga-based LMs endorse their usage as a phase change heat storage material [5, 17].

7.2.2 Alloying Chemistry of Ga-Based Eutectic Alloys

Ga-based alloys are mainly used to modify the surface of negative or positive electrodes in flexible and stretchable energy storage devices. Mixing pure Ga metal with one or more other metal(s) in specific proportions is an effective way to prepare Ga-based alloys with a melting point lower than that of the individual metals. Such Ga alloys are known as eutectic Ga alloys and are very beneficial in developing room-temperature rechargeable battery technologies [5]. Numerous binary, ternary, and quaternary Ga-based alloys have been developed to realize the design and practical applications of LMBs. The Ga alloys' melting point depends on the individual metals' mixing proportions, as shown in Table 7.2. The Ga metal mixed with 8 wt% tin (Sn) melts at 20 °C, which is lower than the melting points of Ga (28.8 °C) and Sn (231.9 °C). Increasing the Sn proportion to 12 wt% further decreases the melting point of GaSn alloy to 17 °C [11]. The ternary Ga alloy GaInSn, with the mixing ratio of 62:25:13 wt% for the respective individual metals, melts at 5 °C. In contrast, the melting point of the quaternary alloy, GaInSnZn (71:15:13:1 wt%) lowered to 3 °C after introducing Zn metal [11].

The appropriate individual metals and their optimal mixing proportions are crucial for designing the required eutectic alloy [18]. The composition ratios could be

TABLE 7.2
Low-Melting Gallium Alloys and Their Mixing Elemental Compositions

Ga-based alloys	Melting point (°C)	Ga-based alloys	Melting point (°C)
$Ga_{95}Zn_5$	25	$Ga_{30}In_{10}Sn_{60}$	12
$Ga_{92}Sn_8$	20	$Ga_{69.8}In_{17.6}Sn_{12.6}$	10.8
$Ga_{88}Sn_{12}$	17	$Ga_{62.5}In_{21.5}Sn_{16}$ (Galinstan)	10.7
$Ga_{72}In_{12}Zn_{16}$	17	$Ga_{62}In_{25}Sn_{13}$	5
$Ga_{75}In_{25}$	16	$Ga_{71}In_{15}Sn_{13}Zn_1$	3
$Ga_{78.6}In_{21.4}$	15	$Ga_{68.5}In_{21.5}Sn_{10}$	−19
$Ga_{67}In_{29}Zn_4$	13		

Source: Adapted with permission [11], Copyright (2007), Higher Education Press and Springer-Verlag.

selected based on the phase diagram of the alloy, as demonstrated in Figure 7.2 for the two typical Ga alloys. A typical eutectic GaIn (EGaIn) alloy with 78.6:21.4 wt% GaIn proportion has a binary phase diagram at 15 °C (Figure 7.2a) [3]. At the intersections of the line at 25 °C, the liquid zone with indium contents of 7.8–23.8 wt% can be determined as a deformable region at room temperature. The melting point of GaIn alloy can be further decreased by introducing Sn metal as an alloy composition. For example, the melting point of the GaInSn (68.5: 21.5:10 wt%) ternary alloy decreased to −19 °C [19]. Figure 7.2b shows the phase diagram of GaInSn ternary alloy, in which several liquid phases were obtained by adjusting the composition ratios [18]. As demonstrated in the phase diagram, various liquid phases were identified and summarized. The e_1, e_2, and e_3 indicate the eutectic points of the InSn, GaSn, and GaIn binary systems, while E represents the eutectic point of the GaInSn ternary system. Similarly, pl and p2 denote the In-Sn peritectic reactions at different compositions and temperatures. The monovariant curves descend from the GaIn eutectic (e3) and the InSn peritectic (pl) to meet at the quasi-peritectic point (p) in the ternary system.

7.2.3 Ga-Based LMBs Configuration and Substrate Effects on LM Wetting

In room-temperature LMBs, the Ga-based LMs can be either negative or positive electrodes. Figure 7.3 shows a schematic of the Ga-based LMB design and LM wettability behavior on various substrates. Ga-based LMs are mainly used to modify the anode electrodes. Figure 7.3a demonstrates the cell configuration and working principle of the Ga-based LMBs. Ga-based LMs enhance the homogeneity of Li nucleation and growth, the stability of the solid electrolyte interface (SEI) layer, self-healing properties, and the wettability of the Ga-based LM-reinforced anode interface. The Ga-based liquid metals can also be used as a positive electrode with organic liquid electrolytes, as illustrated in Figure 7.3b using a Ga-based LM cathode||Na-K anode cell configuration [21]. The solubility of LMs in organic liquid electrolytes is

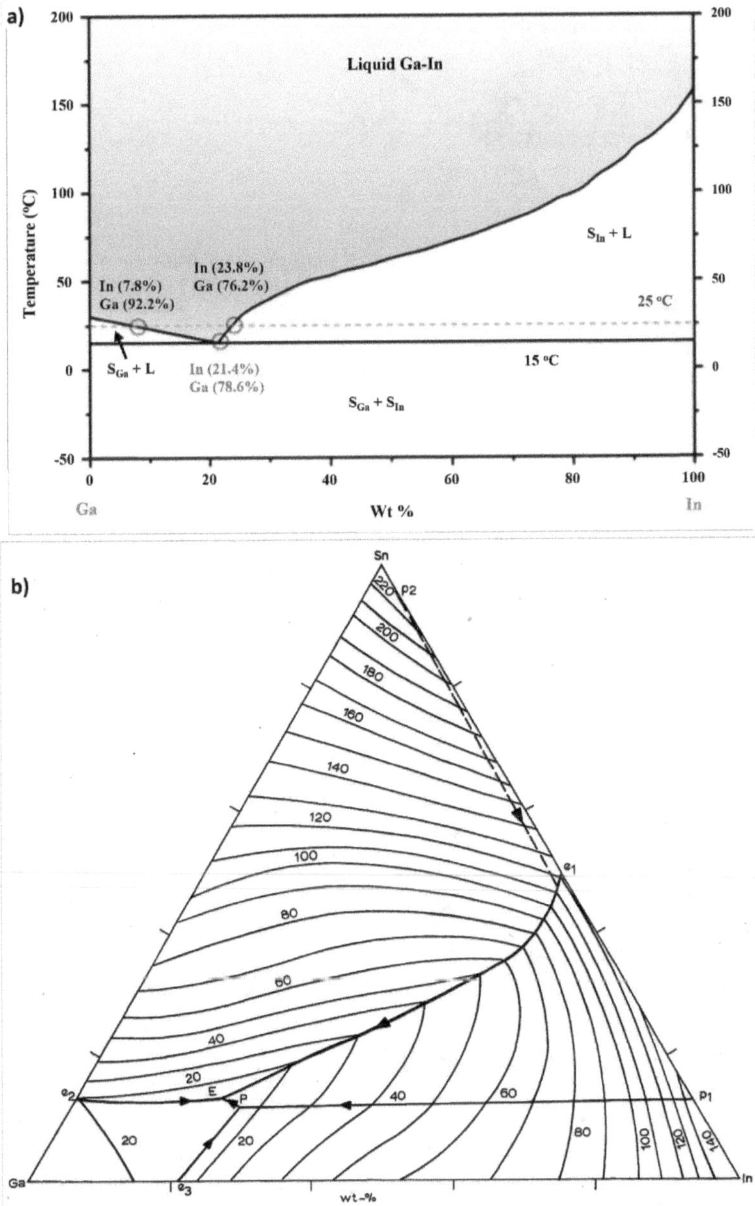

FIGURE 7.2 Phase diagram of eutectic Ga-based alloys. (a) Eutectic GaIn alloy. (b) Eutectic GaInSn alloy.

Sources: (a) Adapted with permission [3], Copyright (2019), The Royal Society of Chemistry. (b) Adapted with permission [20], Copyright (21978), The Metals Society.

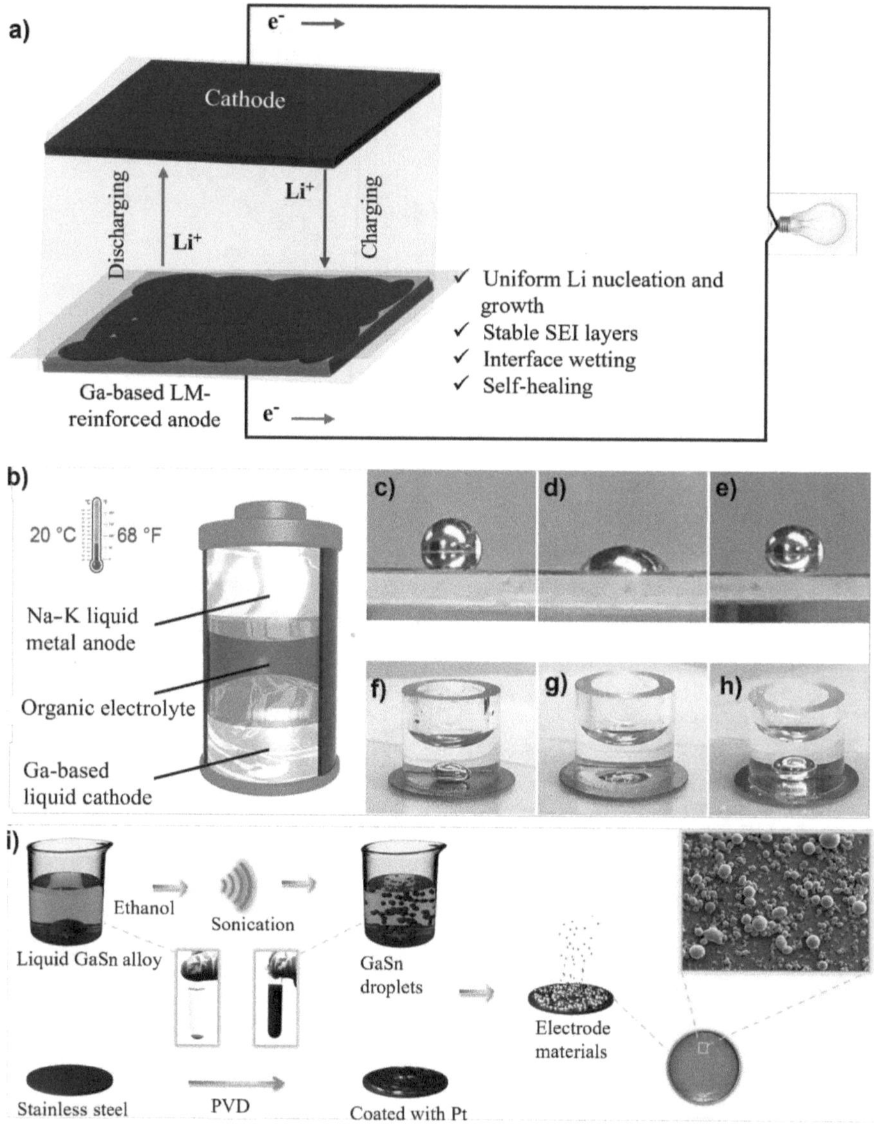

FIGURE 7.3 Configuration of Ga-based LMB and wetting behavior of Ga-based LM on various substrates. (a) Schematic configuration and working principle of the Ga-based LMBs. (b) Room-temperature Ga-based liquid metal cathode||NaK anode cell. (c–e) Wettability of GaIn LM on bare stainless steel, Au-coated stainless steel, and Al-coated stainless steel, respectively. (f–h) The wettability of GaIn on bare stainless steel, Au-coated stainless steel, and Al-coated stainless steel, respectively, immersed in Na electrolyte in the Ar-filled glove box.

Source: Adapted with permission [21], Copyright (2020), Wiley-VCH Verlag GmbH & Co. KGaA, Weinheim. (i) Preparation route for GaSn liquid electrode. Adapted with permission [10], Copyright (2021), Wiley-VCH GmbH.

insignificant at room temperature. Meanwhile, the components of LMBs are self-segregated into three layers. They are not miscible because of differences in component densities, which avoids self-discharge of the LMBs. However, the wetting between Ga-based LMs and current collectors greatly influences the cycling stability and voltage polarization of the Ga-based LMBs. The wettability of Ga-based LMs can be understood by the tendency of LMs to spread on the substrate [5, 22]. Therefore, the wettability of Ga-based LMs varies depending on the nature of the substrates used as a current collector. Figure 7.3c–e shows the wettability of GaIn alloy on bare stainless steel, Au-coated stainless steel, and Al-coated stainless steel, respectively. The GaIn greatly wets Au-coated stainless steel than bare and Al-coated stainless steel in the Ar-filled atmosphere. To understand the effect of electrolytes on the wettability of GaIn on different substrates, the bare stainless steel, Au-coated stainless steel, and Al-coated stainless was directly immersed into Na salt-based organic electrolyte and wettability was tested in the Ar-filled atmosphere, as shown in Figure 7.3f–h. The results show that the GaIn LM has similar wetting behaviors with and without Na salt-based organic electrolyte in the Ar-filled atmosphere [21].

To improve the wettability of GaSn on bare stainless-steel substrates, the GaSn can be coated on the platinum (Pt) film-coated stainless steel without a binder or conducting carbon. Figure 7.3i shows the coating of Pt on the stainless steel via physical vapor deposition (PVD) and synthesis of GaSn alloy droplets by ultrasonication in ethanol. The GaSn droplets are uniformly coated on the Pt-coated stainless steel, indicating the enhancement of witting between GaSn and stainless steel [10].

7.3 UNIQUE FUNCTIONS OF GA-BASED LMS IN RECHARGEABLE BATTERIES

The fluidic property of Ga-based LMs allows them to overcome the problems associated with solid electrode-based batteries in rechargeable batteries. Figure 7.4

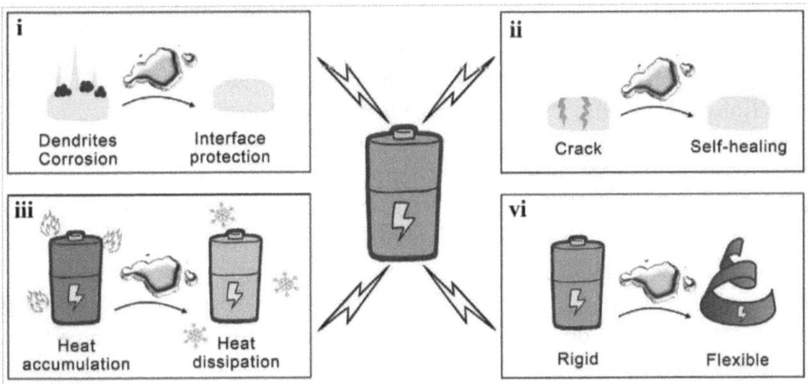

FIGURE 7.4 Schematic of unique functions of Ga-based LMs in rechargeable batteries. (i) Interface protection, (ii) self-healing electrode, (iii) thermal management, (iv) flexible batteries.

Source: Adapted with permission [5], Copyright (2024), Wiley-VCH GmbH.

illustrates the main functions of Ga-based LMs, including i) interface protection to avoid dendrite growth, ii) self-healing of the deformed surface, iii) thermal management, and iv) realizing flexible batteries [5]. Ga-based LMs can repair the deformed surface of the electrodes due to their fluidic and self-healing features, making them unique in many battery applications [3, 5]. They improve the contact at the interface, resulting in them being stable and dendrite-free, and having a long lifespan.

7.3.1 Interface Protection of Electrodes

The quality of the anode/electrolyte interface determines all the interfacial phenomena of the rechargeable batteries. The interface reactions facilitate electrolyte decomposition and result in fast capacity decay [23, 24]. Therefore, interface protection of the anode particularly improves surface contact between the deposited metal and current collector, which is promising for dense metallic deposition and endorses good cycling performance of batteries. Because of their intrinsic unique properties, the Ga-based LMs are an interface modification layer to liquefy the electrode interface [5]. Ga has lithiophilic and sodiophilic properties due to its ability to form GaLi and GaNa alloys, which control the Li^+ and Na^+ plating/stripping processes [14, 25]. Moreover, the Ga-based LMs coated on the anode current collector are good electronic conductors and regulate uniform current distribution over the surface of the current collector. The GaInSnZn alloy is typically used to coat the surface of the Ti_3C_2Tx Mxene film to induce uniform Li nucleation and growth with lower nucleation overpotential [26]. The GaInSnZn has a fluidic feature, which is crucial to cover the originally defective surface of the Ti_3C_2Tx Mxene framework. The GaInSnZn coated on the MXene film can serve as amorphous nucleation seeds, which enables isotropic Li nucleation during the initial plating process and ensures dendrite-free bulk Li deposition. Since the GaInSnZn is lithiophilic, it reduces the nucleation barrier, even at high current density [26].

One could model the formation of the GaLi binary alloy layer on the Li metal anode via a facile in-situ ion exchange reaction by dipping the Li metal anode into 0.1 M solution of $GaCl_3$ in THF. The GaLi layer is a good electronic conductive interface protection layer. It serves as a lithiophilic, which possesses dendrite-free Li deposition with a better discharge capacity of 148.7 mAh g^{-1} after 100 cycles of the NMC811||GaLi-Li cell than bare Li anode (136.4 mAh g^{-1}) [25]. Similarly, in the Na metal batteries, the Na^+ prefers to be deposited on the local swelling sites, resulting in dendritic growth of Na and interface cracks on the bare Na anode. The GaNa alloy layer can be formed by in-situ rolling of Ga metal on the Na metal (NGAL-Na), which avoids Na dendrite growth and improves the electrochemical performance of the NGAL-Na||$Na_3V_2(PO_4)_3$ cell. The GaNa alloy layer regulates Na^+ plating/stripping processes and provides uniform Na deposition, resulting in prolonged cycling of the NGAL-Na anode, in contrast to the bare Na anode [14].

7.3.2 Self-Healing Electrodes in Various Ga-Based LMBs

Due to their fluidity and high surface tension, Ga-based LMs maintain a spherical shape. Consequently, they exhibit self-healing properties, allowing them to return to their initial state after undergoing reversible solid-liquid phase transition processes in various battery technologies. The self-healing property of Ga-based LMs is crucial for ensuring stable and durable batteries [5, 22]. Ga-based LMs can be directly used as an electrode or mixed with electrode compositions such as binders, conductive agents, and active electrode materials. In both cases, Ga-based LMs change their shapes to repair the defects on the electrode surfaces, improve the interface contacts, and accommodate the volume change during the charging/discharging processes. Bulk GaIn and GaIn nanoparticles (GaIn LMNPs) are often employed as a self-healing material to recover the surface defects of the Cu current collector and mitigate the volume expansion during repeated charge/discharge processes in the Li^+ and Na^+ systems, as shown in Figure 7.5a [16]. The confined GaIn LM particles improve the contact between the deposited metals and the Cu current collector, which is crucial to prevent the detachment of deposited metals from the current collector. The GaIn LM nanoparticles disperse on the electrode surface, which facilitates the kinetics of Li^+ or Na^+ and enhances the discharge capacity of Li and Na half-cell systems to 706 mAh g^{-1} and 222.3 mAh g^{-1}, respectively. The Ga-based LMs, such as GaSn, can also be combined with the carbon nanotubes (CNTs) and styrene butadiene rubber (SBR) to coat the Li metal surface as a liquid metal-induced artificial SEI layer [27]. The liquid metal-induced SEI layers are mechanically flexible and mitigate dendrite growth and short-circuiting issues during cycling.

Moreover, Ga-based LMs can be used as self-healing electrodes to recover defects in current collectors and regulate the lithium lithiation/delithiation processes in various metal-ion batteries. Therefore, rather than directly coating Ga-based LMs onto Li metal anodes or electrode current collectors, they can be incorporated into the electrode materials. Figure 7.5b demonstrates the application of Ga-based LMs in Li-ion batteries through mixing with a conductive agent [28]. Due to their fluidic feature, introducing a small amount of Ga-based LMs, such as GaSn, into the electrode components enables the volume expansion to be accommodated. It increases the interface contact between the electrode and the current collector [29]. The Ga-based LMs can be applied for alkali metal batteries and Mg and Al-ion batteries [30, 31]. However, Ga-based LMs like eutectic GaSn (EGaSn) have poor wettability on the stainless-steel mesh substrate, affecting the Mg-ion batteries' electrochemical performance. The $CuGa_2$ layer formation on bare stainless-steel mesh is desirable to enhance surface contact and wettability between EGaSn and stainless steel [32]. The irreversible interfacial phenomena, such as dendrite growth, Al corrosion, and surface cracking of the solid Al negative electrode in the Al-ion battery, can also be resolved using Ga-based LMs as a negative electrode. Figure 7.5c reveals the roles of the Ga negative electrode in stabilizing the cycling performance of the Al-ion batteries [30]. A liquid Ga electrode possesses a Ga-Al alloying/dealloying process without surface defects and dendrite growth (Figure 7.5c–i). The achievement of stable cycling performance in an Al-ion battery with a Ga electrode (Figure 7.5c–ii) is due to the reduced challenges compared to those observed with a solid Al electrode. Due to its self-healing property during cycling, the Ga electrode retains its original spherical shape after the 100th cycle, as shown in Figure 7.5c–iii.

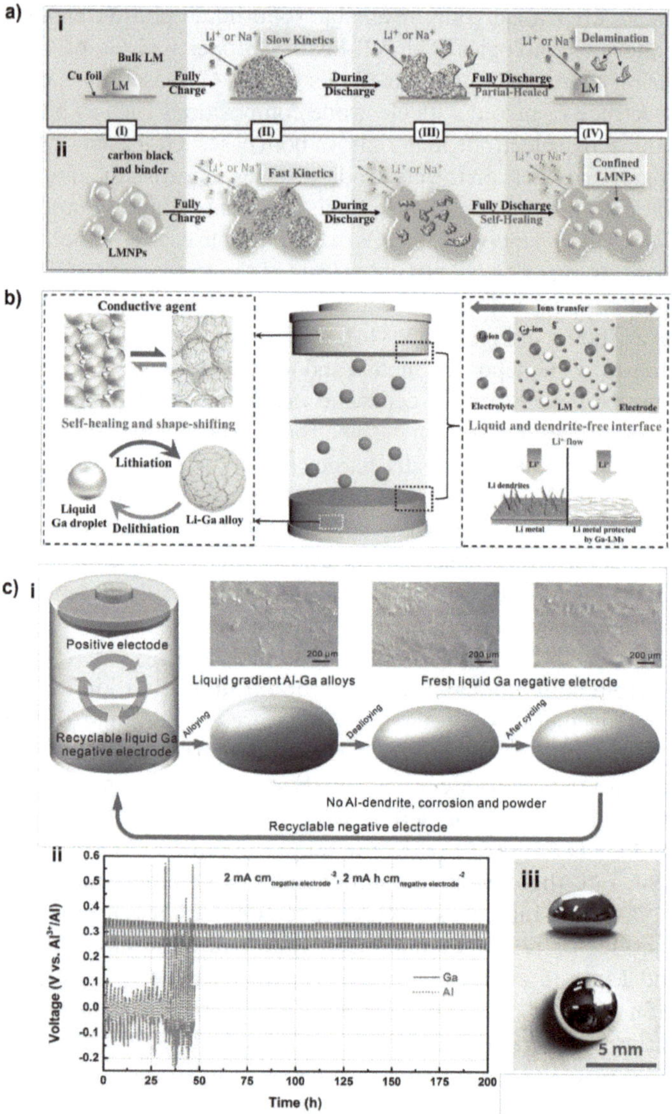

FIGURE 7.5 Schematics of the self-healing electrode design. (a) Comparison of the effects of bulk GaIn LM (a–i) and GaIn LMNPs (a–ii) on the kinetics and self-healing roles on the Cu surface. (b) Introduction of Ga-based LMs to suppress Li dendrite growth in Li-ion batteries. (c) Schematic elucidation of the dendrite-free, corrosion resistant, and recyclable liquid Ga negative electrode-based Al-ion battery (c–i), and cycling performance of symmetric cells for Ga and Al electrodes (c–ii) with a digital photo of liquid Ga negative electrode after 100 cycles (c–iii).

Sources: (a) Adapted with permission [16], Copyright (2018), Wiley-VCH Verlag GmbH & Co. KGaA, Weinheim. (b) Adapted with permission [28], Copyright (2022), The Authors, Interdisciplinary Materials Published by Wuhan University of Technology and John Wiley & Sons Australia, Ltd. (c) Adapted with permission [30], Copyright (2019), Elsevier B.V.

7.3.3 Thermal Management

One concern for many energy storage devices is heat generation during operation. Effective thermal management is essential to address this issue and ensure the steady operation of electronic systems. Ga-based LMs exhibit good thermal stability and can regulate temperature by absorbing and releasing heat through phase changes. This suggests that Ga-based LMs play a crucial role in heat dissipation technologies for the thermal management of rechargeable batteries [5]. Compared to traditional heat conductor materials, Ga-based LMs, such as EGaIn, exhibit superior thermal conductivity (~20 $W m^{-1} K^{-1}$) [33]. However, the higher melting point intermetallic compound Ga_2Cu (32.1 °C) can be easily formed when the Cu current collector is directly in contact with EGaIn alloy. The formation of Ga_2Cu causes instability of EGaIn. It lowers its thermal conductivity, which makes it difficult to use EGaIn for thermal management purposes [34]. Therefore, for a reliable thermal management application, it is crucial to enhance the stability of LMs in electronic systems. Due to its phenolic character, the natural polyphenol is suitable for surface functionalization to prevent the formation of $CuGa_2$. Thus, the tannic acid (TA)-mediated copper-LM composite (EGaIn/Cu@TA) exhibits a superior melting point (66.1 $W m^{-1} K^{-1}$), which makes it a promising material for dissipating heat generated by the integrated electronic devices [35].

In addition, a GaIn-modified polyethylene glycol/GaIn/boron nitride (PEG/GaIn/BN) phase change material (PCM) is preferentially used for thermal management of the LiBs [36]. The deformable nature of GaIn serves as a thermal bridge to reduce the interfacial thermal resistance among adjacent fillers, exhibiting a vertical and in-plane thermal conductivity of 8.8 and 7.6 $W m^{-1} K^{-1}$, respectively. Due to its capability of deformation and having naturally high thermal conductivity, GaIn not only minimizes the interfacial thermal resistance of the PCM but also improves the photothermal conversion capability of the PCM, making it a photoresponsive heater that can be heated in extremely cold temperatures. The heating of batteries can be derived not only in a hot environment but also during charge/discharge at high C-rates [36, 37]. Models show temperature profiles and the corresponding digital photos recorded by the IR camera during charging/discharging of the bare battery pack and battery pack with PEG/GaIn/BN thermal regulator at various high C-rates. Under a 4C charge/discharge rate, the bare battery pack generates a temperature higher than 60 °C. The PEG/GaIn/BN thermal regulator reduces the temperature by over 10 °C. As a result, the PEG/GaIn/BN thermal regulator improves the capacity retentions of the bare 18650 LIB and the 18650 LIB with the PEG/GaIn/BN thermal regulator.

7.3.4 Flexibility

The rigid nature of conventional batteries often leads to the delamination of electrode materials from the current collector during the charge/discharge process, causing significant capacity fading and potential short circuits. Enhancing the adhesion between electrode materials and the current collector without electrode deformation is essential for designing flexible energy storage devices. The

Ga-based LMs are mechanically stable and fluidic, and can be reshaped at a temperature lower than room temperature [18, 38]. Therefore, surface engineering of the Cu current collector and Li metal anode with GaSn, GaIn, GaInSn, GaZn, and other Ga-based LMs improves the flexibility of the anode/electrolyte interface [16, 25]. Moreover, the Ga-based LM nanoparticles (Ga-based LM NPs) can be introduced into the conductive carbon and binder during slurry preparation to enhance the flexibility of the anode material for Li-ion batteries [29]. For example, the GaIn can be used to design flexible energy storage electronic devices with 3D circuits by manipulating the solid-liquid phase change and plasticity of the GaIn [39]. The EGaIn NPs encapsulated in carbon nanofibers/carbon nanotube paper (CNF/CNT@EGaIn NPs) is a 3D free-standing anode, in which the EGaIn NPs accommodate the volume expansion of the electrode by enhancing the flexibility of the electrode and preventing electrode detachment from the current collector [40].

The volume change, polysulfide shuttling, and nonconductive nature of the solid materials, such as sulfur (S) and Li_2S, are the main challenges for alkali metal-sulfur (Li-S, K-S, and Na-S) batteries. The electrode materials should be flexible enough to tolerate mechanical deformation and mitigate the above challenges of S-based electrodes. Due to their conductive and fluidic properties, introducing the liquid metal Ga or Ga-based LMs accommodates the volume expansion. It endues flexible alkali metal-chalcogen batteries, including alkali metal-sulfur (S), alkali metal-selenium (Se), and alkali metal-tellurium (Te) batteries [41]. The liquid Ga can easily interact with the chalcogen (S, Se, and Te) particles, offering a conductive network for the chalcogen cathodes. For example, when Ga is mixed with nano S particles during cathode preparation, a thin S-Ga core-shell structure (S@Ga) can be created, which exhibits a self-adaptive characteristic to afford the volume change of sulfur. The formation of the Ga_2S_3 interlayer between the Ga shell and S core in the S@Ga composite cathode indicates the strong interaction between Ga and S particles. Besides, the S@Ga composite can restrain the polysulfide shuttling and serve as a conductive agent to improve the electronic/ionic conductivity of the cathode, ultimately enhancing the rated capacity of the S@Ga cathode [42]. In the case of silicon (Si)-based anodes, the main challenge is the significant volume change during cycling. The GaInSn (7:2:1 wt.%) in the GaInSn/Si composite anode can rapidly infiltrate gaps or cracks between Si NPs and facilitate spontaneous repairing of the deformation. Therefore, the GaInSn alloy effectively increases the flexibility of the silicon (Si) anode, enabling it to accommodate the volume changes in the Si-based electrodes.

7.4 LIMITATIONS OF GA-BASED LMBs

Despite their advantages, the Ga-based electrodes have various interfacial limitations, which have been the main research aspects of surface chemistry in practical Ga-based energy storage systems.

i) **Limited wettability:** The liquid Ga metal and its alloy, like GaIn, have a high surface tension of ~700 and >590 mN m^{-1} at room temperature, respectively, which results in low interfacial wettability between Ga-based electrodes and solid metal substrates [43]. This increases the resistance and lowers the transfer kinetics of the Ga-based LMBs. For a good wetting behavior between Ga LMs and substrates, the surface tension and viscosity should be reduced regardless of the dimensions of the Ga LMs [44]. The high surface tension nature of Ga-based LMs at normal temperature endows Ga-based LMs with the ability to form spherical droplets, affecting the fluidity of liquid metals and providing the worst wettability, toward not only solid surfaces but also liquid electrolytes. This imposes challenges on employing bare Ga liquid metal during electrode fabrication.

ii) **Corrosive:** Ga has corrosive properties toward various current collectors, so this needs to be considered when proposing Ga-based LMs as a battery component in developing energy storage devices [45]. The eutectic GaIn alloy is regularly used to modify the surface of the Cu current collector. However, the Ga metal in the GaIn alloy forms Ga_2O_3 oxide, which severely corrodes the Cu substrate. Natural polyphenol-based molecules, such as TA, can be employed as surface functionalization to create TA-mediated liquid metal-Cu (GaIn/TA@Cu) composite electrodes. The phenolic hydroxyl groups of TA strongly bind with the Ga_2O_3 layer, protecting it from direct contact with Cu, thereby mitigating Cu corrosion [35].

iii) **Leakage:** Due to their fluidic nature, all LMs, including Ga-based LMs, are generally less viscous than solid current collectors [46]. Thus, a typical Ga-based LM is often prone to leakage problems (short circuit and fire risk), which causes safety concerns for the energy storage systems. To enhance adhesion to the surface and control leakage in the energy storage devices, the nano/micro-sized Ga-based LM droplets are usually employed as dopants by cooperating with polymers [35]. Moreover, hermetic pack sealing technology is crucial to minimize battery leakage.

7.5 SUMMARY AND PERSPECTIVES

In summary, conventional liquid metal batteries require high operating temperatures (>350 °C), leading to temperature-related challenges, including fast chemical reactions and safety concerns. Recently, Ga-based LMBs have shown promise for room-temperature next-generation energy storage devices. Ga alloys liquefy metal electrodes and exhibit high electrical and thermal conductivity, resulting in fast charge-transfer kinetics at the electrode surfaces of Ga-based LMBs. In contrast to conventional LMBs, Ga-based LMBs have minimal uncontrolled chemical reactions, electrode corrosion, and safety concerns due to their low operating temperatures. This chapter briefly discusses the preparation and properties of Ga-based liquid metals, the fabrication of Ga-based LMBs, functional advantages, and their limitations. It also provides prospects for future studies in developing next-generation Ga-based LMBs.

To advance the applications of Ga-based LMBs as next-generation energy storage devices, the following research directions are suggested for further developing dendrite-free, safe, and high-performance Ga-based LMBs.

a) Ga-based LMs liquefy the electrode interfaces. However, the wetting and adhesion strength of various Ga-based LMs on the Li metal anode and other current collectors should be investigated.
b) The Ga-based LMs naturally form spherical droplets on the current collector's surface, making it challenging to achieve uniform coating on the surface. Therefore, designing an appropriate coating method is required to achieve the optimal thickness of the Ga-based LMs layer.
c) The main challenge to using Ga-based LMs in rechargeable batteries is leakage. Thus, the design of hermetic pack sealing technology is required.
d) Ga metal in the Ga alloys typically forms Ga_2O_3 oxide, which corrodes the Cu and Al current collectors when it directly comes in contact with them. Therefore, more efforts are required to employ surface functionalization using a thin artificial layer to mediate the Ga metal from the current collector.

REFERENCES

[1] W. Si, Z. Xiao, W. Ruzhu, L. Tingxian, Progress and perspectives of liquid metal batteries, *Energy Stor. Mater.*, 57 (**2023**) 205–227.
[2] H. Li, H. Yin, K. Wang, S. Cheng, K. Jiang, D.R. Sadoway, Liquid metal electrodes for energy storage batteries, *Adv. Energy Mater.*, 6 (**2016**) 1600483.
[3] X. Guo, L. Zhang, Y. Ding, J.B. Goodenough, G. Yu, Room-temperature liquid metal and alloy systems for energy storage applications, *Energy Environ. Sci.*, 12 (**2019**) 2605–2619.
[4] Y. Ding, X. Guo, G. Yu, Next-generation liquid metal batteries based on the chemistry of fusible alloys, *ACS Cent. Sci.*, 6 (**2020**) 1355–1366.
[5] Z. Zeng, C. Wang, M. Zeng, L. Fu, Gallium-based liquid metals in rechargeable batteries: From properties to applications, *Small*, 20 (**2024**) 2311099.
[6] Y. Lin, J. Genzer, M.D. Dickey, Attributes, fabrication, and applications of gallium-based liquid metal particles, *Adv. Sci.*, 7 (**2020**) 2000192.
[7] S. Liu, K. Sweatman, S. McDonald, K. Nogita, Ga-based alloys in microelectronic interconnects: A review, *Materials (Basel, Switzerland)*, 11 (**2018**) 1384.
[8] K.N. Shitaw, T.M. Tekaligne, S.-K. Jiang, C.-J. Huang, S.-H. Wu, W.-N. Su, B.J. Hwang, Opportunities of liquid metals and liquid metal cations for Li-metal batteries, *Chem. Eng. J.*, 470 (**2023**) 144062.
[9] K.Y. Kwon, V.K. Truong, F. Krisnadi, S. Im, J. Ma, N. Mehrabian, T.-I. Kim, M.D. Dickey, Surface modification of gallium-based liquid metals: Mechanisms and applications in biomedical sensors and soft actuators, *Adv. Intell. Syst.*, 3 (**2021**) 2000159.
[10] K. Wang, J. Hu, T. Chen, J. Tang, Z. Wang, N. Fan, W. Zhang, K. Wang, A high-performance room-temperature Li||Ga–Sn liquid metal battery for grid energy storage, *Energy Technol.*, 9 (**2021**) 2100330.
[11] K. Ma, J. Liu, Liquid metal cooling in thermal management of computer chips, *Front. Energy Power Eng. China*, 1 (**2007**) 384–402.

[12] G. Cao, J. Liang, Z. Guo, K. Yang, G. Wang, H. Wang, X. Wan, Z. Li, Y. Bai, Y. Zhang, J. Liu, Y. Feng, Z. Zheng, C. Lu, G. He, Z. Xiong, Z. Liu, S. Chen, Y. Guo, M. Zeng, J. Lin, L. Fu, Liquid metal for high-entropy alloy nanoparticles synthesis, *Nature*, 619 (**2023**) 73–77.

[13] D.D. Rutooj, L. Juchuan, C. Yang-Tse, W.V. Mark, Liquid metal alloys as self-healing negative electrodes for lithium ion batteries, *J. Electrochem. Soc.*, 158 (**2011**) A845.

[14] X. Lv, F. Tang, Y. Yao, C. Xu, D. Chen, L. Liu, Y. Feng, X. Rui, Y. Yu, Sodium–gallium alloy layer for fast and reversible sodium deposition, *SusMat.*, 2 (**2022**) 699–707.

[15] H. Zhang, P. Chen, H. Xia, G. Xu, Y. Wang, T. Zhang, W. Sun, M. Turgunov, W. Zhang, Z. Sun, An integrated self-healing anode assembled via dynamic encapsulation of liquid metal with a 3D Ti3C2Tx network for enhanced lithium storage, *Energy Environ. Sci.*, 15 (**2022**) 5240–5250.

[16] X. Guo, Y. Ding, L. Xue, L. Zhang, C. Zhang, J.B. Goodenough, G. Yu, A self-healing room-temperature liquid-metal anode for alkali-ion batteries, *Adv. Funct. Mater.*, 28 (**2018**) 1804649.

[17] Y. Zhu, X. Ding, Y. Jiang, Gallium-based liquid metal flexible electronics prepared by solid–liquid phase transition, *J. Electron. Mater.*, 53 (**2024**) 2763–2772.

[18] G. Bo, L. Ren, X. Xu, Y. Du, S. Dou, Recent progress on liquid metals and their applications, *Adv. Phys. X*, 3 (**2018**) 1446359.

[19] S. Koul, Y. Morita, F. Fujisaki, H. Ogasa, Y. Fujiwara, A. Kushima, Effect of liquid metal coating on improved cycle performance of anode-free lithium metal battery, *J. Electrochem. Soc.*, 169 (**2022**) 020542.

[20] D.S. Evans, A. Prince, Thermal analysis of Ga-In-Sn system, *Metal Sci.*, 12 (**1978**) 411–414.

[21] Y. Ding, X. Guo, Y. Qian, L. Xue, A. Dolocan, G. Yu, Room-temperature all-liquid-metal batteries based on fusible alloys with regulated interfacial chemistry and wetting, *Adv. Mater.*, 32 (**2020**) 2002577.

[22] Z. Zhao, S. Soni, T. Lee, C.A. Nijhuis, D. Xiang, Smart eutectic gallium–indium: From properties to applications, *Adv. Mater.*, 35 (**2023**) 2203391.

[23] K.N. Shitaw, S.-C. Yang, S.-K. Jiang, C.-J. Huang, N.A. Sahalie, Y. Nikodimos, H.H. Weldeyohannes, C.-H. Wang, S.-H. Wu, W.-N. Su, B.J. Hwang, Decoupling interfacial reactions at anode and cathode by combining online electrochemical mass spectroscopy with anode-free Li-metal battery, *Adv. Funct. Mater.*, 31 (**2021**) 2006951.

[24] K.N. Shitaw, C.-J. Huang, S.-C. Yang, Y. Nikodimos, N.T. Temesgen, S.K. Merso, S.-K. Jiang, C.-H. Wang, S.-H. Wu, W.-N. Su, B.J. Hwang, Evolution of interfacial phenomena induced by electrolyte formulation and hot cycling of anode-free Li-metal batteries, *ACS Appl. Energy Mater.*, 5 (**2022**) 7770–7783.

[25] Y. Zhou, J. Zhang, K. Zhao, Y. Ma, H. Zhang, D. Song, X. Shi, L. Zhang, Y. Ding, A novel dual-protection interface based on gallium-lithium alloy enables dendrite-free lithium metal anodes, *Energy Stor. Mater.*, 39 (**2021**) 403–411.

[26] W. Chuanliang, F. Huifang, T. Yuan, A. Yongling, G. Huanhuan, F. Jinkui, Q. Yitai, Isotropic Li nucleation and growth achieved by an amorphous liquid metal nucleation seed on MXene framework for dendrite-free Li metal anode, *Energy Stor. Mater.*, 26 (**2020**) 223–233.

[27] Z. Guangzhou, D. Heming, T. Runming, X. Buqiong, H. Tianyu, Y. Song, S. Nima, C. Qi, Z. Weixin, L. Jiyuan, Constructing a liquid-metal based self-healing artificial solid electrolyte interface layer for Li metal anode protection in lithium metal battery, *Mater. Lett.*, 262 (**2020**) 127194.

[28] B.-W. Zhang, L. Ren, Y.-X. Wang, X. Xu, Y. Du, S.-X. Dou, Gallium-based liquid metals for lithium-ion batteries, *Interdiscip. Mater.*, 1 (**2022**) 354–372.

[29] Y. Wu, L. Huang, X. Huang, X. Guo, D. Liu, D. Zheng, X. Zhang, R. Ren, D. Qu, J. Chen, A room-temperature liquid metal-based self-healing anode for lithium-ion batteries with an ultra-long cycle life, *Energy Environ. Sci.*, 10 (**2017**) 1854–1861.

[30] H. Jiao, S. Jiao, S. Li, W.-L. Song, H. Chen, J. Tu, M. Wang, D. Tian, D. Fang, Liquid gallium as long cycle life and recyclable negative electrode for Al-ion batteries, *Chem. Eng. J.*, 391 (**2020**) 123594.

[31] S. Chao, Y. Yuan, G. Dachong, C. Tao, L. Yuping, T. Aitao, W. Liang, L. Dajian, P. Fusheng, The evaluation of Mg–Ga compounds as electrode materials for Mg-Ion batteries via ab initio simulation, *J. Electrochem. Soc.*, 168 (**2021**) 110539.

[32] S. Meijia, Y. Bin, C. Wenrui, Y. Wanfeng, B. Qingguo, Z. Zhonghua, A self-healing room-temperature liquid eutectic GaSn anode with improved wettability for advanced Mg ion batteries, *Chem. Eng. J.*, 435 (**2022**) 134903.

[33] W. Xiaohong, L. Chennan, R. Wei, Liquid metal-based thermal interface materials with a high thermal conductivity for electronic cooling and bioheat-transfer applications, *Appl. Therm. Eng.*, 192 (**2021**) 116937.

[34] A. Francois-Marie, M. Salma, T. Jianbo, Z. Chengchen, M. Andrea, C. Shengxiang, B.G. Mohammad, R. Md Arifur, M. Maxime, L. Sean, Z. Jin, H. Lachlan, M. Mohannad, V.C. Benjamin, S.R. Rodney, K.-Z. Kourosh, Carbonization of low thermal stability polymers at the interface of liquid metals, *Carbon*, 171 (**2021**) 938–945.

[35] C. Zhang, Y. Tang, T. Guo, Y. Sang, D. Li, X. Wang, O.J. Rojas, J. Guo, Liquid metal compartmented by polyphenol-mediated nanointerfaces enables high-performance thermal management on electronic devices, *InfoMat.*, 6 (**2024**) e12466.

[36] C. Guo, L. He, Y. Yao, W. Lin, Y. Zhang, Q. Zhang, K. Wu, Q. Fu, Bifunctional liquid metals allow electrical insulating phase change materials to dual-mode thermal manage the Li-ion batteries, *Nano-Micro Lett.*, 14 (**2022**) 202.

[37] Y. Liu, Y. Zhu, Y. Cui, Challenges and opportunities towards fast-charging battery materials, *Nat. Energy*, 4 (**2019**) 540–550.

[38] M.D. Dickey, Stretchable and soft electronics using liquid metals, *Adv. Mater.*, 29 (**2017**) 1606425.

[39] G. Li, M. Zhang, S. Liu, M. Yuan, J. Wu, M. Yu, L. Teng, Z. Xu, J. Guo, G. Li, Z. Liu, X. Ma, Three-dimensional flexible electronics using solidified liquid metal with regulated plasticity, *Nat. Electron.*, 6 (**2023**) 154–163.

[40] Y. Jiayu, X. Jun, G. Xianggang, X. Gangyi, Z. Heliang, Y. Shuai, C. Luojia, Y. Yang, Z. Shichao, X. Yalan, Y. Puheng, Self-healing liquid metal confined in carbon nanofibers/carbon nanotubes paper as a free-standing anode for flexible lithium-ion batteries, *Electrochim. Acta*, 425 (**2022**) 140721.

[41] L. Ren, B.-W. Zhang, Room temperature liquid metals for flexible alkali metal-chalcogen batteries, *Exploration*, 2 (**2022**) 20210182.

[42] M. Zhu, S. Li, B. Li, S. Yang, A liquid metal-based self-adaptive sulfur–gallium composite for long-cycling lithium–sulfur batteries, *Nanoscale*, 11 (**2019**) 412–417.

[43] S. Zhu, Q. Lin, R. Cao, K. Xie, J. Lai, Interfacial tension of Ga, E-GaIn, galinstan, and GaInSnBiZn high-entropy alloy, *J. Mater. Eng. Perform.*, 33 (**2024**) 2369–2378.

[44] Y. Yuhao, Z. Yuan, O. Hongsheng, H. Jianhan, H. Xiaohong, Experimental investigation on optimization of the performance of gallium-based liquid metal with high thermal conductivity as thermal interface material for efficient electronic cooling, *Int. J. Heat Mass Transf.*, 226 (**2024**) 125455.

[45] Y.-G. Deng, J. Liu, Corrosion development between liquid gallium and four typical metal substrates used in chip cooling device, *Appl. Phys. A*, 95 (**2009**) 907–915.

[46] Y. He, J. You, M.D. Dickey, X. Wang, Controllable flow and manipulation of liquid metals, *Adv. Funct. Mater.* (**2023**) 2309614.

8 K-, Mg-, and Ca-Based Liquid Metal Batteries

Aminur Rahman, Abu Bin Imran, and Md. Abu Bin Hasan Susan

8.1 INTRODUCTION

The urgent need to combat global warming and reduce greenhouse gas emissions has driven substantial progress in renewable and clean energy technologies. Solar and wind power have seen remarkable advancements; however, their inherent intermittency calls for dependable grid-scale energy storage solutions to maintain a consistent energy supply. Liquid metal batteries (LMBs) have emerged as one of the leading contenders to address this challenge [1]. They offer high power density, long cycle life, low cost, self-healing properties, and scalability, making them ideal for sustainable energy storage solutions [2]. The development of LMBs has historical roots in early twentieth-century advancements in electrolytic processes [3]. Initially conceived in a different economic and political context, these batteries have reemerged as cutting-edge technology to tackle modern energy challenges. Unlike traditional lithium-ion batteries, which are primarily designed for portable electronics, LMBs are tailored explicitly for grid-scale applications. Their design and performance meet the demands of contemporary energy systems.

The fundamental structure of an LMB consists of two liquid metal electrodes separated by a molten salt electrolyte. The top electrode typically comprises a highly electropositive metal such as potassium (K), magnesium (Mg), or calcium (Ca), serving as the negative electrode, while the bottom electrode is made of metals such as antimony (Sb), bismuth (Bi), or lead (Pb), functioning as the positive electrode. The selection of these materials is guided by their electronegativity and deposition potentials, ensuring optimal performance. The molten salt electrolyte in LMB operates by separating the liquid metal electrodes and enabling ion conduction. The molten state of the electrolyte ensures that the electrodes remain liquid, allowing for rapid ion transport and fast reaction kinetics, which is particularly advantageous for applications requiring quick charge and discharge cycles. The differences in density between the electrodes and electrolytes naturally segregate these components, eliminating the need for physical separators and simplifying the battery design [4]. Electrochemical reactions in LMBs are driven by differences in electronegativity between the electrodes. During discharge, the negative electrode (located at the top) undergoes oxidation, releasing ions that diffuse through the molten electrolyte and form an alloy with the positive electrode (located at the bottom). During charging, the alloyed metal in the bottom electrode is oxidized, dissolves into the molten

DOI: 10.1201/9781003489726-8

electrolyte, and is then reduced back to its original state at the top negative electrode [5]. These reactions can be summarized as follows:

- Negative electrode reaction: $A \rightarrow A^{n+} + ne^-$; (8.1)
- Positive electrode reaction: $A^{n+} + ne^- + B \rightarrow A \text{ (in B)}$; (8.2)
- Overall cell reaction: $A + B \rightarrow A \text{ (in B)}$. (8.3)

These discharging (alloying) and charging (dealloying) cycles are essential for the efficient operation of LMBs. The discharging and charging cycles of a LMB are schematically presented in Figure 8.1.

Among the metals used in LMBs, K has a relatively low melting point (63.5°C), which facilitates easier maintenance of the molten state during operation, leading to high ionic conductivity and fast ion transport. K-based LMBs exhibit a favorable electrode potential, particularly suitable for grid-scale applications. The lower melting point also alleviates challenges of thermal management, potentially resulting in more cost-effective designs. Mg is another key candidate with a high energy density of up to 5.2 Ah/g and excellent electrochemical stability. However, Mg-based LMBs require higher operating temperatures (approximately 650°C), which can be managed through the internal heat generated during operation. This high temperature ensures the melting of the metal electrodes and the salt electrolyte, enabling rapid ion transport and fast reaction kinetics. K, like Mg, is an earth-abundant element with a high energy density (up to 3.86 Ah/g) and low standard electrode potential, which leads to strong alloying behavior with the positive electrode. Although Ca-based LMBs

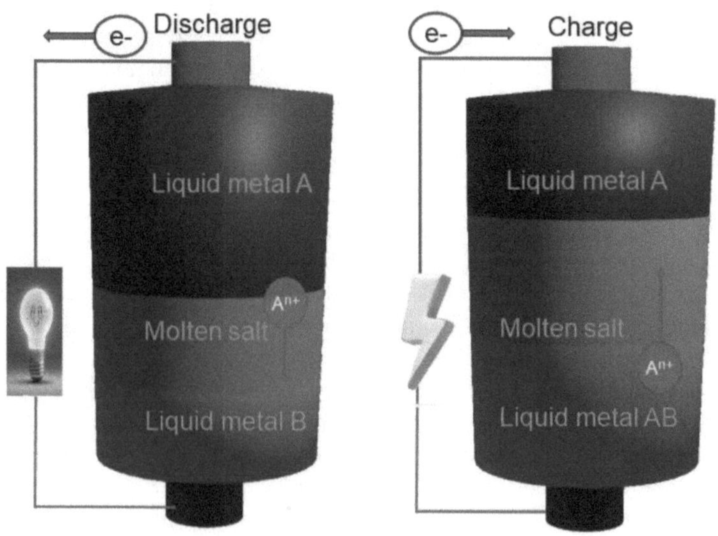

FIGURE 8.1 Schematic illustration of discharging (alloying) and charging (dealloying) cycles of an LMB.

also require higher operating temperatures, typically around 450–600°C, they offer superior electrochemical performance. The abundance and low cost of K make it an ideal candidate for large-scale energy storage applications. Significant advancements have been made in developing K-, Mg-, and Ca-based LMBs. Researchers have optimized electrolyte composition for K-based systems to reduce operating temperatures further while maintaining high ionic conductivity. Innovations in thermal management strategies have also been pivotal, ensuring that K-based LMBs can operate efficiently without excessive external heating. In Mg-based LMBs, the focus has been on improving the alloying behavior of Mg with various positive electrode materials. Research into different molten salt compositions, including $MgCl_2$-based electrolytes, has shown promise in enhancing overall battery efficiency. Developing corrosion-resistant materials for current collectors and other cell components has also been critical for improving battery longevity. K-based LMBs have seen advancements in electrolyte design, with efforts to reduce the melting point of molten salts to lower operating temperatures. Researchers are also working on high-performance cathode materials that can form stable alloys with K, enhancing overall energy density and cycle life.

Despite their potential, several challenges must be addressed to optimize K-, Mg-, and Ca-based LMBs for commercial applications. One primary challenge is mass transfer within the battery. Uncontrolled mass transfer can lead to phase segregation, dendrite formation, and other defects that compromise efficiency and stability, particularly in Mg and Ca-based LMBs, where high operating temperatures exacerbate these problems [6]. Thermal management is another significant challenge. The high temperatures required to maintain the molten state necessitate advanced thermal insulation and heat dissipation strategies. While internal heat generated during battery operation can help sustain the molten state, it can also lead to overheating if not properly managed. Sealing and corrosion must be addressed to ensure the long-term reliability of LMBs. The aggressive thermal environment within these batteries can degrade sealing materials and current collectors, reducing battery lifespan [7]. Developing high-temperature, corrosion-resistant materials for these components is crucial for advancing LMB technology. Scaling up K-, Mg-, and Ca-based LMBs for commercial use presents engineering challenges, including cost-effective manufacturing processes, ensuring the availability of high-purity raw materials, and designing batteries that can withstand mechanical stresses associated with large-scale applications. In parallel, K-, Mg-, and Ca-based metal-ion batteries are emerging as alternatives to traditional lithium-ion batteries, driven by the need for sustainable, cost-effective, and high-performance energy storage solutions [8]. These batteries operate efficiently at room temperature, making them versatile for applications ranging from portable electronics to electric vehicles. The choice of metal ions and electrolyte composition is crucial in determining battery performance, energy density, and overall stability.

K-, Mg-, and Ca-based LMBs thus hold significant promise as next-generation energy storage solutions for grid-scale applications due to their high power density, long cycle life, and low cost. These features make them ideal for integrating renewable energy sources into the grid, reducing reliance on fossil fuels and mitigating climate change. However, challenges such as mass transfer, thermal management, sealing, corrosion, and scalability must be addressed to unlock their full potential. Continued research and development, supported by experimental efforts, are crucial for overcoming these obstacles and promoting the widespread adoption of LMBs in sustainable energy infrastructures. This chapter provides a comprehensive overview of K-, Mg-, and Ca-based LMBs, focusing on advancements in electrode and electrolyte materials, while discussing recent achievements and emerging issues that hinder practical applications. The discussion extends from high-temperature variants to room-temperature options, electrolyte considerations, and perspectives on metal-ion batteries. As the world transitions to renewable energy, LMBs will be vital in providing reliable, cost-effective, high-performance solutions for large-scale energy storage.

8.2 HIGH-TEMPERATURE LIQUID METAL BATTERIES

High-temperature liquid metal batteries (HTLMB) systems are advanced energy storage technologies primarily developed for grid-scale applications. Operating at temperatures typically above 300°C, these batteries utilize molten metals as electrodes and molten salts as electrolytes [9]. The system usually includes a low-density metal, such as Mg or Ca, as the negative electrode, and a high-density metal, such as antimony or lead, as the positive electrode, with a molten salt electrolyte separating them. HTLMBs evolved from metal refining technologies, leveraging the stability of liquid electrodes that maintain their morphology during operation, unlike solid electrodes that degrade over time. The use of abundant and inexpensive materials further enhances their cost-effectiveness. However, the elevated operating temperatures present challenges, such as the need for robust thermal management and specialized corrosion-resistant materials. HTLMBs offer significant advantages, including long cycle life, low cost, and the ability to handle large power fluctuations, making them highly suitable for large-scale energy storage.

8.2.1 Mg-Based HTLMBs

Mg, due to its high volumetric capacity, abundance, and low cost, has emerged as a compelling candidate for use in LMBs, positioning these systems as a potential breakthrough in the field of energy storage. Mg-based LMBs are typically composed of a molten Mg anode, a molten salt electrolyte, and a cathode made of either a metal or a metal compound. One of the primary advantages of Mg-based LMBs is their safety. Unlike lithium, Mg does not form dendrites easily, which reduces the risk of short-circuiting and thermal runaway—a common safety concern in lithium-ion batteries [10]. Furthermore, Mg is non-toxic and more environmentally friendly, aligning with the increasing demand for green and sustainable energy storage technologies. The use of a molten salt electrolyte in Mg-based LMBs also contributes to the system's safety, as these electrolytes typically have high thermal stability and a wide electrochemical window, which further reduces the risk of battery failure.

The development of Mg-based LMBs faces several challenges. One significant challenge is the high operating temperature required to maintain the liquid state of the components, typically above 500°C. This high-temperature operation necessitates robust thermal management systems, which can add complexity and cost to the battery design. Selecting suitable cathode materials that can reversibly intercalate or alloy with Mg ions at these high temperatures is a critical area of research. Identifying cathode materials that provide high energy density, low polarization, and good long-term stability is essential for Mg-based LMBs.

Recent research has focused on optimizing the electrolyte composition to improve ionic conductivity and reduce the operating temperature of Mg-based LMBs. Innovations in electrolyte chemistry, such as developing novel eutectic mixtures and ionic liquid-based systems, lower the melting point of the electrolyte and enhance the overall performance of the battery. Advances in materials science have led to the discovery of new cathode materials and improved understanding of the electrochemical mechanisms governing Mg storage, paving the way for the next generation of Mg-based LMBs.

Bradwell et al. have proposed a high-temperature Mg–antimony (Mg||Sb) LMB operating at 700°C [11]. The cell configuration includes a Mg negative electrode, a molten salt electrolyte ($MgCl_2$–KCl–NaCl), and an antimony (Sb) positive electrode (Figure 8.2a). Due to differences in density and immiscibility, the cell naturally stratifies into three distinct liquid layers: Mg at the top, molten salt in the middle, and Sb at the bottom. During discharge, Mg is oxidized at the negative electrode, releasing Mg^{2+} ions into the electrolyte and electrons into the external circuit. Concurrently, Mg^{2+} ions are reduced at the positive electrode, forming an Mg–Sb alloy. The reverse process occurs during charging, with Mg being electrochemically extracted from the alloy and deposited back onto the negative electrode (Figure 8.2b). Electrochemical performance was assessed through cyclic voltammetry (CV) and electrochemical impedance spectroscopy (EIS). The CV scans exhibited linear current-voltage relationships, and EIS scans showed negligible charge-transfer overpotentials, indicating low resistance and high efficiency. However, mass-transport limitations were observed at lower EIS scan frequencies, likely due to Mg^{2+} ion depletion at electrode-electrolyte interfaces. Cells cycled at 50 mA/cm^2 for a set discharge duration of 10 hours and a cutoff charging voltage of 0.85 V achieved a round-trip coulombic efficiency of 97% and a voltage efficiency of 71%, leading to an overall energy efficiency of 69% (Figure 8.2c). At full discharge, the positive liquid electrode was composed of approximately 12 mol% Mg and 88 mol% Sb. The cells were fully discharged at 50 to 200 mA/cm^2 rates with a 0.05 V cutoff limit (Figure 8.2d). Higher current densities caused increased IR voltage loss and reduced capacity, aligning with observed solution resistance and mass-transport limitations. Efficiency can be improved by reducing electrolyte thickness, lowering current density, or optimizing the current collector design and electrolyte composition to enhance conductivity. The cell performance could be enhanced by optimizing current collector design, electrolyte composition, and operating conditions to reduce mass-transport limitations and improve overall efficiency. Initially, the Mg-Sb cell was not considered viable for practical use due to the high melting points of Mg (648 °C) and Sb (630 °C), as well as the low open-circuit voltage (OCV) of the cell. Moreover, high operating temperatures, low cell voltage, and limited rate capability made Mg-Sb cells unsuitable for large-scale energy storage applications.

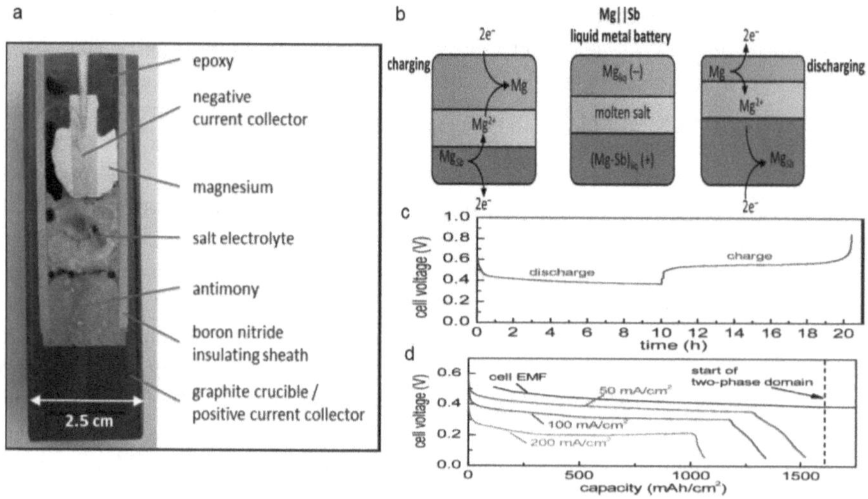

FIGURE 8.2 Depicts the Mg||Sb LMB and its electrochemical performance (a) Upon cooling to room temperature, three distinct liquid layers are formed within the battery. (b) A schematic representation shows the three-layer liquid structure of the battery. During charging, Mg is extracted from the Mg–Sb alloy and deposited as liquid Mg at the top electrode, while during discharging, Mg is removed from the top electrode and incorporated into the Mg–Sb alloy at the bottom electrode. (c) The cell voltage varies in relation to the state of charge during a cycle at 50 mA/cm^2. (d) Results demonstrate deep discharge behavior at different current rates, with the theoretical electromotive force (EMF) calculated based on literature values.

Source: Adapted with permission [11], Copyright (2012), American Chemical Society.

8.2.2 Ca-Based HTLMBs

The abundance, low cost, and high energy density of Ca position them as a sustainable alternative to lithium and sodium for scalable energy storage solutions. The basic structure of a Ca-LMB comprises a molten Ca anode, a molten salt electrolyte, and a cathode made from metal or metal alloy. Operating at high temperatures, these components facilitate rapid ion transport, enabling high current densities and energy efficiencies. The liquid nature of the electrodes and electrolyte helps avoid issues commonly found in solid-state batteries, such as dendrite formation and mechanical stress. The high melting point of Ca (842°C) supports stable operation at elevated temperatures, enhancing the electrochemical performance of the battery and reducing the risk of thermal runaway, a significant safety concern in many battery systems [12]. Furthermore, the high volumetric capacity of Ca contributes to higher energy densities in Ca-LMBs, offering more compact and efficient storage systems. Ca offers several advantages over other alkali and alkaline-earth metal systems. For example, lithium and sodium-based LMBs provide high energy densities, but they are prone to dendrite formation, leading to short-circuiting and catastrophic failure. In contrast, Ca has a lower propensity for dendrite formation, making it inherently safer for high-temperature battery systems. Additionally, widespread availability of

Ca alleviates concerns about resource scarcity and supply chain disruptions, making Ca-LMBs a more sustainable option for large-scale energy storage.

The development of Ca-LMBs is still in its rudimentary stages, with several technical challenges yet to be addressed. A major hurdle is identifying electrolyte compositions that can function efficiently at the high temperatures necessary for molten calcium operation [13]. To ensure efficient operation and long-term stability, the electrolyte must exhibit high ionic conductivity, thermal stability, and compatibility with the Ca anode and the cathode material. Recent research has focused on developing novel molten salt mixtures and ionic liquid-based electrolytes that meet these requirements while lowering the overall operating temperature of the battery. Another challenge is selecting appropriate cathode materials that can effectively intercalate or alloy with Ca ions during battery operation. The cathode must provide high capacity and low polarization while maintaining structural integrity over many cycles to ensure longevity of the battery. Advances in materials science, particularly in understanding Ca-based alloys and intermetallic compounds, are crucial to optimizing the performance of Ca-LMBs.

In the current generation of Ca-LMBs, researchers are exploring Ca-based liquid alloys as electrodes. A typical cell design might include a negative electrode of Ca alloyed with a metal such as Mg, Zn, or Al, and a positive electrode made of Bi, Sb, Sn, or Pb. The electrolyte is a molten mixture of alkali and/or alkaline-earth halide salts. The electrochemical reaction within the cell can be described as Ca(in A) → Ca(in B), where the thermodynamic driving force for this reaction is determined by the change in partial molar Gibbs free energy, which is related to the cell equilibrium voltage by the Nernst equation. Achieving a high equilibrium cell voltage, essential for effective cell discharge, depends on maintaining a low Ca activity in metal B and a high Ca activity in metal A.

Kim et al. studied the Ca-bismuth (Ca-Bi) liquid alloy as a positive electrode material for high-temperature LMBs operating between 500 and 700°C [14]. The research aimed to elucidate the electrochemical behavior, efficiency, and applications of the Ca-Bi system. CV in a LiCl-NaCl-$CaCl_2$ molten salt at 600°C revealed Ca deposition and removal near 0.0 V, with chlorine evolution above 3.5 V. Using a Ca-Bi counter electrode highlighted a distinct faradaic process at approximately 0.78 V related to subvalent Ca ions, a phenomenon absent with tungsten electrodes, though low current densities necessitated further optical spectroscopy validation. The study also observed a significant drop in current density above 1.2 V with a pure bismuth working electrode, indicating mass-transport limitations. Thermodynamic behavior, assessed by coulometric titration, confirmed system stability in the liquid $Ca_{11}Bi_{10}$ two-phase domain, while kinetic analysis showed high coulombic efficiency (>98%) but a sharp electrode potential drop due to $Ca_{11}Bi_{10}$ nucleation at high current densities. Dendrite formation at the electrode-electrolyte interface, voltage inefficiencies, and increasing self-discharge current densities with temperature posed operational challenges. Barium co-deposition was observed during examination, which increased discharge capacities and indicated possible enhancements in performance.

Ca-based systems are a viable candidate for grid-scale energy storage. Ning et al. investigated the Ca-Pb electrode couple in Ca|LiCl-NaCl-CaCl2|Pb LMBs, demonstrating its cost-effectiveness [15]. Their findings revealed that the Ca-Pb alloy exhibited excellent electrochemical properties, including a linear relationship between electrode potential and current density (50–200 mA/cm^2), a high discharge voltage of 0.6 V, and a low self-discharge current density of less than 2 mA/cm^2 at 600°C. The high coulombic efficiency (>98.84%) and uniform intermetallic distribution suggest efficient Ca diffusion, providing valuable insights for the design of high-performance Ca-based LMB electrodes.

Ouchi et al. explored a Ca-Sb LMB using a Ca(in Bi) | LiCl-NaCl-CaCl2 | Ca(in Sb) configuration [16]. Their results indicated a linear electrode potential (0.95–0.75 V vs Ca) across current densities of 50–500 mA/cm^2, achieving a remarkable 100% coulombic efficiency and minimal capacity fade (<0.01% per cycle). The performance enhancements observed with increasing temperature were attributed to improved Ca solubility and diffusivity in Sb, reinforcing the potential of Ca-Sb LMBs for grid-scale applications at a low cost.

The EMF of Ca-based electrodes has provided promising data. For instance, a three-electrode cell designed to investigate the thermodynamic properties of Ca-Bi alloy yielded an average OCV of approximately 0.8 V at temperatures between 600 and 800°C. Comparatively, studies of the Ca-Sb system indicated an average OCV of around 1.0 V, 0.2 V higher than that of Ca-Bi, suggesting that Ca can deliver a relatively high OCV suitable for large-scale energy storage.

Zhou et al. propose a dual-cation (Ca^{2+} and Li^{+}) LMB to address the limitations of single-cation systems, such as low energy density in Ca-based LMBs, high cost in Li-based LMBs, and short lifespan in Na-based LMBs. Their hybrid design integrates traditional methods, first-principles calculations, and machine learning to stabilize the dual-cation system. Adding Mg as an inert additive enhances cycling performance and reveals that specific ion ratios are crucial for optimal dual-cation chemistry. These insights are expected to advance LMB technology, improving performance and expanding its application. One effective strategy to enhance the performance of Ca-based LMBs involves lowering the melting temperature of Ca-based electrodes through alloying. Newhouse et al. demonstrated that forming a low-melting Ca-Mg alloy (with a eutectic temperature of 440°C) could significantly reduce the operating temperature [5]. Using a Ca-Mg alloy with a high Ca concentration would decrease the cell voltage by less than 30 mV compared to pure Ca, while alloys with lower Ca concentrations resulted in a voltage drop of around 200 mV. Ca-Sb LMBs with a LiCl-NaCl-$CaCl_2$ electrolyte achieved near 100% coulombic efficiency and confirmed that Ca dissolution from a Ca-Mg alloy in molten salt electrolytes can be minimized through alloying and the use of multi-cation electrolytes [16]. Ca-Sb cells exhibited a higher discharge cell voltage (approximately 0.1 V higher than Ca-Bi cells), superior cycling performance, and lower costs for electrode and electrolyte materials. Overall, the development of Ca-based LMBs, particularly through the use of eutectic Ca-Mg alloys, has shown significant progress. These advancements enable the operation of liquid Ca-based electrodes at lower temperatures, reduce Ca solubility in molten salt electrolytes, and achieve favorable cell performance, positioning Ca-based LMBs as promising candidates for future energy storage solutions.

8.2.3 K-Based HTLMBs

K-based LMBs are challenging to form at high temperatures primarily because K has a low melting point of about 63.5°C (146.3°F) [17]. K can easily vaporize or react with other materials at high temperatures to maintain a stable liquid metal state. In addition, the low melting point limits the temperature range over which K can remain in a liquid state, complicating the design and stability of the battery under high-temperature conditions.

8.3 ROOM-TEMPERATURE LIQUID METAL BATTERIES

Operating batteries at safer temperatures while maintaining high performance, such as achieving round-trip energy efficiencies above 70% and lifespans over 4000 cycles, is advantageous for both safety and cost. Developing room-temperature liquid metal batteries (RTLMBs) that function between 0 and 40°C aligns with the conditions used in widely adopted lithium-ion batteries, allowing for adapting established chemistries and technologies. Elements and alloys with melting points at or below room temperature, such as mercury, gallium, rubidium, cesium, francium, and NaK alloys, are known. Mercury, with a melting point of 234 K, was initially used in HTLMBs but is limited due to its toxicity. Rubidium, cesium, and francium are less feasible due to radiation hazards. Gallium is favorable for RTLMBs due to its safety and moderate potential relative to lithium [18]. In lithium-ion batteries, Li_2Ga has a theoretical capacity of 769 mAh/g, while sodium-ion batteries with Na_2Ga_3 achieve around 217 mAh/g. NaK alloys, liquid at temperatures down to −12.6°C with capacities of 629 mAh/g for sodium and 579 mAh/g for K, offer a low-cost alternative due to cheaper raw materials and simple synthesis. The liquid-solid interface with NaK enhances contact with solid-state electrolytes. The following section provides a comprehensive analysis of application of RTLMBs in stabilizing metal anodes, focusing on K-, Mg-, and Ca-based systems, and discusses recent progress and future directions in this field.

8.3.1 K-Based RTLMBs

K has a melting point of just 64 °C and a very low electronegativity, similar to lithium. Due to its low operating temperature, liquid mercury (Hg) is well-suited as a positive electrode in combination with a liquid K electrode. In 1960, General Motors introduced one of the earliest LMB models, using K and Hg as the respective negative and positive electrodes, with KOH-KBr-KI as the electrolyte [5]. The K-Hg cells successfully cycled at a current density of 94 mA/cm^2 at 300 °C, achieving a coulombic efficiency of 90–95%. The system operated for 60 hours and achieved a power density of 48 mW/cm^2 with a configuration of three cells. However, the impracticality of using mercury due to its toxicity and high vapor pressure limits the feasibility of K-Hg batteries for large-scale energy storage.

8.3.2 Na-K Alloy-Based RTLMBs

Alkali metal alloys, particularly Na-K, have garnered significant attention for use in RTLMBs due to their low melting points and superior abundance compared to lithium, which is more than a thousand times less available. The eutectic composition of Na-K alloys, which remains liquid below room temperature (−12.6°C), presents a promising alternative to traditional solid metal anodes [19]. This liquid state inherently prevents dendrite formation, a persistent issue in lithium-based batteries. The K-Na phase diagram further highlights the distinct zones of dendrite-free and dendrite-forming regions for creating stable, high-capacity rechargeable batteries.

Utilizing liquid Na-K alloys lies in their immiscibility with most hydrocarbon liquids and their high surface tension, which inhibits wetting and penetration of porous separators [9]. While beneficial in preventing short circuits, this characteristic complicates the direct use of Na-K alloys as anodes. When the alloy is simply placed on a separator, it remains a droplet, rendering it unsuitable for conventional battery designs. However, when absorbed into a porous substrate, Na-K alloys can form a stable, sheet-like electrode. This approach forms a liquid-liquid anode-electrolyte interface, essential for maintaining electrode integrity during cycling.

The fluid nature of Na-K alloys and their high surface tension necessitate using scaffolds for effective encapsulation. Carbon-based materials are particularly suitable for this purpose due to their tunable porosity and low density. Zhang et. al. developed a liquid Na-K alloy-based anode by incorporating the alloy into an in situ-formed graphite intercalation compound (GIC) framework (Figure 8.3a) [20]. Initially, they immersed carbon paper (CP) into the liquid Na-K alloy (Figure 8.3b). The K in the alloy diffused into the CP more rapidly than sodium, creating a GIC (KC8) framework with a gold hue. This framework exhibited excellent wettability with the Na-K alloy. Consequently, the liquid alloy was absorbed into the GIC structure, forming the NaK-GIC-carbon (NaK-G-C) electrode. This design, which incorporates a self-healing property to prevent dendrite formation, allows the GIC networks and liquid Na-K to efficiently facilitate electron and mass transport (Figure 8.3c). This NaK-G-C electrode exhibits remarkable cyclic stability, maintaining consistent performance for over 5000 hours at 20 mA/cm^2 (Figure 8.3d). The electrode also performed well at extreme current densities (50–80 mA/cm^2) and high capacities (16–20 mAh/cm^2) (Figure 8.3d), highlighting the effectiveness of the GIC network in preventing dendrite formation and outperforming other alkali metal anodes (Figure 8.3f).

Other scaffolds, such as metal foams [21], carbon fiber cloths [22], and graphene-based materials [23], have also been employed to encapsulate liquid Na-K alloys, further broadening their applicability in achieving stable, dendrite-free metal anodes. Na-K alloys have also been used to enhance the surface stability of Na metal anodes through coating techniques, improving the overall interfacial stability between solid electrodes and metal anodes.

The low melting point and favorable electrochemical properties of Na-K alloys, including their reduction potentials (–2.71 V for Na and –2.92 V for K vs. standard hydrogen electrode), make them ideal candidates for dendrite-free battery applications. In 2016, Goodenough proposed the use of liquid Na-K alloys as anodes in rechargeable batteries, leveraging their self-healing and deformable nature to block dendrite formation. Subsequent research has focused on the electrochemical behaviors of these alloys in various rechargeable battery systems. The stripping/plating processes of Na or K within the alloy depend on the ion selectivity of the cathode or the type of electrolyte used. For instance, Na^+-based electrolytes favor Na stripping/plating, while K^+-based electrolytes support K processes. A well-designed Na^+/K^+

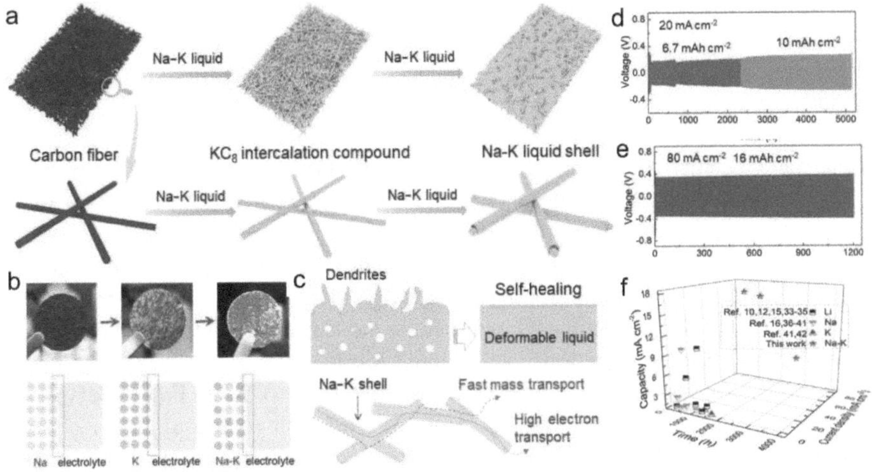

FIGURE 8.3 (a) Schematic of the NaK-G-C fabrication process. (b) Photographs showing the transformation from CP to NaK-G-C. (c) Schematic illustration of the functional mechanism of the NaK-G-C electrode. (d) Displays the long-term cycling performance of a symmetric cell at 20 mA/cm^2 in a 2 M KFSI EC/DEC electrolyte. (e) Highlights the galvanostatic cycling performance at 80 mA/cm^2. (f) Compares the cycling performance of symmetric cells using NaK-G-C with various Li-, Na-, and K-metal electrodes across different current densities and capacities.

Source: Adapted with permission [20], Copyright (2019), Royal Society of Chemistry.

composite electrolyte, with appropriate SEI formation, can enable the stripping/plating of both metals, offering versatility in battery design.

8.3.3 Mg-Ga Alloy-Based RTLMBs

Magnesium-gallium (Mg-Ga) alloy-based LMBs are promising candidates for room-temperature energy storage, leveraging the unique properties of the Mg-Ga alloy [9]. The low melting point of the alloy enables the battery to function as a liquid metal system at ambient conditions, providing a conductive and self-healing liquid metal electrode that mitigates dendrite formation—a prevalent issue in solid metal anodes. This liquid state ensures uniform current distribution, reduces short-circuit risks, and enhances battery safety and longevity. The combination of the high energy density of Mg with the low viscosity of gallium promotes efficient ion transport, making Mg-Ga alloy-based LMBs an attractive option for sustainable energy storage.

A significant contribution to this field was made by Wei et al., who identified the formation of Mg dendrites in Mg metal anodes [24]. These dendrites, typically below 100 nm in diameter, are smaller than those of Li and Na, and can penetrate separators, leading to internal short circuits in batteries. Wei et. al. proposed applying a liquid metal Ga layer on Mg foil, which spontaneously alloys with Mg to form a stable, corrosion-resistant Ga_5Mg_2 layer [25]. This alloy layer ensures a reversible, stable, and dendrite-free Mg metal anode, enhancing the electrochemical performance of both symmetric and Mg-S full cells. Their research demonstrated the preparation of Ga_5Mg_2-Mg by applying liquid Ga to polished Mg foil, allowing it to alloy at room temperature spontaneously. With a melting point of 29.8°C, metallic Ga easily liquefies and spreads on the polished Mg foil. The unpolished Mg foil initially appears dull due to a natural passivation layer, which is removed by polishing to reveal a bright, metallic surface. When first applied, liquid Ga forms spherical droplets due to its high surface tension but can be evenly distributed using a brush. After 24 hours, the surface turns gray, indicating the formation of the Ga_5Mg_2 alloy layer, which remains firmly attached even under bending. XRD analysis confirmed the gradual conversion of metallic Ga into Ga_5Mg_2 alloy, with complete transformation observed after 470 minutes. Contact angle measurements demonstrated that the liquid electrolyte fully wets both the polished Mg foil and the Ga_5Mg_2-Mg surface. Further studies assessed the stability of Ga_5Mg_2-Mg electrodes in liquid electrolytes. The stability of Ga_5Mg_2-Mg was tested by immersing it in a liquid electrolyte for 72 hours. While the Mg electrode became rough and porous due to corrosion, the Ga_5Mg_2-Mg layer remained unaffected, showing its corrosion resistance. In symmetric cell tests, Mg cells experienced short-circuiting after about 27 hours due to dendrite formation, whereas Ga_5Mg_2-Mg cells maintained stability for 500 hours. Mg cells exhibit higher impedance due to corrosion, whereas Ga_5Mg_2-Mg shows lower impedance.

8.4 ELECTROLYTES IN K-, Mg-, AND Ca-BASED LMBs

LMBs are an emerging energy storage technology, valued for their low cost, high energy density, and long cycle life, with electrolyte selection playing a critical role in their performance. Operating at temperatures above 300°C, LMBs typically use either molten salts or solid ceramic electrolytes, with efforts focused on lowering the operating temperature while maintaining high performance. In K-based LMBs, ceramic electrolytes such as β-Al_2O_3 and garnet structures are being explored for their stability, though the large ionic radius of K complicates ionic conductivity [26]. Molten salts, such as mixtures of K halides, offer lower melting points and good conductivity, but stability remains challenging. For Ca-based LMBs, solid electrolytes such as CaF_2 and Ca-doped garnets are studied for high Ca^{2+} conductivity [9]. Mg-based LMBs benefit from molten salts such as $MgCl_2$-$AlCl_3$ for high ionic conductivity, though solid electrolytes are explored despite challenges like passivation and high temperature requirements [27].

8.5 SEALING AND CORROSION OF K-, Mg-, AND Ca-BASED LMBs

In K-, Ca-, and Mg-based LMBs, the high operating temperature of molten salt electrolytes introduces significant challenges related to sealing and corrosion, which are critical to maintaining electrochemical efficiency [28]. The aggressive thermal environment leads to material degradation over time, primarily affecting electrode current collectors, battery casings, insulation, and sealing components. As a result, ensuring robust sealing is essential for long-term battery operation. However, conventional sealing materials often fail under extreme conditions due to their inadequate thermal stability, insulation, and chemical resistance [29]. The high activity of molten salt electrolytes at elevated temperatures exacerbates corrosion, undermining battery stability and reducing lifespan. While progress has been made in understanding corrosion in other applications, the complexity of the processes necessitates further research, particularly on the corrosion mechanisms specific to LMBs. Developing high-temperature, corrosion-resistant sealing materials is critical for advancing LMB technology. While various electrolytes have been explored, there is a lack of focused research on their specific corrosive effects on sealing materials and current collectors. Addressing these issues is essential for the future advancement of LMBs.

8.6 CURRENT COLLECTOR OF K-, Mg-, AND Ca-BASED LMBs

In K-, Ca-, and Mg-based LMBs, the current collector ensures efficient electron transport between the electrodes and the external circuit. The high operating temperatures required for these batteries pose significant challenges for current collector materials, which must exhibit excellent electrical conductivity, mechanical strength, and corrosion resistance under harsh conditions [30]. Common materials like stainless steel, nickel, and copper have been used, but they often suffer from degradation due to the corrosive nature of molten salt electrolytes at elevated temperatures [31]. Therefore, the selection of current collector materials must balance conductivity with durability and resistance to chemical attack. The interface between the current collector and the molten metal anode or cathode is vital; any instability can lead to increased resistance, reduced efficiency, and premature battery failure. Ongoing research focuses on developing and optimizing materials and coatings to withstand the extreme conditions inside LMBs, enhancing performance and extending battery lifespan. Improved current collectors are vital for the reliable operation of K-, Ca-, and Mg-based LMBs, directly impacting their commercialization and widespread application in energy storage systems.

8.7 K-, Mg-, AND Ca-BASED METAL-Ion BATTERIES

K-, Mg-, and Ca-based metal-ion batteries are emerging alternatives to traditional lithium-ion batteries [13]. These are room temperature batteries designed to operate efficiently at ambient temperatures. The ability to function at room temperature makes them highly versatile and suitable for various applications, from portable electronics to electric vehicles. The choice of metal ions and electrolyte composition is crucial in determining the performance, energy density, and overall stability at room temperature of the battery. K-, Mg-, and Ca-ion batteries utilize distinct materials for their anodes, cathodes, and electrolytes. In K-ion batteries, common anode materials include carbon-based compounds and transition metal oxides. At the same time, cathodes often consist of K manganese oxide ($KMnO_2$) and K iron phosphate ($KFePO_4$), with electrolytes typically being K salt solutions in organic solvents [32]. Mg-ion batteries generally feature graphite or Mg metal as anodes, magnesium manganese oxide ($MgMn_2O_4$) as cathodes, and Mg salts in organic solvents as electrolytes [33]. For Ca-ion batteries, anodes can include carbon-based materials or Ca metal, with cathodes commonly being calcium manganese oxide ($CaMnO_3$) or calcium nickel oxide ($CaNiO_3$), and electrolytes usually being Ca salts dissolved in organic solvents [34]. Each battery type is under research to optimize performance and enhance energy storage capabilities.

Metal-ion batteries shuttle metal cations between the anode and cathode through an electrolyte (Figure 8.4). During discharge, metal ions (e.g., Li^+, Na^+, K^+, Mg^{2+}, Ca^{2+}) move from the anode to the cathode, while electrons flow through an external circuit, providing electrical energy. The general discharge reaction is as follows:

Anode: $M \rightarrow M^{n+} + ne-$;
Cathode: $Mn^+ + ne- \rightarrow M$.

Recharging reverses this process, driving the ions back to the anode. The choice of metal ions, electrode materials, and electrolyte composition influences the performance of the battery, each tailored for specific energy, power, and stability requirements.

Each of these metal-ion batteries offers distinct advantages and faces unique challenges that influence their development and potential applications, but their working principle is almost the same.

8.7.1 K-Ion Batteries

K-ion batteries (KIBs) have gained attention due to the abundance of K and their electrochemical similarities to lithium, offering advantages such as faster ion diffusion and higher conductivity in certain electrolytes. Although the larger ionic radius

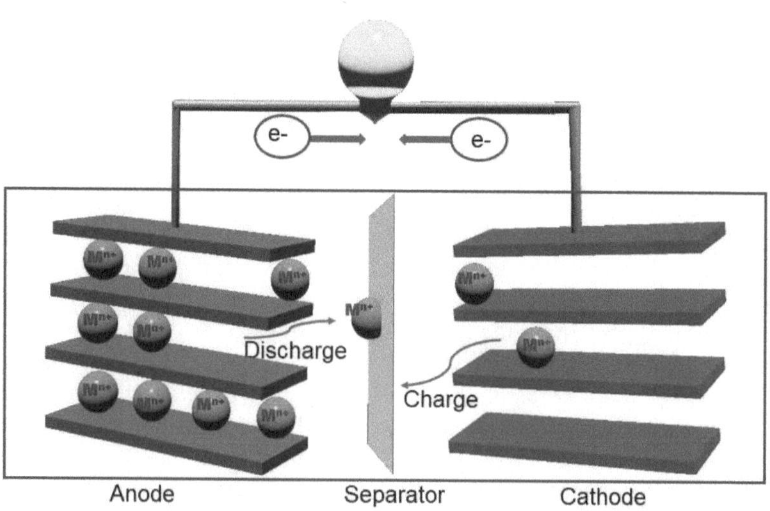

FIGURE 8.4 Schematic illustration of discharging (alloying) and charging (dealloying) cycles of a metal-ion battery (MIB).

and lower reduction potential of K result in lower energy density than lithium-ion batteries, KIBs can utilize graphite as the anode, which is less effective in sodium-ion batteries [35]. This compatibility with existing anode materials reduces production costs and facilitates the transition from lithium to K systems. However, the larger ionic radius of K leads to significant volume expansion during cycling, causing electrode degradation and reduced battery life. To mitigate these challenges, researchers are exploring various anode materials, including carbon-based and alloying options. Eftekhari et. al. developed K-ion batteries offering lower costs for high energy and power densities [36]. The availability and affordability of K resources, along with its comparable redox potential to lithium, make KIBs an appealing alternative for large-scale energy storage. K ions (K^+) exhibit less Lewis acidity than lithium ions (Li^+), attributed to the lower Stokes radius of K^+ and faster diffusion rates. However, the larger size of K^+ (1.38 Å) compared to Li^+ (0.76 Å) can lead to undesirable volume expansions and phase changes in conventional inorganic cathodes during K^+ extraction and insertion, which challenges cycle stability and limits the use of KIBs.

8.7.2 Mg-Ion Batteries

Mg-ion batteries (MgIBs) are gaining attention due to their high theoretical volumetric capacity, as Mg can transfer two electrons per ion compared to one of lithium, making them potentially more efficient. The abundance and lower reactivity of Mg in ambient conditions enhance the safety and cost-effectiveness of MIBs [37]. However, significant challenges remain, particularly in identifying suitable electrolytes that facilitate efficient Mg-ion transport while preventing side reactions and the formation of passivation layers on the anode. Traditional electrolytes used in lithium-ion batteries often prove incompatible with Mg, prompting the need for novel formulations. The divalent nature of Mg ions also leads to stronger ionic interactions, which can hinder diffusion and reduce overall ionic conductivity [38]. Ongoing research is focused on developing cathode materials capable of accommodating Mg ions without significant structural degradation. Dai et. al. published important studies in the early 2010s that advanced the understanding of suitable electrode materials and electrolytes for MIBs, facilitating more effective intercalation and de-intercalation of Mg ions [39].

8.7.3 Ca-Ion Batteries

Ca-ion batteries (CIBs) have the potential to surpass lithium and Mg-ion batteries in energy density and safety. The divalent nature of Ca, like Mg, allows for the transfer of two electrons per ion, potentially doubling the charge capacity [40]. Moreover, Ca is abundant and poses fewer environmental risks than lithium. Challenges include electrolyte compatibility and ion diffusion due to the larger, doubly charged Ca ions with stronger electrostatic interactions. Stable electrolytes that prevent passivation layers on the Ca metal anode are crucial. Recent electrolyte chemistry and material science advancements have reignited interest in CIBs for large-scale energy storage. Studies have focused on optimizing electrolytes for improved ionic conductivity and reducing side reactions [41]. Developing novel electrode materials such as Ca-based compounds remains a key research area, and significant progress has been made in advancing CIB technology.

8.8 CONCLUSION

K-, Mg-, and Ca-based LMBs represent a transformative approach to addressing the pressing need for efficient, grid-scale energy storage solutions in the context of renewable energy integration. Their unique properties, such as high power density, long cycle life, and cost-effectiveness, position them as viable alternatives to traditional lithium-ion batteries. This chapter has explored the electrochemical mechanisms, advancements, and challenges associated with LMBs, highlighting their potential to mitigate the intermittency issues of renewable energy sources. While significant progress has been made in optimizing electrode and electrolyte materials, challenges related to mass transfer, thermal management, sealing, and corrosion remain critical hurdles for commercial viability. Ongoing research and development efforts are essential to refine these technologies and enhance their performance and scalability. By addressing these challenges, K-, Mg-, and Ca-based LMBs can significantly contribute to a sustainable energy future, facilitating the transition from fossil fuels and supporting the global shift toward cleaner energy systems. The insights provided in this chapter lay the groundwork for future innovations in LMB technology.

8.9 ACKNOWLEDGMENTS

ABI greatly acknowledges the research support received from the Internal Research Grant, Research and Innovation Centre for Science and Engineering — RISE, Bangladesh University of Engineering and Technology, Dhaka, Bangladesh (Grant No. 2023-02-024).

REFERENCES

[1] S. Wu, X. Zhang, R. Wang, and T. Li, Progress and perspectives of liquid metal batteries, *Energy Storage Mater.* 57(2023) 205–227.

[2] D. Agarwal, R. Potnuru, C. Kaushik, V. R. Darla, K. Kulkarni, A. Garg, R. K. Gupta, N. Tiwari, and K. S.Nalwa, Recent advances in the modeling of fundamental processes in liquid metal batteries, *Renew. Sustain. Energy Rev.* 158(2022) 112167.

[3] H. Kim, D. A. Boysen, J. M. Newhouse, B. L. Spatocco, B. Chung, P. J. Burke, D. J. Bradwell, K. Jiang, A. A. Tomaszowska, K. Wang, W. Wei, L. A. Ortiz, S.A. Barriga, S. M. Poizeau, and D. R. Sadoway, Liquid metal batteries: Past, present, and future, *Chem. Rev.* 113(2013) 2075–2099.

[4] O. A. Kuznetsov, S. Mohanty, E. Pigos, G. Chen, W. Cai, and A. R. Harutyunyan, High energy density flexible and ecofriendly lithium-ion smart battery, *Energy Storage Mater.* 54(2023) 266–275.

[5] H. Li, H. Yin, K. Wang, S. Cheng, K. Jiang, and D. R. Sadoway, Liquid metal electrodes for energy storage batteries, *Adv. Energy Mater.* 6(2016) 1–19.

[6] W. Cao, Q. Li, X. Yu, and H. Li, Controlling Li deposition below the interface, *eScience.* 2(2022) 47–78.

[7] Y. Yang, R. Wang, Z. Shen, Q. Yu, R. Xiong, and W. Shen, Towards a safer lithium-ion batteries: A critical review on cause, characteristics, warning and disposal strategy for thermal runaway, *Adv. Appl. Energy.* 11(2023) 100146.

[8] L. Yan, W. Yang, H. Yu, L. Zhang, and J. Shu, Recent progress in rechargeable calcium-ion batteries for high-efficiency energy storage, *Energy Storage Mater.* 60(2023) 102822.

[9] S. Zhang, Y. Liu, Q. Fan, C. Zhang, T. Zhou, K. K. Zadeh, and Z. Guo, Liquid metal batteries for future energy storage, *Energy Environ. Sci.* 14(2021) 4177–4202.
[10] J. Verma and D. Kumar, Metal-ion batteries for electric vehicles: Current state of the technology, issues and future perspectives, *Nanoscale Adv.* 3(2021) 3384–3394.
[11] D. J. Bradwell, H. Kim, A. H. C. Sirk, and D. R. Sadoway, Magnesium-antimony liquid metal battery for stationary energy storage, *J. Am. Chem. Soc.* 134(2012) 1895–1897.
[12] L. Zhang, C. Zhao, Y. Liu, J. Xu, J. Sun, and Q. Wang, Electrochemical performance and thermal stability of lithium ion batteries after immersion, *Corros. Sci.* 184(2021) 109384.
[13] L. Stievano, I. de Meatza, J. Bitenc, C. Cavallo, S. Brutti, and M. A. Navarra, Emerging calcium batteries, *J. Power Sources.* 482(2020) 228875.
[14] H. Kim, D. A. Boysen, T. Ouchi, and D. R. Sadoway, Calcium-bismuth electrodes for large-scale energy storage (liquid metal batteries), *J. Power Sources.* 241(2013) 239–248.
[15] X. Ning, S. Phadke, B. Chung, H. Yin, P. Burke, and D. R. Sadoway, Self-healing Li-Bi liquid metal battery for grid-scale energy storage, *J. Power Sources.* 275(2015) 370–376.
[16] T. Ouchi, H. Kim, X. Ning, and D. R. Sadoway, Calcium-antimony alloys as electrodes for liquid metal batteries, *J. Electrochem. Soc.* 161(2014) 1898–1904.
[17] H. Sun, P. Liang, G. Zhu, and H. Dai, A high-performance potassium metal battery using safe ionic liquid electrolyte, *Proc. Natl. Acad. Sci. U. S. A.* 117(2020) 27847–27853.
[18] J. Saint, M. Morcrette, D. Larcher, and J. M. Tarascon, Exploring the Li-Ga room temperature phase diagram and the electrochemical performances of the LixGay alloys vs. Li, *Solid State Ionics.* 176(2005) 189–197.
[19] Z. Tai, Y. Liu, L. Zhao, Y. Ding, Z. Lu, Z. Peng, L. Meng, G. Yu, and L. Liu, Novel quasi-liquid K-na alloy as a promising dendrite-free anode for rechargeable potassium metal batteries, *Adv. Sci.* 8(2021) 1–7.
[20] L. Zhang, S. Peng, Y. Ding, X. Guo, Y. Qian, H. Celio, G. He, and G. Yu, A graphite intercalation compound associated with liquid Na-K towards ultra-stable and high-capacity alkali metal anodes, *Energy Environ. Sci.* 12(2019) 1989–1998.
[21] Z. Wang, Z. Du, Y. Liu, C. E. Knapp, Y. Dai, J. Li, W. Zhang, R.Chen, F. Guo, W. Zong, X. Gao, J. Zhu, C. Wei, and G. He, Metal–organic frameworks and their derivatives for optimizing lithium metal anodes, *eScience.* 4(2024) 100189.
[22] L. Xue, H. Gao, W. Zhou, S. Xin, K. Park, Y. Li, and J.B. Goodenough, Liquid K–Na alloy anode enables dendrite-free potassium batteries, *Adv. Mater.* 28(2016) 9608–9612.
[23] G. Li, B. Huang, Z. Pan, X. Su, Z. Shao, and L. An, Advances in three-dimensional graphene-based materials: Configurations, preparation and application in secondary metal (Li, Na, K, Mg, Al)-ion batteries, *Energy Environ. Sci.* 12(2019) 2030–2053.
[24] C. Wei, L. Tan, Y. Zhang, Z. Wang, J. Feng, and Y. Qian, Towards better Mg metal anodes in rechargeable Mg batteries: Challenges, strategies, and perspectives, *Energy Storage Mater.* 52(2022) 299–319.
[25] C. Wei, L. Tan, Y. Zhang, B. Xi, S. Xiong, J. Feng, and Y. Qian, Highly reversible Mg metal anodes enabled by interfacial liquid metal engineering for high-energy Mg-S batteries, *Energy Storage Mater.* 48(2022) 447–457.
[26] A. Kim, S. Woo, M. Kang, H. Park, and B. Kang, Research progresses of garnet-type solid electrolytes for developing all-solid-state li batteries, *Front. Chem.* 8(2020) 1–13.
[27] Y. Man, P. Jaumaux, Y. Xu, Y. Fei, X. Mo, G. Wang, and X. Zhou, Research development on electrolytes for magnesium-ion batteries, *Sci. Bull.* 68(2023) 1819–1842.
[28] X. Zhou, S. Yan, X. He, H. Zhou, J. Ning, H. Li, K. Wang, and K. Jiang, Low-temperature, high cycling stability, and high Coulombic efficiency liquid metal batteries enabled by lithium halide-potassium halide molten salt electrolytes, *Energy Storage Mater.* 61(2023) 102889.

[29] A. Gharehghani, M. Rabiei, S. Mehranfar, S. Saeedipour, A. M. Andwari, A. García, and C. M. Reche, Progress in battery thermal management systems technologies for electric vehicles, *Renew. Sustain. Energy Rev.* 202(2024) 114654.

[30] T. Masese, G. M. Kanyolo, G. M. Kanyolo, and T. Masese, Advancements in cathode materials for potassium-ion batteries: Current landscape, obstacles, and prospects, *Energy Adv.* 3(2023) 60–107.

[31] P. Wang, K. Du, H. Yin, and D. Wang, Corrosion and protection of metallic materials in molten carbonates for concentrating solar power and molten carbonate electrolysis applications, *Corros. Commun.* 11(2023) 58–71.

[32] Z. Wu, J. Zou, S. Chen, X. Niu, J. Liu, and L. Wang, Potassium-ion battery cathodes: Past, present, and prospects, *J. Power Sources.* 484(2021) 229307.

[33] J. Yin, A. B. Brady, E. S. Takeuchi, A. C. Marschilok, and K. J. Takeuchi, Magnesium-ion battery-relevant electrochemistry of $MgMn_2O_4$: Crystallite size effects and the notable role of electrolyte water content, *Chem. Commun.* 53(2017) 3665–3668.

[34] P. A. Chando, S. Chen, J. M. Shellhamer, E. Wall, X. Wang, R. Schuarca, M. Smeu, and I. D. Hosein, Exploring calcium manganese oxide as a promising cathode material for calcium-ion batteries, *Chem. Mater.* 35(2023) 8371–8381.

[35] S. Dhir, S. Wheeler, I. Capone, and M. Pasta, Outlook on K-ion batteries, *Chem.* 6(2020) 2442–2460.

[36] S. Noreen, M. B. Tahir, A. Hussain, T. Nawaz, J. U. Rehman, A. Dahshan, M. Alzaid, and H. Alrobei, Emerging 2D-Nanostructured materials for electrochemical and sensing application: A review, *Int. J. Hydrogen Energy.* 47(2022) 1371–1389.

[37] M. Javed, A. Shah, J. Nisar, S. Shahzad, A. Haleem, and I. Shah, Nanostructured design cathode materials for magnesium-ion batteries, *ACS Omega.* 9(2024) 4229–4245.

[38] P. Panigrahi, K. Alhameedi, A. Karton, R. Ahuja, and T. Hussain, Improved adsorption and migration of divalent ions over C4N nanosheets: Potential anode for divalent batteries, *Surf. Interfac.* 21(2020) 100758.

[39] A. Medina, C. Pérez-Vicente, and R. Alcántara, Advancing towards a practical magnesium ion battery, *Materials (Basel).* 14(2021) 23.

[40] Z. L. Xu, J. Park, J. Wang, H. Moon, G. Yoon, J. Lim, Y. J. Ko, S. P.Cho, S. Y. Lee, and K. Kang, A new high-voltage calcium intercalation host for ultra-stable and high-power calcium rechargeable batteries, *Nat. Commun.* 12(2021) 1–9.

[41] L. Z. Mohamed, A.A. Lfatah, A. M. Selima, A. Elhamid, M. A. Elhamid, Y. Redac, S. M. El-Raghya, and R. A. Karim, Recent advancements in li-ion batteries electrolytes: A review, *Indian J. Pure Appl. Phys.* 62(2024) 397–411.

9 Chalcogen-Based Liquid Metal Batteries

Ntakadzeni Madima, Siphelo Ngqoloda, Mpfunzeni Raphulu and Bulelwa Ntsendwana

9.1 INTRODUCTION

Energy is a crucial component of modern economies and society. However, the world is currently concerned about the future of energy due to the ever-increasing demand for advancements in contemporary human life [1]. The increasing energy demand has resulted in the depletion of non-renewable hydrocarbon fuels and increased environmental pollution, mainly through the emission of toxic and greenhouse gases [1, 2]. Consequently, achieving global sustainability and a sustainable low-carbon footprint economy depends on integrating renewable energy sources such as solar and wind into the energy grid systems [3]. Unfortunately, the current energy grid systems cannot handle the mass-scale integration of renewable energy sources without causing significant imbalance. The consensus is that a grid system may become significantly unstable if intermittent renewable energy sources contribute more than 20% of its total energy [4]. Thus, a greater reliance on renewable energy sources may lead to more fluctuations than the current grid can manage. Consequently, integrating renewable energy sources into the grid requires low-cost and large-scale energy storage technologies that can potentially reduce inefficiencies and shortcomings in the grid system. Energy storage technologies have the potential to boost grid stability, facilitate the full integration of renewable energy sources, and efficiently regulate power generation [5]. Therefore, energy storage systems are critical for clean and sustainable energy production in the future.

In energy storage systems, rechargeable batteries have been widely recognized for over a decade as one of the most effective solutions for managing the intermittent nature of renewable energy sources and stabilizing power fluctuations in the grid. This is due to their high round-trip efficiency, flexible power output, low maintenance requirements, and pollution-free operation [3]. However, the high cost of existing battery technologies, safety concerns due to thermal runaway, and relatively short cycle life are the most significant barriers to grid energy storage [6]. Nonetheless, following the invention of liquid metal batteries (LMBs) by Donald R. Sadoway and his team, most of the above problems have been resolved, leading to renewed interest in developing battery technologies for grid systems [7]. LMB has a three-liquid-layer electrochemical cell comprising two liquid metal electrodes (a liquid metal negative electrode and a liquid metal positive electrode) separated by a molten salt electrolyte, which self-segregates into three layers based on density and immiscibility [8]. The

DOI: 10.1201/9781003489726-9

electrochemical processes of LMB are driven by the disparate electronegativity of the two liquid metal electrodes, and those liquid metal electrode materials significantly affect battery performance, manufacturing, and overall cost. These batteries are considered ideal candidates for large-scale grid energy storage due to their inherent advantages of low cost, facile cell fabrication, and unprecedented cycle life, benefiting from cheap materials and the amorphous liquidity of electrodes and electrolytes [9].

To date, various types of LMBs have been developed, such as potassium-, magnesium-, and calcium-based, as well as lithium-based and sodium-based, and integrated into the grid system [9]. However, despite their potential, these LMBs face challenges such as high operating temperatures, low discharge voltages, and high cost. Furthermore, the complex chemical reactions of parasites at high temperatures have inhibited practical applications of these LMBs due to safety concerns [7]. To address these challenges, researchers are exploring new ways to develop LMBs operating at room or intermediate temperatures. Moreover, electrode materials play a significant role in influencing the overall energy efficiency of batteries. Therefore, optimizing the electrodes and electrolyte interfaces, searching for highly reactive electrode materials, and adjusting the fundamental electrochemical properties of current existing electrode materials become more crucial for driving innovation and enhancing the electrochemical performance of LMBs [9, 10].

Recently, chalcogenide-based materials have emerged as electrode materials of choice for the fabrication of LMBs owing to their high reversible specific capacities, together with their abundant resources, low cost, low melting points, flexibility, intriguing electrochemical performance, high electrical conductivity, thermal stability, and environmental friendliness [11, 12]. Chalcogenides-based electrode materials contain one or more chalcogen elements, such as sulphides, selenides, or tellurides, as anions and electropositive cations [11, 13]. Consequently, integrating chalcogenide-based materials into LMB electrodes can provide batteries with high energy density, extended cycle life, and cost-effectiveness. This makes chalcogenides-based LMBs a promising choice for large-scale energy storage applications required for integrating renewable energy and stabilizing the grid [13]. In this context, this chapter highlights the application of chalcogen-based materials as alternative electrode materials for high-energy-density battery systems.

9.2 CLASSIFICATION OF BATTERY SYSTEMS

To date, various types of batteries have been developed and are available on the market, each using different materials for the electrodes and electrolytes. On the basis of their electrochemistry, those batteries can be classified as either primary or secondary batteries. Primary batteries, also known as non-rechargeable batteries, are types of batteries that undergo chemical reactions that cannot be reversed. During the discharge process, the electrode material and electrolyte of these batteries break down, preventing the use of an electric current to reverse the chemical reaction [14]. On the other hand, secondary batteries, also known as rechargeable batteries, are electrical batteries that can be charged, discharged, and recharged many times before degradation processes limit their capacity [15]. Some examples of primary and secondary batteries are given in Table 9.1 together with their applications.

TABLE 9.1
Example of Primary and Secondary Batteries Together with Their Possible Applications

Categories	Example	Applications
Primary batteries	Alkaline batteries	Commonly used in household items such as remote controls and flashlights.
	Lithium batteries	Used in small devices such as watches and calculators.
	Zinc-carbon batteries	Used in basic electronics.
	Silver oxide batteries	High-energy-density cells for watches and medical devices.
Secondary batteries	Lithium-ion batteries	Used in portable electronics, electric vehicles, and energy storage systems.
	Lead-acid batteries	Standard in automotive applications and uninterruptible power supplies.
	Nickel-metal hydride batteries	Used in hybrid cars and some portable devices.
	Nickel-cadmium batteries	Utilized in power tools and emergency lighting.
	Lithium-polymer batteries	Found in mobile devices, drones, and electric vehicles.

9.3 FABRICATION PROCESS OF LIQUID METAL BATTERIES

LMBs consist of components including the liquid metal anode, the liquid metal cathode, and the molten salt electrolyte. The anode is typically a low-density metal with high conductivity and electrochemical activity [16]. On the other hand, the cathode is a denser metal, often chosen based on electrochemical properties. Additionally, the electrolyte (molten salt) separates the two liquid metal electrodes and facilitates the transportation of ions [8]. The choice of electrolyte is based on its stability, ionic conductivity, and melting point. During operation, electrons move from one liquid metal electrode to another through the external circuit and the molten electrolyte, connecting these electrodes through an external electrical circuit (Figure 9.1) [17].

In principle, liquid metal batteries operate as follows (Figure 9.1) [8, 9]:

- During the discharge process (alloying), the metal A in the negative electrode is oxidized to A^{n+} (Equation 9.1), which diffuses through the molten electrolyte and creates an alloy with the bottom electrode (metal B).

$$A \rightarrow A^{n+} + ne^{-}. \tag{9.1}$$

- During charging (de-alloying), metal A, present in the bottom electrode as an alloy, is oxidized to A^{n+}, which dissolves in the molten electrolyte and is subsequently reduced to metal A in the top negative electrode (Equation 9.2).

$$A^{n+} + ne^{-} + B \rightarrow A(in\,B). \tag{9.2}$$

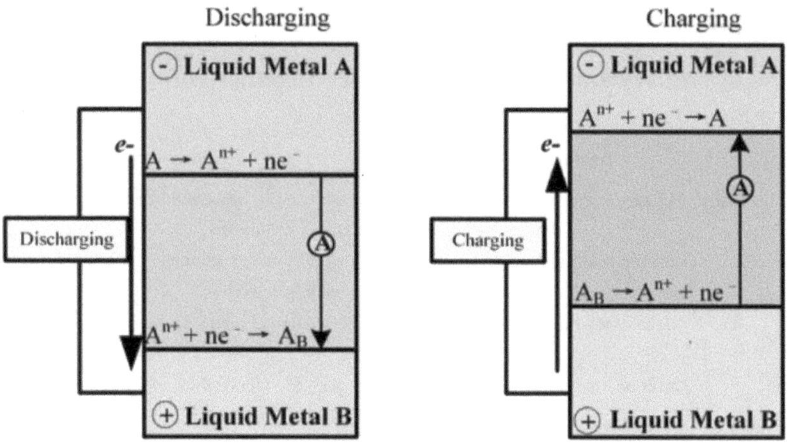

FIGURE 9.1 Principle of a liquid metal battery.

Source: Adapted with permission from ref [18]. Copyright @ The Authors, some rights reserved; exclusive licensee [MDPI, Basel, Switzerland]. Distributed under a Creative Commons Attribution License 4.0 (CC BY).

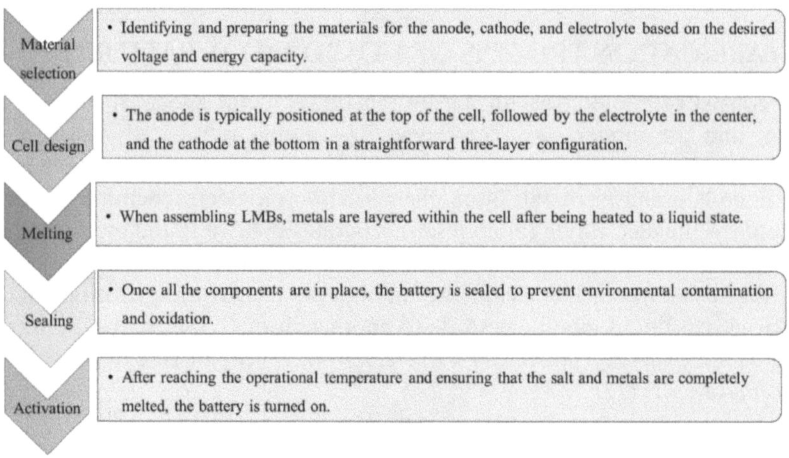

FIGURE 9.2 Schematic representation of the liquid metal battery fabrication process.

- In summary, two metal electrodes are alloyed to transform chemical energy into electrical energy and vice versa. The overall cell reaction (Equation 9.3) is as follows [19]:

$$A + B \rightarrow A(in\,B). \qquad (9.3)$$

In terms of the fabrication process, several steps are involved in the development of LMBs, and those steps include material selection, cell design, melting, sealing, and activation as highlighted in Figure 9.2 [20].

9.4 ELECTRODE MATERIALS FOR LMBs

The widespread use of batteries and their performance depend mainly on electrode materials and the understanding of their chemistry. Metal electrodes are most frequently utilized and typically exist in solid and liquid states. Therefore, selecting ideal electrode materials that optimize performance and provide measurable benefits is fundamental to the design of any battery system. Liquid metal electrodes offer advantages over solid metal electrodes in terms of the absence of dendrite formation during electrodeposition, the absence of phase deformation during charge and discharge, and the simplicity of developing stable electrode/electrolyte interfaces for cell fabrication [19]. From the standpoint of battery operation, the electrode materials' reaction kinetics significantly influence the batteries' voltage polarization and energy density [9].

For decades, various electrode materials have been developed and utilized in LMBs. These electrode materials can be categorized as positive or negative electrodes depending on their electronegativity or deposition potential. For instance, more electronegative metals serve as positive electrodes, and more electropositive metals serve as negative electrodes [16]. Moreover, electrode materials with more than −2.0 V deposition potential are classified as negative electrodes, whereas electrode materials with deposition potential greater than −1.0 V are regarded as positive electrodes [7, 19]. However, despite substantial progress in developing liquid electrode materials for LMBs, several obstacles remain, including severe reactivity and volume expansion during cycling. To address such issues, extensive attempts have been made to improve existing electrode materials and search for novel electrode materials. It has been documented that the ideal liquid electrode materials of LMBs must possess characteristics such as the following [9, 10]:

- Being liquid at ordinary temperatures (the melting point at normal pressure should be less than 1000 °C, and the boiling point should be above 25 °C);
- Electrically conductive (having a minimum electronic conductivity higher than the ionic conductivity of a typical molten salt electrolyte);
- Nonradioactive (existing as a naturally occurring and stable isotope).

Remarkably, chalcogen-based materials, especially those based on sulphur (S), selenium (Se), and tellurium (Te), have become the preferred electrode materials in developing LMBs due to their high reversible specific capacities, stable cycling behavior, abundance in nature, low cost, high electrical conductivity, thermal stability, appropriate melting points that enable them to remain liquid during battery operation, and environmental friendliness [21]. So, integrating chalcogenide-based electrode materials into LMBs is crucial for providing battery technology with a long cycle life, high energy density, and scalability, making LMBs an attractive option for large-scale energy storage [11, 12]. The physicochemical parameters of S, Se, and Te are presented in Table 9.2. All those parameters position chalcogenide-based materials as ideal electrode materials for developing high-performance LMBs for grid systems.

TABLE 9.2
Physicochemical Properties of Chalcogen Elements

Element	Melting point (°C)	Electronic conductivity [S m^{-1}]	Theoretical specific capacity [mA h g^{-1}]	Theoretical volumetric capacity [mA h cm^{-3}]	Mass density [g cm^{-3}]
S	112.8	5×10^{-28}	1675	3461	2.07
Se	220.0	1×10^{-3}	675	3253	4.81
Te	449.5	2×10^{2}	420	2621	6.24

9.5 SYNTHETIC ROUTES OF CHALCOGEN-BASED MATERIALS

Chalcogen-based materials play a crucial role in the evolution of LMB technologies, offering high energy densities, cost-effectiveness, and sustainability. The ongoing research and development in this field aim to overcome existing challenges and unlock the full potential of chalcogen-based batteries for future energy storage solutions. Various synthesis methods have been developed to optimize the performance of chalcogen-based materials, including hydrothermal, solvothermal, sol-gel, microwave-assisted, and chemical vapor deposition, among others. The advantages and challenges of these synthesis methods are summarized in Table 9.3 [22, 23].

9.5.1 Hydrothermal/Solvothermal Synthesis

Hydrothermal/solvothermal synthesis of chalcogenides has emerged as a pivotal method for developing materials suitable for battery applications, particularly in the field of LMBs. This synthesis technique offers several advantages, including the ability to produce materials with controlled morphology and composition, which are crucial for optimizing electrochemical performance. Hydrothermal/solvothermal synthesis is a widely used method for preparing a variety of nanomaterials, including chalcogenides, for their applications in batteries. This technique involves using a sealed vessel (autoclave) to conduct chemical reactions in aqueous or organic solvent solutions at elevated temperatures and pressures, often above the boiling point of these solvents. The high-temperature and high-pressure environment accelerates reactions, promotes crystal growth, and improves the crystallinity of the resulting materials [24]. Typical hydrothermal/solvothermal synthesis procedure usually follows the dissolution of metal salts (e.g., metal nitrates, chlorides, or sulphates) and chalcogen sources (e.g., elemental sulphur, selenium powder, sodium sulphide, or telluride compound) in deionized water or another solvent to form a homogeneous solution through stirring.

In solvothermal synthesis, non-aqueous solvents such as alcohols, ethylene glycol, or organic amines are used instead of de-ionized water, which can provide additional control over reaction dynamics and the properties of the resulting materials. The solution will then be transferred to a Teflon-lined autoclave, sealed, and heated to the desired temperature (usually between 100 °C and 300 °C). The pressure inside the

TABLE 9.3
Advantages and Challenges of Various Synthesis Methods of Chalcogenides

Synthesis method	Advantages	Challenges
Hydrothermal/ solvothermal synthesis	Produces nanostructures with a large surface area and controlled morphology Low cost Low temperature and pressure requirements Eco-friendly and scalable	Complex Reaction Mechanisms Poor material purity Poor material quality during upscaling
Sol-gel synthesis	Good control over material morphology Produces homogeneous materials with consistent electrochemical performance Environmentally friendly, scalable, and versatile Low cost	Poor control over the material's crystallinity, particle size, and surface area Produces materials with poor conductivity and cycling performance Produces non-uniform morphology during upscaling
Microwave-assisted synthesis	Reduced reaction times Lower energy consumption Uniform heating Improved material properties	It requires careful control of synthesis parameters May produce unwanted phases or non-uniform products High initial setup costs
CVD synthesis	Provides precise thickness control up to a monolayer Produces high-quality thin films and nanostructures Highly scalability Versatility in materials synthesis	High-temperature requirements (typically 600–1000 °C) The high cost of CVD reactors and the high cost of high-purity precursors Thin film quality is limited to small thicknesses

autoclave increases as the temperature rises, reaching several tens of atmospheres, promoting the reaction of different compounds. The high temperature and pressure further promote the dissolution and recrystallization of materials, resulting in the formation of crystalline chalcogenide nanostructures. Reaction times can range from several hours to days, depending on the desired product morphology and structure. After cooling to room temperature, the resulting product is washed, dried, and possibly annealed to improve crystallinity further. The schematic illustration of the hydrothermal/solvothermal steps is provided in Figure 9.3 [25].

9.5.2 Sol-Gel Synthesis

The sol-gel synthesis method has emerged as one of the versatile and efficient techniques for producing chalcogen-based materials for their application in batteries. This method involves the transformation of a sol (a colloidal solution) into a gel-like network through hydrolysis and poly-condensation of metal and chalcogen

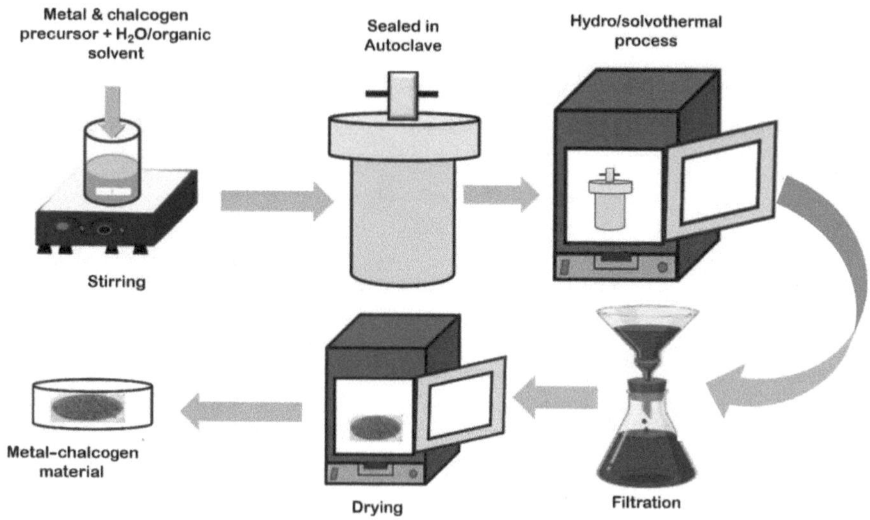

FIGURE 9.3 Schematic representation of the steps involved during hydrothermal/solvothermal synthesis.

precursors. The process follows these steps: a precursor, typically a metal alkoxide or metal salt plus a chalcogen precursor are dissolved in an appropriate solvent. This step is followed by hydrolysis and condensation, where water or other reagents initiate the hydrolysis and condensation reactions, causing the metal ions to form a network of metal-chalcogen bonds. For chalcogenides, this results in the formation of a sol with dispersed metal-chalcogen particles. The sol gradually forms a gel-like network as the particles aggregate. Lastly, the gel is dried to remove solvents, and then heat treatment (annealing) is applied to promote crystallization and improve the material's structural and electrochemical properties [26].

The sol-gel process is advantageous due to its ability to produce materials at lower temperatures than traditional solid-state synthesis, which is crucial for maintaining the integrity of sensitive chalcogen compounds. The final properties of the synthesized materials, such as crystallinity and morphology, can be finely tuned through subsequent heat treatment, allowing for the creation of nanoparticles, thin films, or bulk ceramics that are essential for battery applications. The sol-gel method facilitates the incorporation of chalcogens into composite structures, enhancing their electrochemical performance [27].

9.5.3 Microwave-Assisted Synthesis

Microwave-assisted synthesis has also emerged as a significant synthesis method for the preparation of chalcogenide-based materials for their application in batteries. This technique offers several advantages over other synthesis methods discussed above, including reduced reaction times, lower energy consumption, and improved material properties. These advantages offered by microwave-assisted synthesis stem from the rapid and uniform heating it provides, which can lead to the formation of nanocrystalline materials with enhanced electrochemical properties. The method is energy efficient due to the direct interaction of the microwaves with the reactants, hence reducing energy consumption. It promotes uniform nucleation and growth of crystals, which can enhance the structural and chemical homogeneity of the chalcogenide material. By adjusting the microwave power and reaction time, it is possible to control particle size and morphology, which are critical for battery performance. This synthesis method can further be scaled up, making it essential for industrial applications in energy storage [28].

The microwave-assisted approach also allows for the synthesis of new crystalline chalcogenides by utilizing soluble chalcogenides as precursors. This method could produce uniform jellylike precursors, preventing the formation of thermodynamically stable binary phases and promoting the crystallization of desired cluster compounds. This capability is crucial for developing advanced materials for energy storage applications, where specific structural characteristics can significantly influence performance. While microwave-assisted synthesis offers many benefits, it requires careful control of parameters such as power, time, and solvent selection to avoid unwanted phases or non-uniform products. Initial setup costs for microwave reactors can be high, though they are offset by faster synthesis times and improved material performance. Overall, this synthesis method presents a promising alternative for the development of chalcogenides for battery applications. Its ability to produce high-quality materials rapidly and efficiently, while enhancing their electrochemical properties, positions it as a valuable technique in the field of materials science [29].

9.5.4 Chemical Vapor Deposition Synthesis

Chemical vapor deposition (CVD) is a versatile and widely used method for synthesizing chalcogenides, especially 2D transition-metal chalcogenides such as MoS_2, WS_2, etc., which are critical for various battery technologies. CVD methods, especially thermal chemical vapor deposition (TCVD) and plasma-enhanced CVD (PECVD), offer unique advantages in producing high-quality thin films and nanostructures of chalcogenides, which are critical for enhancing the performance of energy storage devices. CVD techniques enable the bottom-up growth of chalcogenide nanostructures, allowing for the fabrication of materials with single-crystal quality, which is often unachievable with other synthesis methods. CVD produces highly crystalline chalcogenides with few defects, essential for enhancing the electrochemical properties in batteries. It is also a mature technique allowing large-scale production and deposition of chalcogenides, which is necessary for the industrial production of battery materials. CVD can produce thin films of 2D metal chalcogenides with precisely controlled thickness (down to monolayers), essential for battery electrodes with improved charge/discharge rates [30].

CVD involves the chemical reaction of volatile precursors in a gas phase, which are transported to heated substrates by employing inert carrier gases such as argon or nitrogen. The heat initiates the decomposition or reaction of the precursors, resulting in the deposition of a thin film of chalcogenide material on the substrate. For chalcogenides, the metal (e.g., Mo, W, Sn) and chalcogen (S, Se, Te) precursors react at elevated temperatures to form a layered compound. The versatility of the CVD synthesis is particularly beneficial in the context of battery applications, where the interface between the electrode and electrolyte plays a crucial role in device performance. Recent advancements in CVD methods have led to the successful synthesis of 2D transition metal dichalcogenides such as MoS_2 and WS_2, which exhibit remarkable electronic properties. During the deposition of these layered 2D materials, CVD offers control over their thicknesses and may even produce monolayer thicknesses, which are highly desirable due to enhanced surface area and conductivity, improving battery capacity and cycling stability. These materials have been shown to enhance charge transport and storage capabilities, making them suitable candidates for next-generation batteries [30, 31].

9.6 CHARACTERISTICS OF CHALCOGEN-BASED MATERIALS

9.6.1 Crystal Structure

Investigating the crystal structure of the material is crucial to tailor its properties during synthesis, for example, investigating the differences in crystallinity of bulk and nanosheets of Bi_2Se_3. X-ray diffraction (XRD) studies are often conducted on synthesized chalcogen-based materials after synthesis to optimize their crystal structure to suit their application as battery electrodes. Chalcogen compounds, such as sulphides, selenides, and tellurides, are being explored for their unique electrochemical properties and structural characteristics that can be effectively analyzed using XRD techniques. The application of XRD in studying chalcogen-based materials is crucial for understanding their structural transformations during battery operation, that is, before and after cycling. X-ray diffraction has been shown to elucidate phase changes in transition-metal dichalcogenides, revealing the structural distortions that occur during intercalation processes. This understanding is essential for optimizing the performance of these materials in energy storage applications [32].

The development of advanced X-ray techniques, such as operando XRD, allows for real-time monitoring of structural changes in battery materials during charge and discharge cycles. The importance of in situ synchrotron-based X-ray powder diffraction for gaining insights into the structure-chemistry-function relationships of active battery materials, which is vital for the design of high-performance batteries, is often emphasized. This approach enables researchers to observe the dynamic changes in crystal structures, providing a clearer picture of the mechanisms underlying battery performance. Furthermore, the integration of X-ray microscopy techniques enhances the ability to visualize the morphological and chemical structures of battery materials at the nanoscale. This level of detail is particularly beneficial for understanding the behavior of chalcogen-based materials, as it allows for the identification of phase changes and the distribution of elements within the materials during electrochemical cycling [33].

In addition to structural analysis, XRD studies also facilitate the exploration of the electronic properties of chalcogen compounds. The geometry and local electronic environment of chalcogen bonds can be probed using diffraction techniques, which can inform the design of new materials with tailored properties for battery applications. This multifaceted approach, combining XRD with other analytical techniques, is essential for advancing the field of chalcogen-based battery materials. X-ray diffraction studies of chalcogen-based materials play a pivotal role in the development of next-generation batteries. By providing critical insights into the structural and electronic properties of these materials, XRD techniques enable researchers to optimize their performance and address challenges such as capacity fading and structural instability during cycling. The ongoing advancements in X-ray analytical methods will undoubtedly contribute to the evolution of chalcogen-based battery technologies [32, 33].

9.6.2 Morphology

The morphology of chalcogen-based materials is a critical factor influencing their performance in battery applications. Scanning Electron Microscopy (SEM) and Transmission Electron Microscopy (TEM) are often employed to provide detailed images of the material's surface and morphology, including changes during cycling. The SEM does not always provide the best image of the powder material, and therefore, the TEM is employed to reveal more details. Chalcogen compounds, particularly transition metal dichalcogenides, exhibit unique structural properties that can be tailored for enhanced electrochemical performance. The synthesis methods and resulting morphologies of these materials significantly affect their electrochemical behavior, including capacity, cycling stability, and rate capability. Recent studies have demonstrated that the morphology of chalcogen-based materials can be effectively controlled through various synthesis techniques. The ability to manipulate the atomic arrangement and morphology of these materials is crucial for optimizing their performance in batteries, as it directly influences the surface area and electrochemical kinetics [34].

The dimensionality of chalcogen nanostructures also plays a significant role in their electrochemical applications. To further probe the microstructure of these nanomaterials, high-angle annular dark-field scanning transmission electron microscopy (HAADF-STEM) images are used. This mode of the TEM also provides information about the detailed chemical composition and its distribution along the nanostructured material through mapping. These microscopy studies help control the morphology of chalcogen-based materials to engineer them to improve performance metrics such as specific capacity and charge/discharge rates. Moreover, the interaction between chalcogen atoms and metal ions can lead to complex morphologies that enhance battery performance. These atomic arrangements can further be studied through high-resolution TEM (HRTEM), which provides details about the lattice configuration and the crystal nature of the nanomaterial [35, 36].

9.7 PERFORMANCE EVALUATION OF CHALCOGEN-BASED ELECTRODE MATERIALS

To develop chalcogen-based LMBs, numerous sophisticated electrode materials composed of chalcogen hybridized with other metals have been developed. As previously mentioned, chalcogen-containing electrode materials have exceptional properties, making them ideal materials for deployment in LMBs. When utilized as cathodes, they can limit the dissolution of polysulphides and polyselenides, while when used as anodes, they can offer anode materials devoid of dendrites [37]. As such, this section highlights the performance of chalcogen-based electrode materials in battery technologies.

9.7.1 Sulphur-Based

Sulphur-based materials have garnered a lot of interest as a promising electrode material due to the abundance of sulphur on Earth, being harmless and inexpensive, and having a high theoretical capacity. With that, a significant number of research has been conducted on developing sulphur-based batteries such as sodium-sulphur (Na||S) batteries, lithium-sulphur (Li||S) batteries, aluminum-sulphur (Al||S) batteries, potassium-sulphur (K||S) batteries, and others. These sulphur-based batteries represent numerous energy storage solutions to meet various application needs [38]. However, despite the breakthrough in their development, the practical commercialization of these batteries is still confronting many technical challenges. These obstacles include the production of soluble polysulphides that can diffuse to the electrolyte from the sulphur cathode, which would lead to low sulphur utilization and unstable cycling. In addition, poor Coulombic efficiency and quick capacity depreciation are prompted by significant sodium polysulfide shutdown and slow reaction kinetics [39]. Therefore, more research has been undertaken to overcome some of the challenges faced by sulphur-based batteries.

For instance, a study by Zhang *et al.* [40] demonstrates that intrinsically altering the reaction pathway of a sulphur-based cathode with designed additional redox activities could simultaneously suppress polysulfide shuttling and enhance energy density in Na||S batteries. Through adjusting the electrochemical reaction pathways of sulphur chemistry with oxygen a new hybrid sulphur-oxygen chemistry was discovered. This new hybrid sulphur-oxygen cathode achieves a high reversible capacity of over 1400 mA h g^{-1}, low overpotential (~250 mV), and stable cycling performance. The intermediate oxy-sulphur species served as an effective mediator to immobilize the polysulfide species and unlock a high specific capacity from the hybrid cathode. In another study, Yang *et al.* [41] demonstrate the methods for improving the poor conductivity and Li-ion diffusion of lithium sulphide (Li_2S) cathode material in Li||S batteries by coating Li_2S particles with Cobalt (Co) nanomaterial-decorated porous carbon. Following 500 cycles, the developed cathode catalyst (Li_2S/Co@C) maintains a high reversible capacity of 335 mA h g^{-1} at 0.1 °C. It also displays a low activation potential of 3.12 V and an initial reversible capacity of 1006 mA h g^{-1}.

9.7.2 Selenium-Based

Selenium-based electrode materials offer an intriguing possibility for rechargeable battery systems. Selenium can exhibit quicker charge carrier movement during the discharge-charge process, which could lead to improved electrode utilization, electrochemical activity, and rate capability. To enhance the electrochemical performance, Se can be effectively immobilized both physically and chemically to prevent polyselenide dissolution and the side reaction between Se anions and electrolytes. Nevertheless, the low cycle life, Coulombic efficiency, and relatively high cost of selenium, coupled with the easy dissolution of polyselenides with severe cases of the shuttle phenomenon, hinder the applications of Se-based rechargeable metal batteries. For more than a decade, scientists have suggested several approaches to address these issues [42]. For example, to synergistically diminish the shuttle effect and inhibit the formation of polyselenides, Liu *et al.* [43] developed a high-performance potassium-selenium (K-Se) battery, using confined selenium/carbonized-polyacrylonitrile (c-PAN) composite (c-PAN-Se) as cathode and metallic potassium as an anode. The developed K-Se battery shows a reversible capacity of 1904 mAh cm^{-3} after 100 cycles at 0.2 C and a rate retention of 89% from 0.1 to 2 C.

Another work by Zheng *et al.* [44] demonstrated the importance of liquid metal alloys containing WSe_2 by coordination, electrostatic adsorption, and other bonding techniques. The alloy improved the electrochemical performance over repeated cycles of lithium-ion insertion and extraction by promoting the repair of fractured surface areas in the electrode material and reducing internal oxidation-reduction processes. This composite electrode maintained a capacity of 224.5 mAh g^{-1} after 500 cycles at a current density of 1 A g^{-1}. It was impressive that after 50 cycles at a current density of 0.2 A g^{-1}, it could maintain a capacity of 290 mAh g^{-1} at a low temperature of 10 °C. A recent study by Tau *et al.* [45] addresses the problem of the shuttle effect and slow reaction kinetics during cycling by putting forth a novel design idea for low-temperature electro-deoxidation in a neutral NaCl-$AlCl_3$ molten salt system to fabricate uniform Se/C hollow microspheres with controllable morphologies. The Se/C hollow microspheres in their prepared state have a high reversible capacity of 720.1 mAh g^{-1} at 500 mA g^{-1}. Se/C provides long-term stability over 1100 cycles, a high Coulombic efficiency of 98.6%, and a high discharge capacity of 564.0 mAh g^{-1} even at the high current density of 1000 mA g^{-1}. This work offers important new insights into the value-added energy storage applications and short-process recovery of improved Se-containing materials.

9.7.3 Tellurium-Based

Metal tellurium (Te) is a special class of metal chalcogenides that have also shown unique characteristics and consequently make excellent electrode candidates for LMBs. These characteristics include high melting points, appropriate electronegativity, high specific volumetric capacity, and superior electrical conductivity. Unfortunately, the poor cycling stability of Te due to chemical and electrochemical disintegration limits its applications. Nonetheless, to enhance the electrochemical performance of Te-based materials, various studies have been conducted. For instance, a study by Yan *et al.* [46] designed a positive electrode of a metalloid dual-active material by alloying Te with Sb in an effort to increase the energy density of traditional LMB material systems. This alloying strategy effectively provides new insights into the design of multi-active-component cathodes for high-performance LMBs. The designed Li||Sb-Te battery system exhibits excellent cycling performance, high discharge voltage (1.02 V), and ultrahigh energy density (459 Wh kg^{-1}) due to the meticulous design.

In another study, Zhou *et al.* [47] propose a strategy for the operando formation of Li_2Te with the multi-channel structure on the surface of the positive electrode as an effort to improve the rate capability of LMBs to meet the demand for practical applications. The designed positive electrode structure exhibits enhanced ion transport during cycling, which leads to better rate capability. The designed Li||Sb-Bi-Tes5 cell exhibits superior rate performance with a capacity retention of 84.4% and an energy density of 143 Wh kg^{-1}. This facile strategy guides the design of novel electrode materials with benign energy density and rate capability. Furthermore, the high solubility of tellurium in molten salt also poses a challenge to its utilization in LMBs. To tackle this, Li *et al.* [48] conducted a study demonstrating a high-voltage LMB with high energy density by alloying metalloid Te with metallic Tin (Sn). Through alloying, the solubility of Te in molten salt electrolytes was suppressed, and the electronic conductivity was dramatically enhanced. The designed Li||Te-Sn system exhibits the highest discharge voltage of 1.6 V and energy density of 495 Wh kg^{-1}.

9.8 CHALLENGES AND LIMITATIONS

Although liquid metal batteries—including chalcogen-based ones—present promising alternatives for energy storage, particularly for grid-scale applications, due to their unique properties, they also face several significant challenges and limitations that must be addressed to realize their full potential. One of the primary challenges is the operational temperature required for LMBs. These batteries typically operate at elevated temperatures to maintain the liquid state of the metals and the molten salt electrolyte, which can complicate system design and increase energy losses due to thermal management issues. The high operating temperatures, often exceeding 300 °C, can lead to thermal stability concerns and corrosion of the materials used in the battery construction. Additionally, the sensitivity of liquid metal batteries to liquid motion can result in flow instabilities, which may disrupt the electrochemical processes and affect the overall performance of the battery. Another significant limitation is the relatively low voltage output of these technologies compared to conventional lithium-ion batteries. This lower voltage can limit the energy density and overall efficiency of these batteries, making them less competitive in certain applications where higher voltages are required. Moreover, the compatibility of structural materials with liquid metals poses a challenge. Liquid metals can react with many materials, leading to corrosion and degradation over time. This issue necessitates careful selection of materials that can withstand the corrosive environment of LMBs while maintaining structural integrity and performance.

9.9 CONCLUSION

The advancement of synthetic routes for metal chalcogen certainly opens avenues for the fabrication of liquid metal-based batteries. This necessitates further optimization of all the components during battery assembly in order to reach acceptable temperature management. Additionally, advanced fabrication methods through the utilization of flexible thin films have been shown to exhibit better performance. In order to realize these systems in real applications, upscaling approaches from materials, electrode fabrication, and system assembly should be investigated to ensure technology advancement.

REFERENCES

[1] O. Krishan, S. Suhag, An updated review of energy storage systems: Classification and applications in distributed generation power systems incorporating renewable energy resources, *Int J Energy Res* 43 (2019) 6171–6210. https://doi.org/10.1002/er.4285.

[2] A.G. Olabi, M.A. Abdelkareem, M.S. Mahmoud, K. Elsaid, K. Obaideen, H. Rezk, T. Wilberforce, T. Eisa, K.J. Chae, E.T. Sayed, Green hydrogen: Pathways, roadmap, and role in achieving sustainable development goals, *Process Saf Environ Prot* 177 (2023) 664–687. https://doi.org/10.1016/j.psep.2023.06.069.

[3] S.O. Rey, J.A. Romero, L.T. Romero, À.F. Martínez, X.S. Roger, M.A. Qamar, J.L. Domínguez-García, L. Gevorkov, Powering the future: A comprehensive review of battery energy storage systems, *Energies (Basel)* 16 (2023). https://doi.org/10.3390/en16176344.

[4] K.M. Tan, T.S. Babu, V.K. Ramachandaramurthy, P. Kasinathan, S.G. Solanki, S.K. Raveendran, Empowering smart grid: A comprehensive review of energy storage technology and application with renewable energy integration, *J Energy Storage* 39 (2021). https://doi.org/10.1016/j.est.2021.102591.

[5] R. Hou, B. Liu, Y. Sun, L. Liu, J. Meng, M.D. Levi, H. Ji, X. Yan, Recent advances in dual-carbon based electrochemical energy storage devices, *Nano Energy* 72 (2020) 104728. https://doi.org/10.1016/j.nanoen.2020.104728.

[6] M. Wu, S. Han, S. Liu, J. Zhao, W. Xie, Fire-safe polymer electrolyte strategies for lithium batteries, *Energy Storage Mater* 66 (2024) 103174. https://doi.org/10.1016/j.ensm.2024.103174.

[7] H. Kim, D.A. Boysen, J.M. Newhouse, B.L. Spatocco, B. Chung, P.J. Burke, D.J. Bradwell, K. Jiang, A.A. Tomaszowska, K. Wang, W. Wei, L.A. Ortiz, S.A. Barriga, S.M. Poizeau, D.R. Sadoway, Liquid metal batteries: Past, present, and future, *Chem Rev* 113 (2013) 2075–2099. https://doi.org/10.1021/cr300205k.

[8] D. Agarwal, R. Potnuru, C. Kaushik, V.R. Darla, K. Kulkarni, A. Garg, R.K. Gupta, N. Tiwari, K.S. Nalwa, Recent advances in the modeling of fundamental processes in liquid metal batteries, *Renew Sust Energy Rev* 158 (2022) 112167. https://doi.org/10.1016/j.rser.2022.112167.

[9] S. Wu, X. Zhang, R. Wang, T. Li, Progress and perspectives of liquid metal batteries, *Energy Storage Mater* 57 (2023) 205–227. https://doi.org/10.1016/j.ensm.2023.02.021.

[10] H. Ponnuru, I. Marriam, I. Rambukwella, J.C. Zheng, C. Yan, Recent advances in liquid metals for rechargeable batteries, *Adv Funct Mater* 2309706 (2023) 1–23. https://doi.org/10.1002/adfm.202309706.

[11] S. Palchoudhury, K. Ramasamy, J. Han, P. Chen, A. Gupta, Transition metal chalcogenides for next-generation energy storage, *Nanoscale Adv* 5 (2023) 2724–2742. https://doi.org/10.1039/d2na00944g.

[12] K. Mensah-Darkwa, D.N. Ampong, E. Agyekum, F.M. de Souza, R.K. Gupta, Recent advancements in chalcogenides for electrochemical energy storage applications, *Energies (Basel)* 15 (2022) 1–62. https://doi.org/10.3390/en15114052.

[13] S. Yadav, S.R. Yashas, H.P. Shivaraju, Transitional metal chalcogenide nanostructures for remediation and energy: A review, *Environ Chem Lett* 19 (2021) 3683–3700. https://doi.org/10.1007/s10311-021-01269-w.

[14] S. Ha, C. Lim, Y.S. Lee, Fluorination methods and the properties of fluorinated carbon materials for use as lithium primary battery cathode materials, *J Ind Eng Chem* 111 (2022) 1–17. https://doi.org/10.1016/j.jiec.2022.03.044.

[15] S. Hemavathi, S. Srirama, A.S. Prakash, Present and future generation of secondary batteries: A review, *ChemBioEng Rev* 10 (2023) 1123–1145. https://doi.org/10.1002/cben.202200040.
[16] C. Wei, L. Tan, Y. Zhang, Z. Wang, B. Xi, S. Xiong, J. Feng, Y. Qian, Review of room-temperature liquid metals for advanced metal anodes in rechargeable batteries, *Energy Storage Mater* 50 (2022) 473–494. https://doi.org/10.1016/j.ensm.2022.05.024.
[17] Q. Zeng, Z. Lv, S. Li, B. Yang, J. He, J. Song, Electrolytes for liquid metal batteries, *Mater Res Bull* 170 (2024) 112586. https://doi.org/10.1016/j.materresbull.2023.112586.
[18] X. Wang, Z. Song, K. Yang, X. Yin, Y. Geng, J. Wang, State of charge estimation for lithium-bismuth liquid metal batteries, *Energies (Basel)* 12 (2019). https://doi.org/10.3390/en12010183.
[19] H. Li, H. Yin, K. Wang, S. Cheng, K. Jiang, D.R. Sadoway, Liquid metal electrodes for energy storage batteries, *Adv Energy Mater* 6 (2016) 1–19. https://doi.org/10.1002/aenm.201600483.
[20] X. Zhou, K. Jiang, K. Wang, Enhancing the discharge performance of liquid metal batteries through external magnetic fields, *Energy Procedia* 41 (2024). https://doi.org/10.46855/energy-proceedings-10981.
[21] E.S. Sowbakkiyavathi, S.P. Arunachala Kumar, D.K. Maurya, B. Balakrishnan, J.Z. Guo, A. Subramania, Research progress in the development of transition metal chalcogenides and their composite-based electrode materials for supercapacitors, *Adv Compos Hybrid Mater* 7 (2024) 130. https://doi.org/10.1007/s42114-024-00913-7.
[22] P. Muthu, S. Rajagopal, D. Saju, V. Kesavan, A. Dellus, L. Sadhasivam, N. Chandrasekaran, Review of transition metal chalcogenides and halides as electrode materials for thermal batteries and secondary energy storage systems, *ACS Omega* 9 (2024). https://doi.org/10.1021/acsomega.3c08809.
[23] K. Mensah-Darkwa, D.N. Ampong, E. Agyekum, F.M. de Souza, R.K. Gupta, Recent advancements in chalcogenides for electrochemical energy storage applications, *Energies (Basel)* 15 (2022). https://doi.org/10.3390/en15114052.
[24] J. Cui, H. Zheng, K. He, In situ TEM study on conversion-type electrodes for rechargeable ion batteries, *Adv Mater* 33 (2021). https://doi.org/10.1002/adma.202000699.
[25] J. Huang, Z. Wei, J. Liao, W. Ni, C. Wang, J. Ma, Molybdenum and tungsten chalcogenides for lithium/sodium-ion batteries: Beyond MoS2, *J Energy Chem* 33 (2019) 100–124. https://doi.org/10.1016/j.jechem.2018.09.001.
[26] S. Kim, J. Yang, D. Liu, Y. Huang, Y. Lee, Bio-derived materials achieving high performance in alkali metal–chalcogen batteries, *Adv Funct Mater* 31 (2021). https://doi.org/10.1002/adfm.202008354.
[27] M. Liu, Y. Zhang, Z. Xu, X. Han, W. Gou, Y. Sun, C. Ming Li, Highly microporous carbon from enteromorpha for high-performance aqueous zinc-chalcogen (S, SeS 2) batteries, *Batter Supercaps* 6 (2023). https://doi.org/10.1002/batt.202300145.
[28] K.L. Salvatore, J. Fang, C.R. Tang, E.S. Takeuchi, A.C. Marschilok, K.J. Takeuchi, S.S. Wong, Microwave-assisted fabrication of high energy density binary metal sulfides for enhanced performance in battery applications, *Nanomaterials* 13 (2023) 1599. https://doi.org/10.3390/nano13101599.
[29] J. Tang, X. Zhou, B. Jin, J. Luo, X. Ning, L. Zhan, Z. Wu, X. Fan, G. Wang, X. Zhou, Microwave-assisted synthesis of a ternary MoS2/carbon-FeOx composites with 3D hierarchical nanostructure for lithium-ion battery application, *J Mater Sci Mater Electron* 34 (2023). https://doi.org/10.1007/s10854-022-09553-3.
[30] B. Tang, D. Hu, X. Zhao, X. Wang, Z. Liu, Recent developments in chemical vapor deposition of 2D magnetic transition metal chalcogenides, *ACS Appl Electron Mater* 4 (2022) 3303–3324. https://doi.org/10.1021/acsaelm.2c00425.

[31] K. Zhang, T. Zhang, J. You, X. Zheng, M. Zhao, L. Zhang, J. Kong, Z. Luo, S. Huang, Low-temperature vapor-phase growth of 2D metal chalcogenides, *Small* 20 (2024). https://doi.org/10.1002/smll.202307587.

[32] Y. Zhao, Y. Lu, H. Li, Y. Zhu, Y. Meng, N. Li, D. Wang, F. Jiang, F. Mo, C. Long, Y. Guo, X. Li, Z. Huang, Q. Li, J.C. Ho, J. Fan, M. Sui, F. Chen, W. Zhu, W. Liu, C. Zhi, Few-layer bismuth selenide cathode for low-temperature quasi-solid-state aqueous zinc metal batteries, *Nat Commun* 13 (2022). https://doi.org/10.1038/s41467-022-28380-y.

[33] H. Ji, S. Hu, S. Shi, B. Guo, W. Hou, G. Yang, Rapid microwave-hydrothermal preparation of few-layer MoS2/C nanocomposite as anode for highly reversible lithium storage properties, *J Mater Sci* 53 (2018) 14548–14558. https://doi.org/10.1007/s10853-018-2631-7.

[34] X. Liu, T. Najam, G. Yasin, M. Kumar, M. Wang, One-pot synthesis of high-performance tin chalcogenides/C anodes for li-ion batteries, *ACS Omega* 6 (2021) 17391–17399. https://doi.org/10.1021/acsomega.1c01647.

[35] Y. Liu, X. Hu, J. Li, G. Zhong, J. Yuan, H. Zhan, Y. Tang, Z. Wen, Carbon-coated MoS1.5Te0.5 nanocables for efficient sodium-ion storage in non-aqueous dual-ion batteries, *Nat Commun* 13 (2022). https://doi.org/10.1038/s41467-022-28176-0.

[36] N. Kapuria, B. Nan, T.E. Adegoke, U. Bangert, A. Cabot, S. Singh, K.M. Ryan, Colloidal synthesis of multinary alkali-metal chalcogenides containing Bi and Sb: An emerging class of I-V-VI2 nanocrystals with tunable composition and interesting properties, *Chem Mater* 35 (2023) 4810–4820. https://doi.org/10.1021/acs.chemmater.3c00673.

[37] L. Ren, B.W. Zhang, Room temperature liquid metals for flexible alkali metal-chalcogen batteries, *Exploration* 2 (2022). https://doi.org/10.1002/EXP.20210182.

[38] C. Ye, H. Li, Y. Chen, J. Hao, J. Liu, J. Shan, S.Z. Qiao, The role of electrocatalytic materials for developing post-lithium metal||sulfur batteries, *Nat Commun* 15 (2024) 15–18. https://doi.org/10.1038/s41467-024-49164-6.

[39] T. Yang, B. Guo, W. Du, M.K. Aslam, M. Tao, W. Zhong, Y. Chen, S.J. Bao, X. Zhang, M. Xu, Design and construction of sodium polysulfides defense system for room-temperature Na–S battery, *Adv Sci* 6 (2019). https://doi.org/10.1002/advs.201901557.

[40] S. Zhang, T.P. Pollard, X. Feng, O. Borodin, K. Xu, Z. Li, Altering the electrochemical pathway of sulfur chemistry with oxygen for high energy density and low shuttling in a Na/S battery, *ACS Energy Lett* 5 (2020) 1070–1076. https://doi.org/10.1021/acsenergylett.9b02746.

[41] H. Yang, Y. Lei, Q. Yang, B.W. Zhang, Q. Gu, Y.X. Wang, S. Chou, H.K. Liu, S.X. Dou, Cobalt-induced highly-electroactive Li2S heterostructured cathode for Li-S batteries, *Electrochim Acta* 439 (2023) 141652. https://doi.org/10.1016/j.electacta.2022.141652.

[42] J. Tu, C. Chang, J. Wang, H. Lei, R.V. Kumar, M. Wang, S. Jiao, Fundamental understanding on selenium electrochemistry: From electrolytic cell to advanced energy storage, *Energy Environ Mater* 7 (2024) 1–21. https://doi.org/10.1002/eem2.12664.

[43] Y. Liu, Z. Tai, Q. Zhang, H. Wang, W.K. Pang, H.K. Liu, K. Konstantinov, Z. Guo, A new energy storage system: Rechargeable potassium-selenium battery, *Nano Energy* 35 (2017) 36–43. https://doi.org/10.1016/j.nanoen.2017.03.029.

[44] X. Zheng, L. Guo, C. Zhu, T. Hu, X. Gong, C. Wu, G. Wang, Y. Hou, Preparation and electrochemical performance study of a self-healing electrode composite material with WSe2/liquid metal Galinstan for lithium-ion batteries, *J Alloys Compd* 969 (2023) 172304. https://doi.org/10.1016/j.jallcom.2023.172304.

[45] J. Tu, Z. Huang, C. Chang, H. Lei, S. Wang, S. Jiao, An efficient molten-salt electrodeoxidation strategy enabling fast-kinetics and long-life aluminum–selenium batteries, *SusMat* 4 (2024) 126–139. https://doi.org/10.1002/sus2.183.

[46] S. Yan, X. Zhou, L. Fan, X. He, W. Zhang, H. Zhou, L. Yuan, K. Jiang, B. Li, H. Li, K. Wang, Tellurium-antimony electrodes with multistep discharge mechanisms for high-energy-density liquid metal batteries, *ACS Sustain Chem Eng* 12 (2024) 11831–11838. https://doi.org/10.1021/acssuschemeng.4c04926.

[47] Y. Zhou, G. Li, B. Li, X. Ning, Operando formation of multi-channel positive electrode achieved via tellurium alloying in liquid metal battery, *Energy Storage Mater* 53 (2022) 927–936. https://doi.org/10.1016/j.ensm.2022.10.015.

[48] H. Li, K. Wang, H. Zhou, X. Guo, S. Cheng, K. Jiang, Tellurium-tin based electrodes enabling liquid metal batteries for high specific energy storage applications, *Energy Storage Mater* 14 (2018) 267–271. https://doi.org/10.1016/j.ensm.2018.04.017.

10 Transition Metal-Based Liquid Metal Batteries

Yamini Gupta, Poonam Siwatch, S. K. Tripathi, and Kriti Sharma

10.1 INTRODUCTION

As a result of the rapid growth of a low-carbon economy, which has expedited the revolutionary shift away from hydrocarbon fuels and toward sustainable and renewable energy sources, there has been a considerable change in global energy consumption [1]. Presently, there are restricted methods for storing a substantial quantity of energy at the grid scale, including electrochemical energy storage systems (ESS), pumped hydro, and compressed air. There are geographical limitations accompanying pumped hydro and compressed air systems; as a result, efforts are being made to find solutions for ESS, with electrochemical ESS eventually becoming the technology of choice [2]. Nowadays, lead-acid batteries and lithium-ion batteries (LIBs) are among the most prevalent ESS. In microgrids and other small-scale applications, lead-acid batteries are frequently employed; however, they are limited by their low energy density (E_d) and calendar aging [3]. Lithium-ion-based batteries are employed in the majority of grid-level ESS that are currently operational or will be constructed using ESS. On the other hand, these are a serious cause for concern in terms of safety because their electrolyte contains flammable organic solvents, which puts them in danger of starting a fire. Additionally, the cost of ESS is elevated to \$150/KWh due to the limited availability of lithium reserves, which are also exceedingly costly. Therefore, it is imperative to develop battery technologies that combine high cycle life with the use of earth-abundant materials [4]. Similar to batteries, ESS is essential for permitting intermittent but sustainable energy harvesting from sources including solar, wind, and geothermal energy. Up to this point, numerous technologies for rechargeable batteries have been established so far for the persistence of high-efficiency ESS.

Liquid metal batteries (LMBs) are a viable choice for large-scale power storage technology. These batteries are built on liquid metal (LM) electrodes and molten salt electrolytes, making them one of the most favorable options. LMBs are gaining importance due to their all-liquid structure's merits of low cost, extended life, and faster reaction kinetics [5]. In LMBs, there are three layers one above the other, comprising a lighter-density anode, a medium-density electrolyte, and a heavy-density cathode [6]. The temperatures of operation are slightly elevated from the active materials' greatest melting points, which are typically between 275 and 700°C [7]. The operating temperature of LMBs is connected to the screening of electrode materials and electrolytes,

DOI: 10.1201/9781003489726-10

the solubility of electrodes, the wettability, the E_d, the energy efficiency, and other factors. So, LMBs can be classified into three types (based on operating temperature): 1) high-temperature LMBs (HTLMBs) that operate at temperatures higher than 350°C, with molten salt electrolyte and two different types of LM or alloy electrodes; 2) medium-temperature LMBs (MTLMBs) that operate in the 100 to 350°C temperature window, with solid or liquid electrolytes and at least one LM or alloy electrode; and 3) room-temperature LMBs (RTLMBs) that operate at or near room temperature, with solid or liquid electrolytes and at least one LM or alloy electrode. In numerous reported LMBs, the bottom layer is composed of a high-density metal/alloy with high electronegativity (e.g., Pb, Hg, Bi), while the upper layer is composed of a low-density and low electronegativity metal (e.g., Li, Mg, K, and Na). For LMBs, electrolytes with a low melting temperature and high ionic conductivity, as well as minimal metal solubility, are mostly preferred. These electrolytes also provide the transport pathways for ions, thereby preventing short circuits.

The excellence of electric vehicles is increasingly being assessed based on their E_d. Lithium metal batteries (LiMBs) can offer a significantly higher E_d than conventional LiBs due to their relatively low electrode potential (–3.04 V vs. SHE) and high theoretical specific capacity (3860 mAh g^{-1}). Consequently, they are suitable candidates for the next generation of batteries. However, the lifetime of LiMBs is typically less than a few hundred cycles since lithium dendrites penetrate the polymer separator and short-circuit the batteries when liquid electrolytes are used [8]. The underlying physical mechanism of LMB has also been the subject of extensive research and development efforts, in addition to the battery chemistries. Several complex multiphysical processes take place inside LMBs, in contrast to solid-state batteries. These activities include the distribution of electric potential, the flow of electro-vortex, the flow of thermal convection, the movement of mass, and the transfer of heat, among others. To maximize the performance of the battery in real-world operations and to speed up the deployment of LMBs on a broad scale, it is of the utmost importance to have a comprehensive grasp of the electrochemistry interaction regime. More work needs to be done to keep up with the advancement of battery chemistry to produce better LMBs, even though there are a variety of models that can be used to analyze the processes that occur inside the LMBs. These models offer insights into the optimization of cell design and operating conditions [9].

Till now a lot of work has been done on Li-, Na-, Ga-, K-, Mg-, and Ca-based LMBs. Wei et al. [10] passivated the surface of the Li-metal anode with GaInSnZn LM, and the stability of the anode is increased. At ambient temperature, a spontaneous reaction between metallic Li and LM results in the formation of a framework for an Li-based alloy with grains on the submicron scale. This framework is created on the surface of Li metal. Even when bent and folded, the framework of the lithium-based alloy is securely connected to the lithium metal, preventing any exfoliation or material separation. Even though lithium-based batteries are currently the most prevalent in the ESS market, their implementation in large-scale grid-scale ESS is hindered by the high cost and the irregular geological distribution of lithium sources.

Na-based batteries have also been shown to be a promising option for stationary and mobile ESS. In addition, sodium is an element that is abundant, inexpensive, and safe for the environment, which is what drives the research and development of sodium-based batteries. A low redox potential (2.714 V vs. SHE) and a large theoretical capacity (1166 mAh g^{-1}) are some of the characteristics of the sodium anode. These batteries can effectively inhibit dendrite growth and can be charged and discharged in a short amount of time. However, the high operating temperature of sodium batteries, low reliability, and short life limit their availability for wider applications. In addition, in comparison to Na-ion batteries, Na-metal batteries have a higher E_d and a greater potential for use. Numerous Na-metal battery systems, including Na-CO_2, Na-O_2, and Na-S, have garnered a great deal of attention lately.

In recent times, it has been discovered that exchanging Na metal for K can be relatively fascinating due to the substantial amount and low melting point of K. Even though research on this topic is still in its infancy, a K-BASE electrolyte was used in K-based HTLMBs. This electrolyte showed promise in terms of solving both the Na^+ exchange in Na-BASE and the large K solubility in the molten electrolyte [11]. Although gallium-based LMBs are among the most promising of these, their further deployment is limited because of the unpredictable volume variation and phase transitions that occur during electrochemical reactions [12]. Because magnesium in its liquid electrolyte has an extraordinarily low solubility, the discovery of magnesium-based LMBs gave a more appealing perspective. This is because just half as many ions would be required to attain the same capacity as Li. The standard Mg|NaCl-KCl-$MgCl_2$|Sb cell was discovered to work at temperatures higher than 700°C, which is the primary reason why it has been mainly eliminated from the market. As the fifth most prevalent element on the Earth's surface, calcium holds great promise due to its non-toxic nature, abundance, and low electronegativity. It possesses a high theoretical capacity of 2060 mA h cm^{-3}. Ca was initially employed as a negative electrode in thermal batteries from the 1950s to the 1980s. After the concept of LMBs was introduced, Ca's low cost (0.14 \$ mol^{-1}) garnered significant attention. Nevertheless, the continued deployment of Ca-based LMBs is confronted with two primary challenges: the high melting point of 842°C and the high solubility of calcium in electrolytes that contain molten salt. Ning et al. [13] investigated the electrochemical characteristics of a Ca-Pb alloy that was used in a Ca-LiCl-NaCl-$CaCl_2$-Pb cell. The Ca-Pb electrode displayed exceptional features with a discharge voltage of 0.6 V, a self-discharge current density of less than 2 mA cm^{-2} at 600°C, and a high coulombic efficiency of more than 98.84%. However, the Ca-Pb electrode included a significant quantity of dispersed intermetallic $CaPb_3$ near the end of the Ca-Pb alloying process. This may result in an abrupt decrease in the electrode potential that occurs. In spite of this, the intrinsic strong reactivity that involves LM in electrolyte continues to have negative effects on the realization of LMBs that have satisfactory electrochemical reversibility. Furthermore, cycling at high cell voltages along with adequate E_d in pack-level LMBs has not yet been accomplished.

Zinc metal is a material that is low-cost and naturally abundant. Recently, Zn-based LMBs have been reported that utilize solid Zn or liquid Zn-Sn alloy as the negative electrode; Sn, Pb, or Bi as the positive electrode; and $ZnCl_2$-KCl eutectic molten salt as the electrolyte, which can operate at low to intermediate temperatures. These LMBs can have fewer corrosive effects, higher safety, and lower costs than traditional LMBs. However, the charge-discharge performance and coulombic efficiency of these Zn-based LMBs are unsatisfactory and need to be clarified further.

As we have observed there are certain drawbacks of Li-based LMBs (dendrites formation), Na-based LMBs (high operating temperature, small lifetime), Ca-based LMBs (high melting point and high solubility of Ca in molten electrolytes), etc., so transition metals/transition metal-based alloys may be incorporated in these LMBs to improve the performance of these batteries. Also, the issues associated with Zn-based LMBs (charge-discharge performance and coulombic efficiency) need to be addressed. In this chapter, we will discuss the transition metal/transition metal alloys-based LMBs. The difficulties and advancements that have been made in transition metal-based LMB electrodes, including the chemical reactivity of metals, dendrite formation, and unstable interfaces, are discussed. The proposed ways for transition metal-based electrode modification will also be discussed. These strategies include modification based on alloys, electrolytic engineering, separator engineering, and sealing strategies. The optimization of electrolytes, including both liquid and solid electrolytes, is also discussed. Furthermore, a summary and prospectus of the future research directions of transition metal-based LMBs will also be presented.

10.2 WORKING MECHANISM OF LMBs

The negative electrode layer undergoes electrochemical oxidation ($A \rightarrow A^{z+} + ze^-$) during discharge, resulting in a decrease in thickness. The cations are then transported across the molten salt electrolyte to the positive electrode as electrons are released into an external circuit (Figure 10.1). Concurrently, the positive electrode layer increases in thickness as the cations undergo electrochemical reduction to produce a liquid A-B alloy [$A^{z+} + ze^- \rightarrow A(\text{in } B)$] [14]. Charging reverses this process, as illustrated in Figure 10.1.

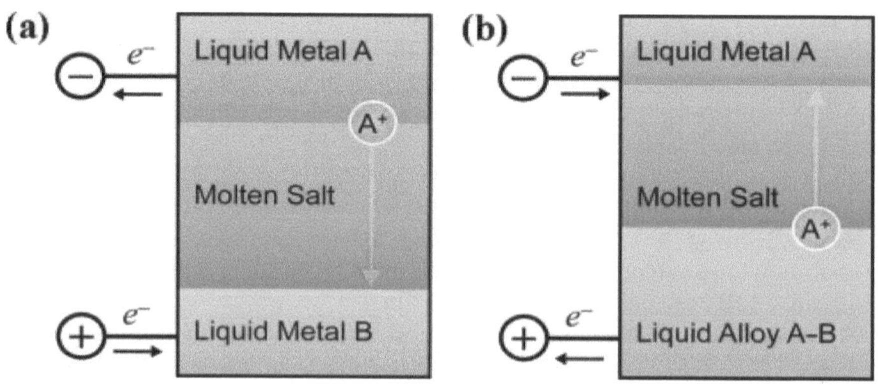

FIGURE 10.1 (a) The discharging and (b) charging process of LMBs.
Source: Adapted with permission [14]. Copyright (2012), ACS.

10.3 TRANSITION METAL-BASED LMBs

The Zn-based LMBs use economical Zn and a low-temperature electrolyte [15]. But, low voltage plateau and coulombic efficiency hinder their advancement. Zhao et al. [316] reported Zn-based LMB, which uses Zn_5Sn_5 as the negative electrode, $ZnCl_2$-LiCl-KCl as the electrolyte, and Bi as the positive electrode. This battery can operate at a temperature of 375°C, possesses a high voltage plateau of 0.8–0.9 V, and achieves coulombic efficiency of ~ 96%. Adding Sn to the negative electrode can resolve the challenge of Zn dendrites. The transition metal alloys and metal halides can be utilized in the construction of LMBs, in addition to the cathodes that are based on chalcogenides, due to the merits of their feasibility and low cost. Wu et al. [217] fabricated Fe_2O_3/PEO/Li (FPL) cells that delivered a high discharge capacity of 990 mAh g^{-1} at 0.01 A g^{-1}, when measured at 60 °C, which is analogous to the Fe_2O_3 liquid cell. One could model the *coulombic* efficiency and capacity retention of the FPL cell at 60 °C for 71 cycles and 15 cycles, the charging and discharging plots, and the demonstration of the FPL cell for powering LEDs. Chu et al. [18] fabricated a novel lithiation method shown by the $Sb_{64}Cu_{36}$ cathode. The Li||$Sb_{64}Cu_{36}$ cell exhibits a significantly superior electrochemical performance, with an E_d of 353 Wh kg^{-1} at 0.4 A cm^{-2}, in addition to a significantly lower cost for ESS, which is only 38.45 \$ kW h^{-1}.

In contrast to Li, a single Mg atom has a higher gravimetric capacity due to its ability to liberate two electrons when oxidized (Mg^{2+}) [19]. Chen et al. [20] proposed crystallizing deep eutectic solvents composed of urea and crystallizing eutectic $Mg(TFSI)_2$. The results concluded that the rapid Mg^{2+} migration has been facilitated by the coordination structure of anion-rich Mg^{2+}/urea. The full cells Mg//V_2O_5 when tested delivered a high reversible capacity of 172 mAh g^{-1}. Chu et al. [21] utilized the method of chemical vapor deposition in the preparation of the graphite layer-coated Ni-Fe foam. With capacity retention of 98.10% and high coulombic efficiencies of 98.3% at 0.4 A cm^{-2} over 100 cycles at 500°C, the built-up Li||Bi batteries, which utilize Graphite@Ni-Fe foam as the negative current collector, demonstrate a stable cycling performance. When the current density is changed from 0.2 to 2.0 A cm^{-2}, it is seen that nearly no capacity degradation occurs, which is a significant achievement with exceptional rate capability.

To further study the transition metal-based LMBs, Xie et al. [22] reported a novel dual-active Sb-Cd cathode that exhibits a new ternary alloying discharge process. As a result of the exceptional capacity of Cd to store lithium, the $Sb_{80}Cd_{20}$ cathode can achieve an impressive capacity of more than 500 mAh g^{-1}, surpassing the performance of all LMB cathodes that have been reported. This fast electron and lithium diffusion network for the subsequent discharge significantly accelerates the electrode reaction kinetics. Because of this, the $Sb_{80}Cd_{20}$ electrode can produce a remarkable combination of high E_d (398.4 Wh kg^{-1} at 200 mA cm^{-2}) and better rate capability (542.5 W kg^{-1} at 2400 mA cm^{-2}).

Thus, we have observed that the incorporation of transition metals/transition metal-based alloys in Li, Na, K, Mg, and Ca-based LMBs has produced LMBs at lower costs and with high discharge capacity, stable cycling performance, and high E_d. However, they have also faced some challenges leading to the deterioration of high-performance LMBs. In the next section, we will discuss the challenges faced by these LMBs and their modification strategies, which will help speed up the implementation of LMBs for grid ESS.

10.4 CHALLENGES FACED BY LMBs

LMBs face several significant issues, including high-temperature operation, toxicity, chemical reactivity with electrolytes, high volume change, corrosion, dendrite formation, etc.

10.4.1 High-Temperature Operation

Low interface resistance and rapid kinetics of electrode reaction are typical features of the LMB with an all-liquid structure, which facilitates the battery's operation at a high rate. As can be observed, the battery operations can function at temperatures that are higher than their respective melting points. This is necessary to enable the formation of interfaces between liquid-liquid electrodes and electrolytes. Therefore, high-temperature LMBs can operate at high voltage efficiencies because of the rapid mass movement of reactants and products, as well as the reduced ohmic losses that are caused by highly conductive molten electrolytes. One of the most important characteristics of high-temperature LMBs is that their rapid charge-transfer kinetics are extremely dependent on the elevated operating temperature.

High operating temperatures are one of the obstacles that prevent the widespread deployment of high-temperature LMBs. This is because high temperatures cause insulating seals to deteriorate, necessitate rigid thermal control, and provide an environment that is highly reactive to chemicals. Specifically, the high reactivity causes the electrode species to become metallically soluble in the molten salt electrolyte. This, in turn, causes the cell to experience high rates of self-discharge, which ultimately results in the cell's performance becoming irreversible. All liquid components considerably restrict the mobility of the battery, but they provide a possible risk if there is a leakage of the liquid phase. Because of this, there is a possibility that the cell will experience rapid heat generation as well as short circuits. Because of the external energy input and the necessities of hermetic sealing, the cost load is readily increased, which makes HTLMBs uncompetitive in comparison to other technologies that are currently available in both the grid-scale and portable markets. It is important to emphasize that the high temperature also involves multiple cost-based challenges.

The kinetics of the reactions and the high performance that occurs at elevated temperatures are negatively affected when the operating temperature of HTLMBs is lowered. However, this allows for the possibility of using polymeric sealing materials that have improved mechanical strength and reduced heat loss, which ultimately results in an improvement in the ability to resist corrosion and simplifies the packing process.

10.4.2 Toxicity

There are limitations imposed on the use of LM as anode materials in conventional battery systems due to the hazardous effects that they have on human health and environmental damage. Some transition metals may require extensive mining and processing, leading to environmental degradation. Within the category of LM, metallic mercury (Hg) is a poisonous substance, which limits its utilization in the field of battery technology. Therefore, the disposal of trash with caution is required for Hg-based batteries. Ga and its alloys are non-toxic, are reasonably stable with battery components, and have a stronger self-healing property to cure faulty anodes. They are also relatively stable with battery components [20].

10.4.3 Corrosiveness

When LM reacts chemically with liquid electrolytes, the result is corrosion of the LM that can be classified as either chemical or electrochemical. Conditions such as the operating temperature, the electrolyte, and the oxygen level can have a significant impact on the degree of corrosion that occurs in LMs.

10.4.4 Low Energy and Power Density

There is a necessity for energy and power densities for portable devices, and grid-scale markets. It is clear that the E_d and P_d are influenced by several elements, the most important of which are the transference kinetics between the hydrolysis solution and electrode, the efficiency of conduction of ions in the electrolytic solution, and the electrode couple [17]. When it comes to ESS that involves high-temperature LMBs, the energy densities consistently go below 200 Wh kg^{-1}. After bringing their more elevated operational temperatures down to intermediate zones, these values would experience a significant reduction. To meet the increased demand for energy-related applications, there has been a significant movement in the focus toward higher-energy-density HTLMB devices (>200 Wh kg^{-1}) in recent years. This shift has been particularly focused on high-temperature LMBs that operate at high current densities but at low temperatures.

10.5 MODIFICATION STRATEGIES

Transition metal-based LMBs present unique challenges, but several modification strategies can be employed to address these issues. Below are some of the key strategies to overcome them.

10.5.1 Alloy-Based

The technique of alloying was an efficient method for overcoming the high melting point of some metals. In comparison to the melting points of the separate metallic components that make up negative electrodes, eutectic alloys have demonstrated that they can significantly lower the melting points of positive electrodes. When it comes to alloy cathodes, there are an infinite number of combinations that can be implemented when taking into account the chemical species (such as Ni, Mn, Co, Zn, Sn) and the composite percentage. Beyond binary and ternary alloy systems, the battery chemistry of LM should be extended to multi-element fusible alloys to further reduce the operating temperature of LMBs. When compared to pure metals, the multi-element alloy system offers greater versatility, allowing for a more extensive selection and rational screening of cost-effective and fusible metallic electrodes to design novel LMBs. In addition, the utilization of the low-cost metal component for the alloy electrode provides the opportunity to contribute to the overall cost reduction. To achieve a high level of energy efficiency, it is obvious that the electrical conductivity of the LM or alloy electrode, as well as its wettability on the surface of the current collector and solid electrolyte, should be improved to the greatest extent possible.

Li et al. [18] fabricated Sb-Sn alloys for the very first time as being able to function as positive electrodes for LMBs. The Li||Sb-Sn LMB demonstrates exceptional rate performance, with 13% capacity loss from 100 mA cm^{-2} to 1 A cm^{-2} of current densities. Additionally, it has a high E_d (200.4 Wh kg^{-1}) at a lowered operating temperature. Guo and his colleagues [23] came up with the idea for a ternary hybrid-cation battery after being inspired by the charge selection mechanism investigations themselves. Only lithium ions were able to intercalate into the cathode side, which was regulated by size, and only potassium ions were able to strip-deposit on the anode in the Na-free electrolyte. The commercially available lithium-ion cathode, $Li_2Ni_{1/3}Mn_{1/3}Co_{1/3}O_2$, was paired with an Na-K anode in the hybrid electrolyte that included lithium and potassium ions. Through the examination of the chemical distribution on the anode surface and the examination of the cyclic voltammograms, they were able to establish that the ion distribution in the solid electrolyte interphase (SEI) layer was the cause of this charge selection behavior.

10.5.2 Electrolyte Designing

An essential component of the LMB is the molten salt electrolyte, which serves a dual function by conducting lithium ions for the electrode reaction and segregating anode and cathode materials. The following criteria must be satisfied by a fluid salt electrolyte that is suitable: (1) low metal solubility; (2) no side reaction with electrode materials in the operating electrochemical window; (3) intermediate density between cathode and anode; (4) low cost; (5) superior ionic conductivity; (6) adequate melting temperature [24].

At high temperatures, however, it is unavoidable that ions will be produced as a result of the fact that the molten electrolyte contains a significant amount of metallic solubility. This leads to an increase in the rate of self-discharge, which in turn leads to a reduction in energy efficiency. This is also an obstacle in terms of maintaining the reversibility. It is also considered that the maximal metal solubility can only be found in molten halides that include the cation of their respective metal. To advance the development of molten salt electrolytes, it would be necessary to conduct reasonable screenings of chemical species and to optimize the ratios of active to inactive components [25].

In addition, it is important to note that pursuing a low operating temperature by designing molten salt electrolytes with low melting points is not a good idea from the perspective of the overall performance of LMB. This is because an excessively low operating temperature results in a decreased lithium-ion diffusivity in both the electrolyte and the electrode, which in turn leads to sluggish electrode reaction kinetics [26]. A cost-effective electrolyte that has high conductivity and a moderate melting point is strongly desired to couple with the electrode materials that are currently available. This electrolyte is expected to break the dilemma that LMB technology faces between long-term operating stability and electrochemical performance as well as material cost with its combination of these two characteristics. The mechanistic linkages between material processing, defect chemistry, transport dynamics, and practical performance for the solid ceramic electrolyte that is employed in MTLMB and RTLMB are still not fully understood [27]. Sun et al. [28] reported an ionic liquid electrolyte composed of sodium bis(trifluoromethanesulfonyl)imide with bis(fluorosulfonyl)imide. Lithium cobalt oxide ($LiCoO_2$) paired with Li anode and $LiNi_{0.8}Co_{0.1}Mn_{0.1}O_2$ exhibits a high specific capacity of 199 mAh g^{-1} and E_d of 765 Wh kg^{-1} with ~99.9% Coulombic efficiencies up to 1200 cycles. Yan et al. [30] generated $Cu(BF_4)_2$ electrolyte and PANI/Cu battery with a high specific capacity of 70 mAh g^{-1} and P_d of 3000 W kg^{-1}. Moreover, the full cell displayed nearly 100% coulombic efficiency over 10,000 cycles of charging/discharging. Some studies have shifted their focus to the solid electrolyte as an alternative solution to the high operating temperature issue of the LMB, rather than addressing the melting temperature limitation of molten salt electrolytes [30], [316]. Chen et al. [32] achieved even deposition of sodium, to moderate the volume change, and to simultaneously construct a stable SEI layer. For this, a porous sodiophilic zinc metal framework that has been modified with a surface layer of tin oxide is introduced on a copper substrate and used as a three-dimensional current collector for sodium metal batteries. The cycling life of a Cu/Zn/SnO_2@Na symmetric cell is greater than 1000 h, which is the long-term rate performance of the cell. Furthermore, the Cu/Zn/SnO_2@Na symmetric cell has the capability to undergo cycling for around 700 h regardless of the current density of 5 mA cm^{-2}, resulting in a total capacity of 5 mAh cm^{-2}. Models could show the capacity retention and coulombic efficiency performance, voltage outlines, and rate performance of the Cu@Na//$Na_3V_2(PO_4)_3$ and Cu/Zn/SnO_2@Na//$Na_3V_2(PO_4)_3$ full cells.

10.5.3 Separator Engineering

The commercial development of LiMBs has been significantly impeded by issues such as low coulomb efficiency, limited cycling life, potential safety hazards associated with lithium dendrites, and different interface reactions. The strategy of developing functional separators is highly effective, as it can suppress the generation of lithium dendrites by modulating the transport of Li^+ ions or utilizing mechanical barriers. Additionally, this approach does not significantly reduce the E_d of the battery. The strategy of suppressing lithium dendrites through separator engineering primarily entails three key components: (1) The mechanical strength of the separator can be enhanced to impede the growth of lithium dendrites through physical action; (2) the flux of Li^+ ions can be redistributed by controlling separator voids; and (3) the nucleation sites on the separators can be increased to induce consistent plating of Li^+ ions [33]. Qu et al. [34] employed graphene oxide (GO) to prevent the formation of lithium dendrites. The cathode was commercial lithium manganate (LMO), the separator was a three-layer film, and the anode was lithium foil. To create LMO/GO/Li batteries, they sandwiched the GO film as an intermediate layer between two polypropylene microporous films to create a three-layer film. They have shown that the modified separator effectively eliminates the lithium dendrite in the LMBs. The LED remains on, and the battery functions normally even when penetrated with a syringe needle. Figure 10.2. shows the microscopy image of the yielded lithium foil after it has been penetrated, the EDS spectra, demonstration of the separator, SEM pictures of the subsequent separator, demonstration of LMO on Al foil, and SEM pictures of the prepared LMO/Al foil. Further, Feng and colleagues optimized separator modification by bonding and combining amorphous SiO particles with polyacrylic acid (PAA). The emergence of dendrites is effectively suppressed, and the cycle stability of LMBs is improved by the formation of a single-sided SiO@PAA-PE on polyethylene (PE) separators. The SiO@PAA-PE separator was employed in stainless steel Li batteries, resulting in closer contact between the SiO@PAA coatings and lithium. The Li-SiO/PAA layer not only mitigates the interfacial side reactions but also aids in the homogenization of the Li^+ flux, thereby effectively impeding the formation of lithium dendrites. The battery's cycle stability is significantly enhanced when the SiO@PAA-PE separator is employed in $LiNi_{0.8}Mn_{0.1}Co_{0.1}O_2Li$ batteries, as evidenced by a high coulombic efficiency of 99% at 5C [35].

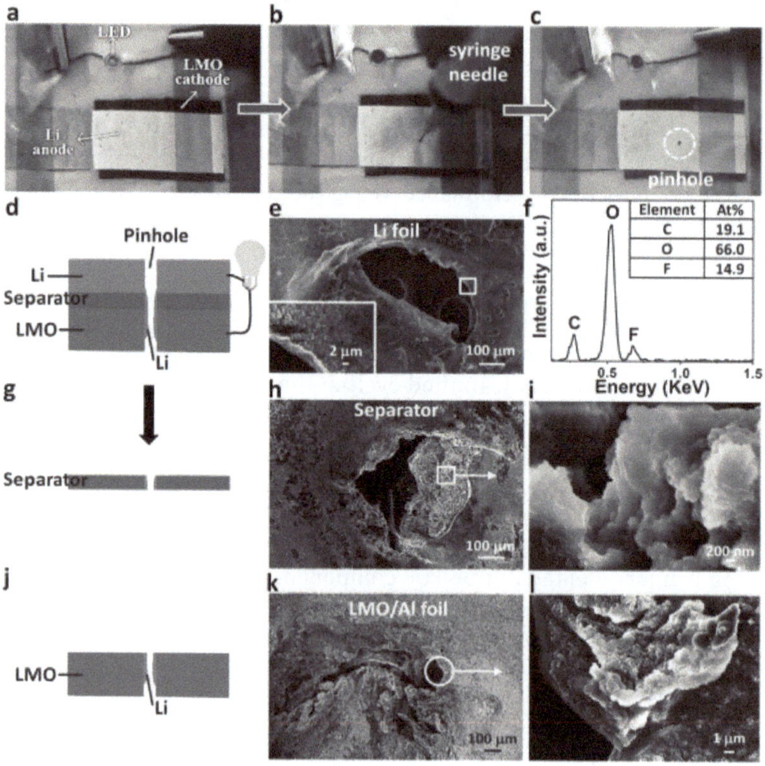

FIGURE 10.2 (a-c) The piercing process, (d) LMO/Li battery experiences a short circuit, (e) Microscopy image, (f) EDS spectra, (g) Separator, (h, i) SEM pictures, (j) LMO on Al foil, and (k, l) SEM pictures.

Source: Adapted with permission [34]. Copyright (2018), Elsevier.

10.5.4 Sealing Strategy

The sealing technique is an essential component since it helps to prevent hazardous secondary reactions and minimize corrosion that can be caused by the introduction of oxygen, nitrogen, and moisture from the surrounding atmosphere. The seal must be able to preserve its integrity in addition to being gastight, electrically insulating, chemically stable, and resistant to thermomechanical forces. For high-temperature battery applications, the engineering of a seal presents a particularly difficult challenge. This is because the requirement for electrical insulation necessitates the combination of materials that are not equivalent to one another and have significantly different coefficients of thermal expansion. The reason for this is the seal needs to be able to withstand prolonged exposure to high temperatures [21].

The two basic categories that can be utilized in the classification of seals are compressive seals and adhesive seals. The space between two or more surfaces is filled with a compressive seal to prevent leakage through the joined items while exposed to a mechanical load. The utilization of metal/metal compressive seals is an effective method for sealing at high temperatures. For applications that include temperatures of up to about 200°C, compressive elastomeric linings and O-ring seals are employed. But when it comes to their long-term durability and high heat losses, elastomeric seals may cause complications.

The intermolecular forces that are responsible for the combining of two materials by an adhesive seal include dispersive bonding, chemical bonding, and diffusive bonding. The thermal expansion coefficient of the seal, on the other hand, needs to be carefully designed to match the joining materials. This is because dispersive bindings are often weak. Chemical bonding and diffusion bonding, on the other hand, can produce insulating seals that are more robust and long-lasting. As a consequence, these bonding techniques are more promising for applications that make use of LMBs [14]. Figure 10.3 demonstrates the schematic of various modification strategies for challenges faced by LMBs.

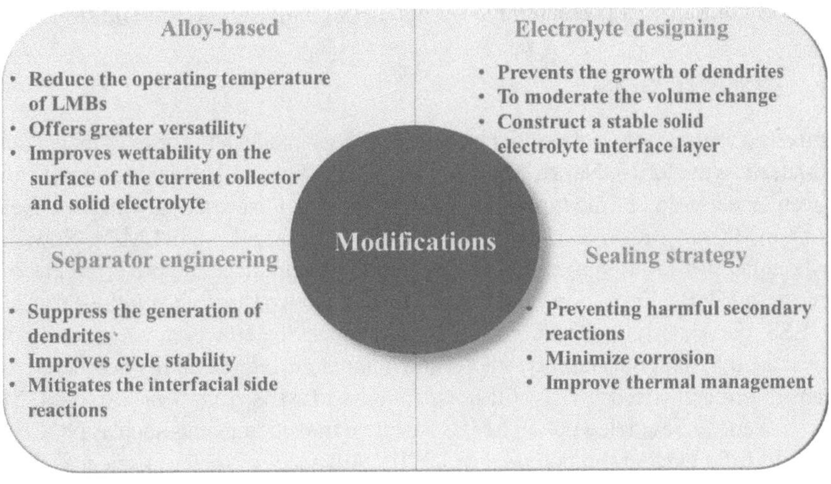

FIGURE 10.3 Modification strategies.

10.6 CURRENT CHALLENGES AND FUTURE PROSPECTS

To ensure that transition metal-based LMBs are commercially viable, one of the primary obstacles is to reduce production costs and increase manufacturing capacity. It is essential to refine the manufacturing processes, as the materials used, such as LM and ions, can be costly. To endure multiple charge/discharge cycles in contact with molten compounds, transition metals and their alloys must be capable of resisting corrosion and degradation. This challenge necessitates the careful selection of protective coatings and materials. The challenge of enhancing E_d and efficiency persists. Although LMBs possess a high P_d, it is imperative to increase their E_d to compete with other battery technologies, such as lithium-ion, in order to gain broader adoption. Further, it is essential to maintain the operating temperature of LMBs to ensure their efficient operation. Stable operation in a variety of environmental conditions without sacrificing longevity is a technical challenge. Throughout numerous charge-discharge cycles, LM and compounds must maintain a stable and non-reactive state within the battery. A persistent challenge is preventing the degradation or contamination of the electrolyte by materials. The safety challenges posed by the corrosive and reactive nature of the materials inside LMBs necessitate the implementation of strong monitoring and suppression systems. Grid-scale ESS is well-suited to LMBs, as they are capable of managing high currents and have extended cycle lives. A consistent power output could be provided by them, which could be instrumental in stabilizing renewable energy sources such as solar and wind. LMBs are appealing for infrastructure and utility applications due to their potential for longer lifespans and reduced maintenance requirements, which result from their comparatively fewer degradation issues. Compared to certain other battery formulations, transition metal-based LMBs typically employ a variety of non-toxic and abundant materials, which mitigates environmental concerns. This feature has the potential to facilitate their adoption as a sustainable ESS solution.

In summary, transition metal-based LMBs are a promising prospect for future development, even though they encounter challenges related to performance and safety. Their potential benefits in grid-scale ESS and long-term reliability are particularly inspiring. To overcome current limitations and realize their maximum potential in the ESS landscape, it is essential to engage in ongoing research and innovation.

10.7 CONCLUSION

Over the last few decades, substantial progress has been made in the development of electrodes/electrolytes for Li, Na, K, Mg, Ca, Ga, Zn, etc.-based LMBs. Additionally, there has been an extension of the types of cathode electrochemistry along with the development of anode materials (based on molten metals and their alloys) in LMBs. Nowadays, transition metals/transition metal alloys are gaining attention as the preferred electrode materials in LMBs. Transition metal-based LMBs offer promising solutions for large-scale ESS, particularly in terms of cost, E_d, and longevity. However, overcoming challenges related to material stability, safety, and manufacturing scalability is crucial. Also, in transition metal-based LMBs, till now, Zn-based LMBs have been explored. More research needs to be carried out on LMBs based on transition metals such as Ni, Co, Cu, Fe, etc., to fully explore the potential of LMBs. Advances in material science, thermal management, and electrolyte development will be key to unlocking the full potential of LMBs, making them a viable alternative in the future energy landscape.

REFERENCES

[1] Y. Gupta, P. Siwatch, R. Karwasra, K. Sharma, S.K. Tripathi, Recent progress of layered structured P2- and O3-type transition metal oxides as cathode material for sodium-ion batteries, *Renew. Sustain. Energy Rev.* 192 (2024) 114167.
[2] Poonam, K. Sharma, A. Arora, S.K. Tripathi, Review of supercapacitors: Materials and devices, *J. ESS.* 21 (2019) 801–825.
[3] G. Tatrari, R. An, F.U. Shah, Designed metal-organic framework composites for metal-ion batteries and metal-ion capacitors, *Coord. Chem. Rev.* 512 (2024) 215876.
[4] M.D. Singh, C. Kaushik, J. Nishanth, A.V. Shinde, R. Gupta, N. Tiwari, K.S. Nalwa, Enroute to sub-300 °C operating Sn–Zn||Pb–Bi liquid metal battery by compositional engineering, *J. Power Sources.* 564 (2023) 232855.
[5] H. Zhou, H. Li, S. Yan, X. Zhou, W. Zhang, S. Feng, Y. He, K. Jiang, K. Wang, In situ displacement reactions of molten sodium anode and multi-cationic halide electrolytes enabling high-performance liquid metal batteries, *Chem. Eng. J.* 485 (2024) 149786.
[6] V. Ranawade, R. Verma, N. Tiwari, K.S. Nalwa, A comprehensive multiphysics approach to model solutal buoyancy, thermal convection, and electro-vortex flow phenomena for liquid metal batteries, *J. ESS.* 85 (2024) 111100.
[7] N. Weber, P. Beckstein, W. Herreman, G.M. Horstmann, C. Nore, F. Stefani, T. Weier, Sloshing instability and electrolyte layer rupture in liquid metal batteries, *Phys. Fluids.* 29 (2017).
[8] X. Zhang, J. Wang, D. Hu, W. Du, C. Hou, H. Jiang, Y. Wei, X. Liu, F. Jiang, J. Sun, H. Yuan, X. Huang, High-performance lithium metal batteries based on composite solid-state electrolytes with high ceramic content, *ESS Mater.* 65 (2024) 103089.
[9] H. Li, H. Yin, K. Wang, S. Cheng, K. Jiang, D.R. Sadoway, Liquid metal electrodes for ESS batteries, *Adv. Energy Mater.* 6 (2016) 1–19.
[10] A.A. Nechikott, P.K. Nayak, Electrochemical capacitance properties of pre-sodiated manganese oxide for aqueous Na-ion supercapacitors, *RSC Adv.* 13 (2023) 14139–14149.
[11] A.C. Baclig, G. McConohy, A. Poletayev, A. Michelson, N. Kong, J.H. Lee, W.C. Chueh, J. Rugolo, High-voltage, room-temperature liquid metal flow battery enabled by Na-K|K-β″-alumina stability, *Joule.* 2 (2018) 1287–1296.
[12] B. Zhang, L. Ren, Y. Wang, X. Xu, Y. Du, S. Dou, Gallium-based liquid metals for lithium-ion batteries, *Interdiscip. Mater.* 1 (2022) 354–372.
[13] X. Ning, C. Liao, G. Li, Electrochemical properties of Ca—Pb electrode for calcium-based liquid metal batteries, *Int. J. Miner. Metall. Mater.* 27 (2020) 1723–1729.
[14] H. Kim, D.A. Boysen, J.M. Newhouse, B.L. Spatocco, B. Chung, P.J. Burke, D.J. Bradwell, K. Jiang, A.A. Tomaszowska, K. Wang, W. Wei, L.A. Ortiz, S.A. Barriga, S.M. Poizeau, D.R. Sadoway, Liquid metal batteries: Past, present, and future, *Chem. Rev.* 113 (2013) 2075–2099.
[15] W. Wu, M. Wang, J. Wang, Z. Wei, T. Zhang, S. Sen Chi, C. Wang, Y. Deng, Transition metal oxides as lithium-free cathodes for solid-state lithium metal batteries, *Nano Energy.* 74 (2020) 104867.
[16] L. Gao, J. Chen, Q. Chen, X. Kong, The chemical evolution of solid electrolyte interface in sodium metal batteries, *Sci. Adv.* 8 (2022) 1–7.
[17] S. Zhang, Y. Liu, Q. Fan, C. Zhang, T. Zhou, K. Kalantar-Zadeh, Z. Guo, Liquid metal batteries for future ESS, *Energy Environ. Sci.* 14 (2021) 4177–4202.

[18] H. Li, K. Wang, S. Cheng, K. Jiang, High performance liquid metal battery with environmentally friendly antimony – tin positive electrode, *ACS Appl. Mater. Interfaces* (2016) 6–11.
[19] P. Chu, J. Wang, H. Xie, Q. Zhang, J. Feng, Z. Li, Z. Yang, H. Zhao, Sb-Cu alloy cathode with a novel lithiation mechanism of ternary intermetallic formation: Enabling high energy density and superior rate capability of liquid metal battery, *J. Energy Chem.* 78 (2023) 393–400.
[20] J.R. Zhang, L. Gao, Hydrothermal synthesis of tin oxide nanoparticles, *Wuji Cailiao Xuebao/J. Inorg. Mater.* 19 (2004) 1177–1180.
[21] F. Chen, J. Yu, R. Li, F. Shi, X. Che, K.C. Chan, Y. Sun, W. Xue, Z.L. Xu, Direct crystallization of deep eutectic solvent into solid-state electrolyte for magnesium metal batteries, *J. Power Sources.* 611 (2024) 234780.
[22] P. Chu, H. Zhao, J. Wang, H. Xie, C. Han, Z. Yang, Bifunctional 3D Graphite@Ni-Fe foam negative current collector toward stable liquid metal battery, *J. Alloys Compd.* 903 (2022) 163952.
[23] H. Xie, J. Feng, H. Zhao, Ternary intermediate phase mediated high energy density and kinetically accelerated Sb-Cd electrode for liquid metal batteries, *Energy Storage Mater.* 72 (2024) 103738.
[24] X. Guo, Y. Ding, H. Gao, J.B. Goodenough, G. Yu, A ternary hybrid-cation room-temperature liquid metal battery and interfacial selection mechanism study, *Adv. Mater.* 32 (2020) 1–8.
[25] H. Xie, J. Feng, H. Zhao, Lithium metal batteries with all-solid/full-liquid configurations, *ESS Mater.* 61 (2023) 102918.
[26] Y. Ding, X. Guo, G. Yu, Next-generation liquid metal batteries based on the chemistry of fusible alloys, *ACS Cent. Sci.* 6 (2020) 1355–1366.
[27] M. Hou, Y. Zhou, F. Liang, H. Zhao, D. Ji, D. Zhang, L. Li, Y. Lei, Research progress of solid electrolyte interphase for sodium metal anodes, *Chem. Eng. J.* 475 (2023) 146227.
[28] S. Wang, B. Peng, J. Lu, Y. Jie, X. Li, Y. Pan, Y. Han, R. Cao, D. Xu, S. Jiao, Recent progress in rechargeable sodium metal batteries: A review, *Chem.—A Eur. J.* 29 (2023).
[29] H. Sun, G. Zhu, Y. Zhu, M.C. Lin, H. Chen, Y.Y. Li, W.H. Hung, B. Zhou, X. Wang, Y. Bai, M. Gu, C.L. Huang, H.C. Tai, X. Xu, M. Angell, J.J. Shyue, H. Dai, High-safety and high-energy-density lithium metal batteries in a novel ionic-liquid electrolyte, *Adv. Mater.* 32 (2020) 1–13.
[30] C. Yan, Z. Chen, H. Huang, X. Deng, Spontaneous proton chemistry enables ultralow-temperature and long-life aqueous copper metal batteries, *Angew. Chem.* 135 (2023).
[31] J. Huang, K. Wu, G. Xu, M. Wu, S. Dou, C. Wu, Recent progress and strategic perspectives of inorganic solid electrolytes: Fundamentals, modifications, and applications in sodium metal batteries, *Chem. Soc. Rev.* 52 (2023) 4933–4995.
[32] Q. Chen, B. Liu, L. Zhang, Q. Xie, Y. Zhang, J. Lin, B. Qu, Sodiophilic Zn/SnO 2 porous scaffold to stabilize sodium deposition for sodium metal batteries, *Chem. Eng. J.* 404 (2021) 126469.
[33] Y. Huang, H. Yang, Y. Gao, G. Chen, Y. Li, L. Shi, D. Zhang, Mechanism and solutions of lithium dendrite growth in lithium metal batteries, *Mater. Chem. Front.* 8 (2023) 1282–1299.
[34] M. Ye, Y. Xiao, Z. Cheng, L. Cui, L. Jiang, L. Qu, A smart, anti-piercing and eliminating-dendrite lithium metal battery, *Nano Energy.* 49 (2018) 403–410.
[35] L. Tan, Y. Sun, C. Wei, Y. Tao, Y. Tian, Y. An, Y. Zhang, S. Xiong, J. Feng, Design of robust, lithiophilic, and flexible inorganic-polymer protective layer by separator engineering enables dendrite-free lithium metal batteries with LiNi0.8Mn0.1Co0.1O2 cathode, *Small.* 17 (2021) 1–11.

11 Antimony-Based Liquid Metal Batteries

Akhilash Mohanan Pillai, Patteth S. Salini, Bibin John, and Mercy Thelakkattu Devassy

11.1 INTRODUCTION

Nowadays, large-scale energy storage technologies (ESTs) play a crucial role in overcoming the bottleneck issue of utilizing renewable energies, such as solar and wind [1–4]. Among the various stationary energy storage systems, battery systems gained much attention due to their high efficiency, mechanical simplicity, low maintenance, and pollution-free operation [5–7]. Lithium-ion batteries (LiBs), the most popular rechargeable battery system, have conquered the worldwide battery market owing to their being lightweight, having high energy density and power density, fast charging/discharging capabilities, low self-discharge, high working voltage, and no memory effect. The 2019 Nobel Prize in Chemistry was awarded jointly to John B. Goodenough, Stanley Whittingham, and Akira Yoshino for the development of LiBs, which rely on the intercalation chemistry of cathode and anode, and this indicates the high impact of LiBs in day-to-day life [8–11]. However, recently, many safety concerns have emerged with LiBs, particularly in electric vehicles (EVs) and large-scale energy storage applications, as compared to portable gadgets [12–15], which has prompted the ongoing research on attempts to improve safety, cycle life, and other performance aspects of LiBs. At the same time, researchers are looking at various forms of electrochemical energy storage systems with higher energy density for usage in electric vehicle applications. The liquid metal battery (LMB) is one example of such an electrochemical system. The invention of LMBs by Prof. Donald R. Sadoway has resolved many of the concerns described above, generating a surge in battery technology development [16–18].

11.1.1 Liquid Metal Batteries: Components and Working Mechanism

11.1.1.1 Various Components in LMBs

A typical LMB comprises two liquid metal electrodes separated by an electrolyte in the molten state, which self-segregates into three layers depending on immiscibility and density. Thus, no additional physical separation is required for such types of batteries. Electrode materials are usually classified as positive or negative based on their deposition potential from aqueous solution. However, since LMBs use molten salt electrolytes, these deposition potentials are not strictly comparable [19–21]. Negative electrodes (A metals) have a deposition potential more negative than –2.0 V, while

DOI: 10.1201/9781003489726-11

positive electrodes have a potential more positive than −1.0 V. But aluminum is unique in that it can be either a negative electrode (A metal) or a positive electrode (B metal) under different reaction conditions since its deposition potential lies in between −2 to −1 V. One can classify elements based on their electronegativity: the metals (including the semimetals) with high electronegativity (electron acceptors: Pb, Bi, Sb, Sn, Hg, and Zn) are the candidates for the positive electrode, which are located at the bottom due to their high density. The metals with high electropositivity (electron donors: Li, Na, Ca, Mg, and K) are the candidates for the negative electrode and are on top owing to their low density. The alloying reaction between the positive and negative electrodes in LMBs is driven by the electron-donating/accepting nature and the difference in electronegativities. Though thermodynamically favored, the alloying reaction is electrochemically regulated in LMBs by using a molten electrolyte as a separator to allow ion migration. Apart from the electrode materials, the properties, selection, and performance of molten salt electrolytes are also very important for LMBs [19, 22]. The molten salt electrolytes employed in LMBs should have the following properties: (i) minimal metal solubility; (ii) high conductivity; (iii) low melting temperature; (iv) adequate density to isolate the three liquid layers; and (v) no side reactions with both positive and negative electrodes within the voltage window and operating temperature. On the other hand, metal solubility in molten salt electrolytes may be the most significant issue since it can result in low coulombic efficiency and excessive self-discharge currents [16, 23].

11.1.1.2 Working Mechanism of LMBs

During discharging, the negative electrode at the top (A) is oxidized to A^{n+}, which moves towards the positive electrode at the bottom (B) through the molten salt electrolyte, and at the same time, electrons move in the same direction through the external circuit. The cation A^{n+} is electrochemically reduced to form an alloy with the metal B (A-B alloy). On the other hand, during charging, the metal A in the A-B alloy is oxidized to A^{n+}, moves towards the negative electrode through the molten electrolyte, and is subsequently reduced to A at the top. A schematic diagram of LMB during the charge-discharge process is depicted in Figure 11.1 [22]. The oxidation, reduction, and overall cell reactions during the charge-discharge process are shown below:

During discharging:

At the negative electrode: $A \rightarrow A^{n+} + ne^-;$ (11.1)

At the positive electrode: $A^{n+} + ne^- + B \rightarrow A\ (\text{in B});$ (11.2)

Overall cell reaction: $A + B \rightarrow A\ (\text{in B}).$ (11.3)

During charging:

At the positive electrode: $A\ (\text{in B}) \rightarrow A^{n+} + ne^- + B;$ (11.4)

At the negative electrode: $A^{n+} + ne^- \rightarrow A;$ (11.5)

Overall cell reaction: $A\ (\text{in B}) \rightarrow A + B.$ (11.6)

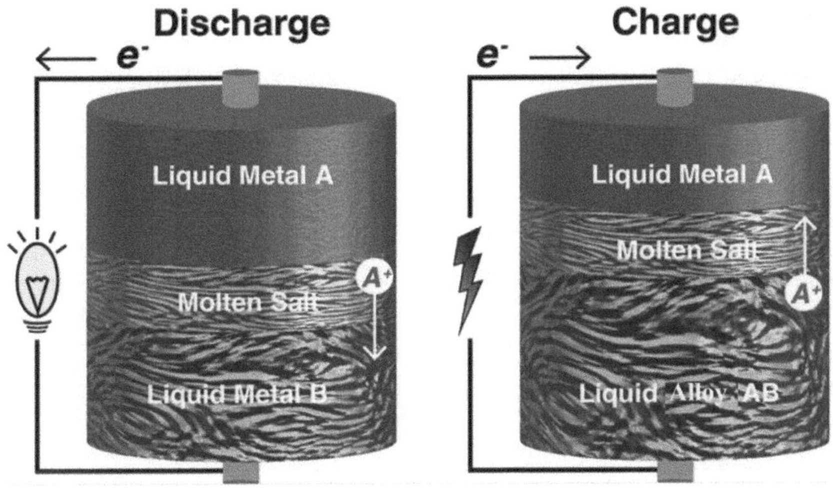

FIGURE 11.1 Schematic representation of LMB upon (a) discharging and (b) charging.
Source: Adapted with permission [22]. Copyright (2022), Elsevier.

11.1.2 Advantages and Disadvantages of LMBs

Metal electrodes are commonly employed in batteries and exist in both solid and liquid states. Liquid metal electrodes have several advantages over solid metal electrodes, including (i) being devoid of dendrite formation during electrodeposition; (ii) exhibiting no grain size change or phase deformation during the charge/discharge process; and (iii) being easy to develop stable electrode/electrolyte interfaces, and also fabricating cells. The liquidity of metal electrodes provides superior transport and kinetic properties for LMBs. The main advantages of LMBs are as follows [16, 19, 24]:

i) LMBs are relatively inexpensive because the majority of the electrode materials used are abundant on earth and cheaper than other materials. Furthermore, the self-segregation of the liquid components allows a simpler manufacturing process and excellent scalability of LMBs when compared to other conventional batteries.
ii) Ultrafast charge transfer kinetics owing to the liquid-liquid electrode-electrolyte interface.
iii) Highly conductive molten salt electrolytes (up to 3 S cm^{-1}) provide low ohmic losses and an excellent rate capability.
iv) The completely liquid electrode and electrolyte system allow LMBs to operate at relatively high current densities.
v) The continuous formation and obliteration of the liquid metal electrodes upon charge-discharge cycling provide unprecedented cycle life to LMBs.

Despite these advantages, LMBs have some disadvantages that make them unsuitable for portable applications. Due to the solubility of metal electrodes in the molten

TABLE 11.1
Properties of Liquid Metals Employed as Negative Electrodes in LMBs

Anode	Melting point (°C)	Specific capacity (mAh g^{-1})	Cost (\$ mol^{-1})	Potential (V vs. SHE)	Solubility in molten salts (mol %)
Li	180	3862	0.43	−3.04	0.5–2.0
Na	92.3	1165	0.057	−2.71	1.6–3.0
K	64	687.2	5.1	−2.93	4.9–19
Ca	842	1340	0.14	−2.87	2.3–9.6
Mg	650	2233	0.069	−2.37	0.2–1.2

Source: Adapted with permission [19]. Copyright (2016), John Wiley and Sons.

TABLE 11.2
Properties of Liquid Metals Employed as Positive Electrodes in LMBs [19]

Cathode	Melting point (°C)	Density (g cm^{-3})	Cost (\$ mol^{-1})	OCV (V vs. Li^+/Li)	Average OCV (V)
Te	449.5	6.24	44	≈1.7	>1.5 V
Bi	271.3	10.05	4.9	≈0.95	<0.8 V
Sb	630.7	6.70	1.8	≈0.92	≈0.9 V
Sn	231.9	5.75–7.28	3.2	≈0.80	≈0.42 V
Pb	327.5	11.34	0.52	≈0.75	≈0.50 V

Source: Adapted with permission [19]. Copyright (2016), John Wiley and Sons.

salt electrolyte, LMBs possess some serious issues, including (i) highly corrosive active cell components, (ii) low theoretical specific energy density (typically <200 Wh kg^{-1}), (iii) high self-discharge rates, (iv) comparatively low equilibrium cell voltages (typically <1.0 V), and (v) elevated operating temperatures (generally >200 °C). The most important threat regarding LMB is that the three-liquid system makes them very sensitive to motion and poses a risk of short-circuiting and rapid heat buildup if the electrodes come into contact.

Tables 11.1 and 11.2 represent the properties of some typical liquid metals used as negative electrodes (Li, Na, K, Ca, and Mg) and positive electrodes (Te, Bi, Sb, Sn, and Pb), respectively, for LMBs.

11.2 ANTIMONY-BASED LIQUID ELECTRODE

Antimony (Sb), a typical semi-metal or metalloid with high electronegativity, is considered as a propitious cathode candidate in liquid metal batteries because of its relatively lighter atomic weight (127.6, about half of Bi), environment-friendliness, earth abundance, low cost (\$ 1.8/mol), high specific capacity (660 mAh g^{-1}), and energy density (above 528 Wh kg^{-1}); it also exhibits high cell voltage when coupled with negative electrodes such as alkali metals (Li, Na, or K) and alkaline earth metals (Mg or Ca) [16]. However, because of its high melting point (630 °C), it requires an operating temperature higher than 630 °C to ensure that the positive electrode remains liquid. Weppner and Huggins [25] demonstrated that Sb and Bi exhibit similar thermodynamic properties based on the electromotive force (EMF) measurements. Alloying with suitable metals such as Pb and Sn is considered an efficient strategy to reduce the melting point of antimony-based materials, and this will eventually decrease the operating temperature of LMBs thermodynamically. However, the attainable energy density is severely limited by the alloying elements (Pb or Sn), which can only lower the melting point of the positive electrode and cannot contribute any capacity to the battery. In spite of this circumstance, Sb-based batteries still continue to surpass Bi-based batteries in terms of energy density, with ~ 200 Wh kg^{-1}. Thus, Sb-based material is considered the most promising positive electrode option available for the LMB at this time. Researchers developed the Sb-based LMB as a potential low-cost competitor for LIB.

11.2.1 Li||Sb-Based Systems

Lithium, the lightest metal with the lowest oxidation potential (–3.04 V vs. SHE), is widely employed as a negative electrode (anode) in many electrochemical storage systems because of its high specific and volumetric energy density, high voltage capability, and facile transport characteristics [26, 27]. The lithium-antimony (Li-Sb) system was first tried to be studied as a high-energy LMB because of its high open circuit voltage (OCV) of 0.92 V. However, because of its high melting point, a battery that uses a pure liquid Sb electrode needs to run at a temperature higher than 630 °C. This ultimately increases the cost of the materials used in secondary batteries and also leads to some corrosion problems. Thus, one efficient method for creating high-performance Sb-based LMBs is to lower the working temperature of batteries while keeping the Sb-based electrodes in the liquid state [19]. In 2014, Wang and his team [28] demonstrated a Li||Sb-Pb battery with a high OCV (0.9 V), using Sb-Pb eutectic alloy as the positive electrode. Due to its low melting point and low cost (0.59 $ mol^{-1}), Pb was chosen as the alloying metal. Furthermore, Pb functions only as a supporting element to lower the melting point of Sb because it has a lower electronegativity than Sb and no intermetallic compound is formed between the two elements. This discovery allowed LMBs to operate at 450 °C, and if the electrolyte compositions were optimized, the temperature could be reduced much lower, to less than 400 °C. In their study, a three-electrode cell was designed to determine the EMF of Sb-Pb alloys and to analyze the utilization of Sb in Sb-Pb alloys using ternary molten salt electrolyte (LiF-LiCl-LiI) at 450 °C. Sb-Pb alloy with various Pb/Sb ratios was employed as the working electrode. The results indicate that even the Sb-Pb alloy with low Sb concentration exhibits a high OCV of 0.87 V against Li, which is only ≈0.05 V lower than the Li-Sb alloy. The OCV of the Sb-Pb alloy is significantly greater than that of pure Pb (0.6 V). This suggests that Sb is the only electrochemically active element present in the electrode materials at a potential higher than 0.6 V, while Pb only acts as a supporting element. The electrochemical performance of 1.9 Ah theoretical capacity Li I LiF-LiCl-LiI ISb-Pb cells (3.16 cm^2 cathode/electrolyte interfacial area) was tested at a current density of 275 mA cm^{-2}. The results showed that the cells consistently delivered 93% of their theoretical capacity with a nominal discharge voltage of 0.73 V. The cycling performance study demonstrates that even after 450 cycles, the Li||Sb-Pb cell still maintains 94% of the initial capacity with 98% coulombic efficiency and a 73% round-trip energy efficiency. It is clear that after the 100th cycle, the capacity fade rate drastically dropped, and from the 100th to the 450th cycle, the fade rate was only 0.004% per cycle. The Li||Sb-Pb cells also exhibit outstanding rate performance under a range of current densities between 100 and 1000 mA cm^{-2}. Even at a high current density of 1000 mA cm^{-2}, the cells retained a capacity of >50% of the theoretical capacity. To demonstrate the scalability of Li||Sb-Pb, Wang et al. constructed cells with a 62 Ah theoretical capacity. At an operating potential of 500 °C, the cell (nominal discharge voltage is ~0.69 V) exhibited negligible capacity fade after 20 cycles with 98% coulombic efficiency and 71% round-trip energy efficiency. Thus, this work proved that without decreasing the cell voltage, the operating temperature of the cell as well as the melting point of the positive electrode can be reduced by alloying Sb with Pb. Further research is necessary to fully understand the underlying mechanism that alloys Sb with Pb without lowering the cell voltage. Furthermore, for large-scale applications, the addition of Pb to an electrode may raise environmental problems.

Later in 2016, Li and his team [29] constructed a high-performance LiǁSb-Sn LMB with enhanced thermodynamic characteristics (>0.9 V of OCV) with an environment-friendly Sb-Sn positive electrode for the first time. Studies revealed that among the various LMBs, the LiǁSb–Sn system exhibited exceptional rate capability, outstanding cycling performance, and unprecedented thermal robustness. Using coulometric titration in a LiCl–KCl electrolyte at 500 °C, the thermodynamic characteristics of the LiǁSb-Sn system were studied. EMF profiles of four compositions of Sb-Sn alloys were demonstrated and compared (Figures 11.2a and 3b). LiǁSn chemistry had a lower EMF of roughly 0.55 V, whereas LiǁSb chemistry displayed the greatest EMF of 0.92 V. However, all Li-Sb-Sn alloys displayed EMF trends that resembled those of the Li–Sb electrode. This indicates that the electrode potential of Li-Sb-Sn systems is primarily determined by the Li-Sb reaction. Thus, these results proved that Sn serves as an inert solvent to reduce the melting point and increase the utilization of the Sb-positive electrode. Furthermore, Li solubility into the positive electrode is also significantly increased by alloying Sb with Sn, which lowers the cost of materials. The rate capability of LiǁSb-Sn cell with a theoretical capacity of 1.5 Ah was investigated at various current densities (100 to 1000 mA cm^{-2}). The results revealed that even at the highest current density of 1000 mA cm^{-2}, the cell retained a capacity of 87 % as compared to that of 100 mA cm^{-2} (Figure 11.2c). Furthermore, the cells exhibit excellent cycling performance at a current density of 300 mA cm^{-2} with a capacity retention of 96.7% after 430 cycles (Figure 11.2d). Interestingly, the Li-Sb-Sn system in this work also demonstrated the simplicity of scaling from 1.4 Ah to 47 Ah for a single cell.

Studies on LiǁSb-Sn LMBs showed that the liquid Sb-Sn cathode undergoes severe corrosion with commonly employed current collectors such as Mo, Ti, and stainless steels. Fortunately, research showed that tungsten (W) is a very stable

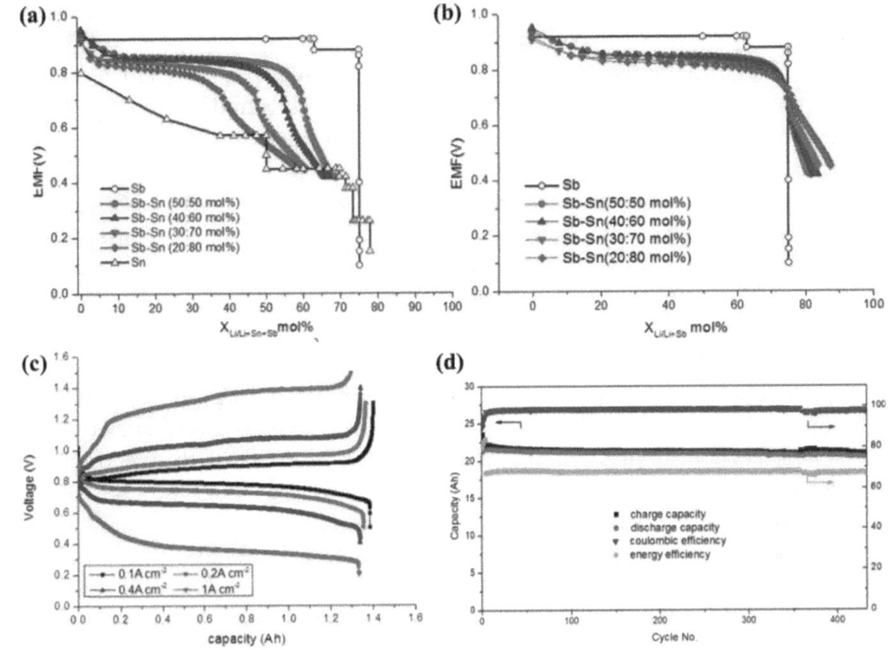

FIGURE 11.2 EMF of Li||Sb-Sn electrodes: (a) EMF as a function of lithium concentration in Sb-Sn alloys, (b) as a function of lithium concentration normalized with respect to Sb, (c) voltage curves of Li||Sb-Sn cell at various current densities (100 to 1000 mA cm^{-2}), and (d) cycling performance of Li||Sb-Sn cell.

Source: Adapted with permission [29]. Copyright (2016), American Chemical Society.

cathode current collector of Li||Sb-Sn LMBs and is considered to act as a protective covering material to lower cost and address the problem of being hard to form. Cui and his team [30] reported a uniform, dense, and low-oxidized W coating by plasma spraying for the first time in the field of LMBs. The results demonstrated that W coating effectively suppresses the cathode corrosion, thereby improving the cycling stability of Li||Sb-Sn LMBs. Moreover, in the present study, Cui et al. designed and fabricated a new LMB that can perform both electrochemical tests as an LMB and non-destructive inspection without needing to dismantle the battery or damage its critical components. Using X-ray micro-computed tomography (micro-CT), researchers were able to observe the corrosion micromorphology of the Sb-Sn cathode and SS304 matrix, both with and without a plasma-sprayed W coating, and the results demonstrated that the W coating effectively protects the SS304 matrix. Additionally, this study also proves that adding a small amount of carbides to W and thickening the W coating can further improve the stability of W coating, thereby inhibiting the cathode corrosion on their current collector at higher temperatures.

The same group [31] designed Li||Sb-Bi-Sn (mol% of Sb-Bi-Sn is 45:45:10) cells by replacing a part of Sn with Bi. They systematically investigated the effect of different operating temperatures on the electrochemical performance of the cells and discovered that the optimum operating temperature was 500 °C. To test the practical applicability of Li||Sb-Bi-Sn LMBs, the energy density, specific capacity, and electrode costs of Li||$Sb_{45}Bi_{45}Sn_{10}$ cells with different capacities (400 and 700 mAh) were calculated. The results showed that both the cells delivered a discharge capacity of 395 mAh g^{-1} with an extremely high energy density of ~ 265 Wh kg^{-1}. The electrode cost of the Li||Sb-Bi-Sn was only 58–60 \$ kWh^{-1}, which is slightly lower than Li||Sb-Pb and Li||Sb-Sn batteries, and much lower than Li||Bi, Li||Te-Sn, and Li||Bi-Pb batteries. The self-healing action of the Sb-Bi-Sn cathode greatly inhibits its corrosion and also maintains a lower melting point. Moreover, the Sb-Bi-Sn cathode increases the overall ratio of active components in the electrode (since the Sn in the Sb-Bi-Sn alloy acts as an inert component), thereby making these types of cells achieve high energy density, discharge capacity, and low costs.

Yang and his team [32] introduced a novel Ti additive to the positive electrode of the Li||Sb-Sn system for the first time. A thorough investigation was carried out on the reaction mechanism of the titanium additive and its effect on the electrochemical performance of Li||Sb-Sn cells (Figure 11.3). During cycling, the high surface tension of the liquid alloy limits the Sb-Sn electrodes from making contact with the graphite collector in non-Ti additive Li||Sb-Sn cells. This reduces the wettability and increases the internal resistance of the cell, which in turn lowers battery efficiency (Figure 11.3a). During the discharge process, the lithium ions move through the molten layer and interact with Sb, and this process increases the wettability of the electrode and the graphite-based current collector (Figure 11.3b). During charging, the discharge product Li_3Sb is converted into Sb-Sn alloy, and the contact between the electrode and current collector again returns to the initial state (Figure 11.3c). Nevertheless, in the case of the Li||Sb-Sn-Ti system, the wettability is improved owing to the lower surface tension of the multicomponent alloy (Figure 11.3d). When Li interacts with Sb, the positive electrode separates into two layers. In the upper layer, Li_3Sb is distributed, whereas in the lower layer, Sn and Ti are found as alloys. Moreover, the interaction of Ti from the Ti-Sn alloy with C results in the formation of a TiC_x transition layer on the graphite surface (Figure 11.3e). This TiC_x layer thickens gradually over time (Figure 11.3f). The presence of the TiC_x transition layer enhances the wettability between the graphite substrate and the positive electrode, ensuring that the electrode maintains good contact throughout the charge-discharge cycles. This layer remains stable during subsequent cycles, contributing to improved wettability and better voltage efficiency of the battery. The results proved that the voltage efficiency of the Li||Sb-Sn-Ti system was 88.4% at 2C, which is 2.8% higher than that of Li||Sb-Sn cells.

Dai and his team [33] developed Li||Sb-Bi LMB with eutectic LiF-LiCl as an electrolyte and thoroughly investigated its electrochemical performances. They employed Sb-Bi alloys as a cathode in LMBs for the first time. The alloying of Sb with Bi not only reduces the melting point of Sb but also allows easy liquefaction of the electrode at operating temperatures. Moreover, both Sb and Bi have similar thermodynamic properties; hence, both elements can contribute to the overall cell capacity. The experimental results showed that the Sb-Bi combination enhanced the discharge voltage of Li||Sb-Bi cells and improved the overall cell capacity and cyclic energy efficiency. In this study, by varying the mole fractions of Sb and Bi, four

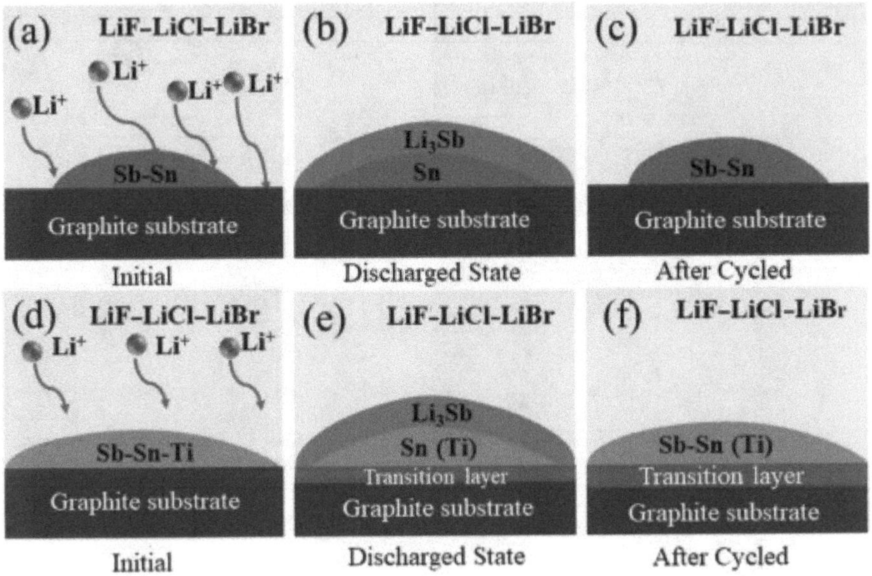

FIGURE 11.3 Schematic representation of microstructural changes of Sb-Sn and Sb-Sn-Ti systems at the interfaces: (a,d) initial state, (b,e) discharged state, and (c,f) charged state after being cycled of Li||Sb-Sn and Li||Sb-Sn-Ti battery systems, respectively.

Source: Adapted with permission [32]. Copyright (2023), American Chemical Society.

alloys were prepared, and among these 60:40 mol % Bi-Sb alloy exhibited the highest discharge voltage. Moreover, a schematic microstructure analysis was carried out on 60:40 mol % Bi-Sb alloy at different levels of discharge. The cross-section micrographs of these alloys prove that the discharge cycle is similar to an extraction process in which Bi and Sb combine to form intermetallics with lithium and then separate into distinct layers (Figure 11.4a–d). Finally, the cycling performance of Li||Sb-Bi cells was investigated by employing a typical two-electrode configuration with 60:40 mol % Bi-Sb alloy as the cathode and Li as the anode. The results showed that the cells exhibited a stable discharge capacity over 160 cycles with a round-trip energy efficiency of 89% (Figure 11.4e).

For LMBs, it seems to be quite challenging to get a high energy density without sacrificing the operating temperature. From the viewpoint of thermodynamics, ternary systems often have a lower eutectic temperature as compared to that of their binary systems, which means that the amount of inactive components required to maintain the liquid state could be lowered. Based on this design strategy, Zhao et al. [34] successfully developed and investigated a Sb-Bi-Sn ternary alloy positive electrode material with excellent rate performance and high energy density. From the Li||Bi and Li||Sb chemistries, the electromotive force (emf) of Li||Sb-Bi-Sn system was determined. High energy density is ensured by this ternary component design, which permits a low percentage of inactive component Sn for the eutectic requirement of a positive electrode operating at a moderate temperature (500 °C).

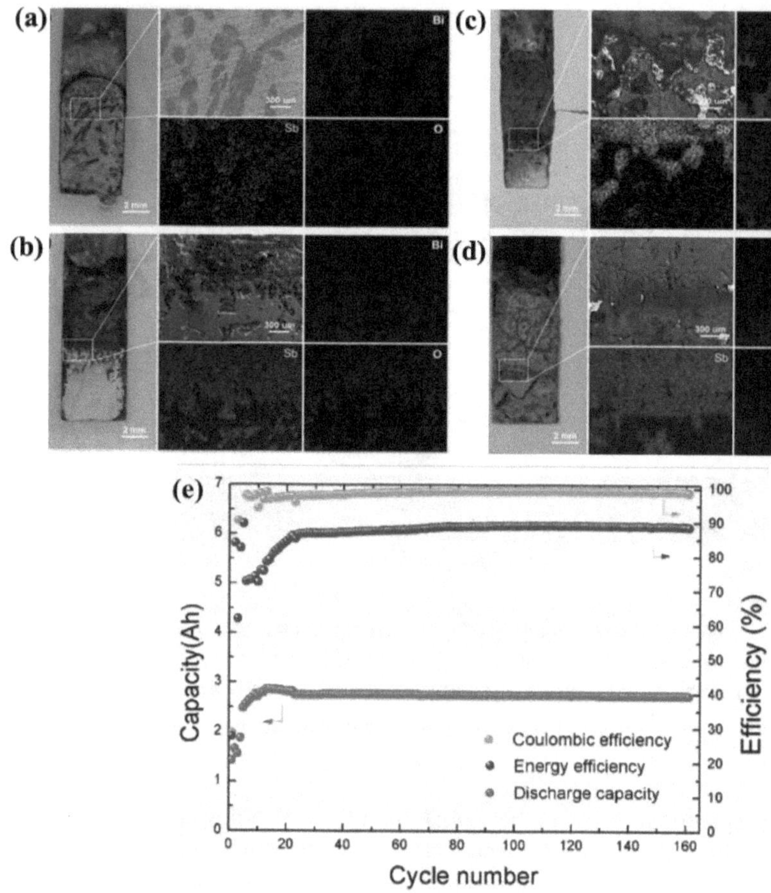

FIGURE 11.4 Cross-sectional micrographs of 60:40 mol % Bi-Sb alloy electrode at different levels of discharge (a) after 25% discharge (b) after 50% discharge (c) after 75% discharge, (d) and after getting fully discharged in the three-electrode cell, and (e) cycling performance of Li||Sb-Bi cells in terms of discharge capacity, energy efficiency, and coulombic efficiency as a function of the number of cycles.

Source: Adapted with permission [33]. Copyright (2018), Elsevier.

The inactive Sn component plays a bi-functional role in this ternary system: firstly, it reduces the melting point of the Sb-Bi-Sn system and thereby facilitates the rapid lithium-ion diffusion in the electrode; secondly, it increases the utilization of active Bi and Sn in the ternary system. A thorough investigation of this ternary system revealed that during the charge-discharge process, the stepwise reaction characteristics of Bi and Sb provide a change in the dynamic microstructure of intermediate compounds, which allows the easy penetration and direct contact of electrolyte with the positive electrode, thereby permitting a fast electrode reaction. Due to the synergic combination of these advantageous characteristics, the present Li||Sb-Bi-Sn system exhibited a very high energy density of 260 Wh kg^{-1} at 346 mA cm^{-2}, which is higher than that of most previously reported LMBs. Moreover, the Li||Sb-Bi-Sn system reduces the material cost ((59 \$ kWh^{-1})) and also exhibits excellent rate capability and superior cycling performance.

11.2.2 Na||Sb-Based Systems

Sodium-based liquid metal batteries (Na-based LMBs) have gained more attention in the field of large-scale energy storage systems than Li-based LMBs owing to the abundance and low melting point of sodium. Moreover, compared with other negative electrodes, Na-based LMBs have economic and environmental advantages. The good compatibility between the Na metal and other sealing components of batteries could ensure the long-term service of Na-based LMBs. Na|NaF-NaCl-NaI|Bi cells were first reported by Argonne National Laboratory in the 1960s. However, this cell requires an operation of 560–585 °C because of the high melting point of the electrolyte; also, the cell exhibits a very low coulombic efficiency of <80 %. Studies revealed that this poor electrochemical performance is owing to the severe dissolution of Na metal in molten electrolyte; this challenge hindered the development of Na-based LMBs for decades. Many strategies have been developed during this period to reduce the dissolution problem of sodium metal [19, 35].

Recently, Fetzer and his team [35] developed Na||$SbBi_9$ LMBs with LiCl-NaCl-KCl (61-3-36 mol%) as a molten salt electrolyte, and their performance is evaluated in different cell designs. In one design, a metal foam hosts the liquid sodium, while in the second design, there is no metal foam. Furthermore, the material employed as the positive current collector and the electrolyte amount were also varied. The detailed studies revealed that the LMBs had a self-discharge current of ~0.35 mA/cm^2, which mainly depends on the electrode distance as well as the operating temperature. Besides, the cell voltage was found to be independent of operating temperature, cell design, and cell age. The long-term cycling tests revealed that both cells retained a stable coulombic efficiency of >99% at the end of 750 cycles. To inhibit the Na dissolution in the molten electrolyte, Zhou et al. [36] recently designed Na||Bi_9Sb cells with a low-melting-point multi-cationic electrolyte (LiCl-NaCl-KCl). Additionally, the compatibility of the molten LiCl-NaCl-KCl inside the batteries with the Na metal electrode was thoroughly investigated and discussed. The process of enhancing battery performance in the Na|LiCl-NaCl-KCl|Sb-Bi batteries was also demonstrated, and the displacement reaction of Na and molten LiCl at high temperatures was disclosed for the first time. Based on the displacement reaction, a Na||Bi_9Sb cell was constructed using a LiCl-KCl (54:46 mol%) electrolyte, without NaCl in the initial state. The cell exhibited an excellent cycling performance (~100% capacity retention after 2500 cycles) at 450 °C, along with a high coulombic efficiency of over 98%.

11.2.3 Mg||Sb-Based Systems

Magnesium, an earth-abundant metal having a melting point of 650 °C, is regarded as a promising negative liquid metal electrode in LMBs. Unlike lithium and sodium, a magnesium atom can release two electrons during oxidation. Hence, they possess a very high gravimetric capacity. The extremely low solubility of Mg in the molten liquid electrolyte makes it an ideal negative electrode candidate for LMBs with high coulombic efficiency and low self-discharge rate [19]. In the year 2012, Bradwell and his team [37] constructed an Mg||Sb cell with liquid Sb as the cathode, liquid Mg as the anode, and NaCl-KCl-$MgCl_2$ as the electrolyte. The electrochemical performance studies showed that under 100 mA cm^{-2}, the Mg||Sb cell had a nominal discharge voltage of 0.35 V. The coulombic and energy efficiencies of the cells were found to be 94% and 73%, respectively. The composition of the positive electrode was determined as $Mg_{12}Sb_{88}$ from the fully charged state, which was then converted into two phases of Sb and Mg_3Sb_2 on cooling to room temperature. Because of the poor OCV of Mg-Sb and the high melting points of Sb (630 °C) and Mg (650 °C), the Mg||Sb cell has not received much attention for practical application initially. However, in terms of the cost of materials (overall cost was estimated as 170 $ kWh^{-1} in this work) and excellent electrochemical performance, the Mg||Sb cell is considered a suitable candidate for LMBs. Later on, with the use of an in-situ pseudo reference electrode (a tungsten wire, sprayed by an insulating layer of boron nitride) at 700 °C, Leung et al. [38] investigated the polarization properties and performance of Mg||Sb-based LMBs. Since the adjacent salt and metal phases are immiscible, the battery is comprised of three separate layers: the electrolyte, the negative electrode, and the positive electrode. To prevent direct electrical contact or short-circuiting with the other electrodes, the reference electrode was immersed in the middle of the electrolyte layer. Polarization studies reveal that the area resistances of the positive and negative electrodes, as well as the overall battery, were approximately 0.65, 0.55, and 1.20 Ω cm^{-2}, respectively. During a typical 1-hour charge/discharge cycle, the battery achieves an average round-trip energy efficiency of approximately 64% at 60 mA cm^{-2}, and this efficiency gradually decreases in the subsequent cycles. Furthermore, half-cell experiments showed that the sprayed layer of boron nitride on the reference electrode maintains its chemical stability and functions as a good electrical insulator during prolonged operation at elevated temperatures.

11.2.4 Ca||Sb-Based Systems

Calcium, another abundant element with low electronegativity, has also been employed as the negative electrode in LMBs. From 1950 to 1980, calcium was used as a negative electrode in thermal batteries owing to its low electronegativity and cost [39, 40]. Presently, their abundance on Earth and low cost make Ca a suitable electrode material for LMBs. However, the Ca-based LMBs face two thorny issues that hinder their commercial application: the high solubility of Ca in the molten salt electrolyte as well as its higher melting point. Poizeau et al. [41] designed a Ca(s)|CaF_2|Ca-Sb cell, and by using EMF measurements, the thermodynamic properties of Ca-Sb alloys were determined over the temperature range 550–830 °C. The average OCV of the Ca-Sb cells was found to be about 1.0 V, which was higher than that of the Ca-Bi cells (~0.8 V). To explain the strong deviation from ideal thermodynamic properties and the interaction of liquid metal particles, two thermodynamic models called the regular associated solution model and the molecular interaction volume model (MIVM) were explored. With an average error of less than 0.6% in the EMF value, the activity coefficient of Ca was predicted at 650 °C based on the interaction parameters optimized by fitting at 800 °C. Ouchi and his team [42] successfully constructed a Ca (in Bi) | LiCl-NaCl-$CaCl_2$ | Ca (in Sb) cell with Ca-Sb alloy as the positive electrode. The EMF measurements showed that the equilibrium potential of the Ca-Sb electrode was found to be 1.2–0.95 V versus Ca. Moreover, the electrochemical performance studies revealed that the discharge capacity of the Ca||Sb cells increases with the increase in operating temperature owing to the higher diffusivity and solubility of Ca in Sb. The Ca-Sb electrode exhibits a good cycling performance of ~100% high coulombic efficiency with < 0.01% $cycle^{-1}$ capacity fade.

11.3 SUMMARY

Liquid metal batteries (LMBs) have made significant advancements in grid-scale energy storage systems owing to their appealing features, such as low cost, ease of scalability, self-healing capability, high power density, extended cyclability, and high efficiency. Moreover, the liquid metal electrodes eliminate the possibility of dendrite formation, a problem that solid electrode batteries constantly face. The last decade has witnessed tremendous progress in the field of LMBs, and various high-power, low-cost, and environmentally friendly electrode materials were developed. Among the various positive electrodes employed in LMBs, antimony (Sb)-based positive electrodes have gained significant attention, especially due to their low cost ($ 1.8/mol), high specific capacity (660 mAh g^{-1}), and energy density (above 528 Wh kg^{-1}). In the present chapter, we have thoroughly reviewed the battery chemistries, the latest advancements, and the challenges of various Sb-based LMBs such as Li||Sb, Na||Sb, Mg||Sb, Ca||Sb, etc. There are a lot of intriguing possibilities for research in the field of liquid metal batteries, especially in the future. By focusing on cost-effective developments and leveraging existing knowledge in the field, there is a great potential to transform the concept of liquid metal batteries into a practical energy storage solution.

11.4 ACKNOWLEDGMENTS

We acknowledge the Director, Vikram Sarabhai Space Centre, Thiruvananthapuram, for granting permission to publish this chapter.

REFERENCES

[1] Z. Yang, J. Zhang, M. C. W. Kintner-Meyer, X. Lu, D. Choi, J. P. Lemmon, J. Liu, Electrochemical energy storage for green grid, *Chem. Rev.* 111 (5) (2011) 3577–3613.
[2] J. B. Goodenough, Electrochemical energy storage in a sustainable modern society, *Energy Environ. Sci.* 7 (2014) 14–18.
[3] M. Pasta, C. D. Wessells, R. A. Huggins, Y. Cui, A high-rate and long cycle life aqueous electrolyte battery for grid-scale energy storage, *Nat. Commun.* 3 (2012) 1149.
[4] Y. Li, Y. Lu, C. Zhao, Y. S. Hu, M. M. Titirici, H. Li, X. Huang, L. Chen, Recent advances of electrode materials for low-cost sodium-ion batteries towards practical application for grid energy storage, *Energy Storage Mater.* 7 (2017) 130–151.
[5] M. Akhilash, P. S. Salini, B. John, T. D. Mercy, The renaissance of high-capacity cathode materials for lithium ion cells, in *Energy Harvesting and Storage*, Editors: M.K. Jayaraj, Aldrin Antony, P.P. Subha, 2022, Springer, Singapore.

[6] M. Akhilash, P. S. Salini, B. John, T. D. Mercy, A journey through layered cathode materials for lithium ion cells-from lithium cobalt oxide to lithium-rich transition metal oxides, *J. Alloys Compd.* 869 (2021) 159239.
[7] A. Manthiram, Y. Fu, Y. S. Su, In charge of the world: Electrochemical energy storage, *J. Phys. Chem. Lett.* 4 (8) (2013) 1295–1297.
[8] Y. Ding, X. Guo, G. Yu, Next-generation liquid metal batteries based on the chemistry of fusible alloys, *ACS Cent. Sci.* 6 (8) (2020) 1355–1366.
[9] D. Lin, Y. Liu, Y. Cui, Reviving the lithium metal anode for high-energy batteries, *Nat. Nanotech.* 12 (2017) 194–206.
[10] A. M. Pillai, P. S. Salini, B. John, T.D. Mercy, Aqueous binders for cathodes: A lodestar for greener lithium ion cells, *Energy Fuels* 36 (10) (2022) 5063–5087.
[11] B. W. Zhang, L. Ren, Y. X. Wang, X. Xu, Y. Du, and S. X. Dou, Gallium-based liquid metals for lithium-ion batteries, *Interdisciplinary Materials* 1 (2022) 354–372.
[12] M. Akhilash, P. S. Salini, B. John, N. Supriya, S. Sujatha, T. D. Mercy, Thermal stability as well as electrochemical performance of Li-rich and Ni-rich cathode materials-a comparative study, *Ionics* 29 (2023) 983–992.
[13] F. M. N. U. Khan, M. G. Rasul, A. S. M. Sayem, N. K. Mandal, Design and optimization of lithium-ion battery as an efficient energy storage device for electric vehicles: A comprehensive review, *J. Energy Storage* 71 (2023) 108033.
[14] Y. Chen, Y. Kang, Y. Zhao, L. Wang, J. Liu, Y. Li, Z. Liang, X. He, X. Li, N. Tavajohi, B. Li, A review of lithium-ion battery safety concerns: The issues, strategies, and testing standards, *J. Energy Chem.* 59 (2021) 83–99.
[15] M. Akhilash, P. S. Salini, B. John, S. Sujatha, T. D. Mercy, Surface modification on nickel rich cathode materials for lithium-ion cells: A mini review, *Chem. Rec.* 23 (2023) e202300132.
[16] H. Kim, D. A. Boysen, J. M. Newhouse, B. L. Spatocco, B. Chung, P. J. Burke, D. J. Bradwell, K. Jiang, A. A. Tomaszowska, K. Wang, W. Wei, L. A. Ortiz, S. A. Barriga, S. M. Poizeau, D. R. Sadoway, Liquid metal batteries: Past, present, and future, *Chem. Rev.* 113 (3) (2013) 2075–2099.
[17] H. Zhou, B. Li, M. Yu, S. Li, G. Fan, X. Ning, Accelerated design of electrodes for liquid metal battery by machine learning, *Energy Storage Mater.* 56 (2023) 205–217.
[18] Y. Ding, X. Guo, G. Yu, Next-generation liquid metal batteries based on the chemistry of fusible alloys, *ACS Cent. Sci.* 6 (8) (2020) 1355–1366.
[19] H. Li, H. Yin, K. Wang, S. Cheng, K. Jiang, D. R. Sadoway, Liquid metal electrodes for energy storage batteries, *Adv. Energy Mater.* 6 (2016) 1600483.
[20] X. Ning, S. Phadke, B. Chung, H. Yin, P. Burke, D. R. Sadoway, Self-healing Li-Bi liquid metal battery for grid-scale energy storage, *J. Power Sources* 275 (2015) 370–376.
[21] M. Irfan, D. Yang, F. Jabbar, A. Elbourne, K. Chiang, T. Daeneke, K. Zuraiqi, Liquid metals for renewable energy synthesis and storage, *Curr. Opin. Green Sustain. Chem.* 49 (2024) 100952.
[22] D. Agarwal, R. Potnuru, C. Kaushik, V. R. Darla, K. Kulkarni, A. Garg, R. K. Gupta, N. Tiwari, K. S. Nalwa, Recent advances in the modeling of fundamental processes in liquid metal batteries, *Renew. Sust. Energy Rev.* 158 (2022) 112167.
[23] H. Liu, X. Zhang, S. He, D. He, Y. Shang, H. Yu, Molten salts for rechargeable batteries, *Mater. Today* 60 (2022) 128–157.
[24] W. Zhao, P. Li, Z. Liu, D. He, K. Han, H. Zhao, X. Qu, High-performance antimony-bismuth-tin positive electrode for liquid metal battery, *Chem. Mater.* 30 (24) (2018) 8739–8746.
[25] W. Weppner, R. A. Huggins, Thermodynamic properties of the intermetallic systems lithium-antimony and lithium-bismuth, *J. Electrochem. Soc.* 125 (1978) 7–14.

[26] M. M. Thackeray, C. Wolverton, E. D. Isaacs, Electrical energy storage for transportation—approaching the limits of, and going beyond, lithium-ion batteries, *Energy Environ. Sci.* 5 (2012) 7854–7863.
[27] Y. Lu, Z. Tu, L. A. Archer, Stable lithium electrodeposition in liquid and nanoporous solid electrolytes, *Nat. Mater.* 13 (2014) 961–969.
[28] K. Wang, K. Jiang, B. Chung, T. Ouchi, P. J. Burke, D. A. Boysen, D. J. Bradwell, H. Kim, U. Muecke, D. R. Sadoway, Lithium-antimony-lead liquid metal battery for grid-level energy storage, *Nature* 514 (2014) 348–350.
[29] H. Li, K. Wang, S. Cheng, K. Jiang, High performance liquid metal battery with environmentally friendly antimony-tin positive electrode, *ACS Appl. Mater. Interfaces.* 8 (20) (2016) 12830–12835.
[30] K. Cui, P. Li, W. Zhao, C. Liu, Q. Wan, S. Li, X. Qu, Advancing knowledge of plasma spraying coatings for Li||Sb-Sn liquid metal batteries by X-ray micro-CT, *Chin. Chem. Lett.* 34 (7) (2023) 107797.
[31] K. Cui, P. Li, W. Zhao, S. Li, C. Liu, X. Qu, Electrochemical performance and application prospect analysis of Li||Sb-Bi-Sn liquid metal batteries, *Mater. Lett.* 338 (2023) 134067.
[32] S. Yan, H. Li, Z. Li, W. Chen, X. Zhou, H. Zhou, W. Zhang, Y. He, K. Jiang, K. Wang, In situ transition layer design based on Ti additive enabling high-performance liquid metal batteries, *ACS Appl. Mater. Interfaces.* 15 (4) (2023) 5265–5272.
[33] T. Dai, Y. Zhao, X. H. Ning, R. Lakshmi Narayan, J. Li, Shan, Z. Wei, Capacity extended bismuth-antimony cathode for high-performance liquid metal battery, *J. Power Sources.* 381 (2018) 38–45.
[34] W. Zhao, P. Li, Z. Liu, D. He, K. Han, H. Zhao, X. Qu, High-performance antimony-bismuth-tin positive electrode for liquid metal battery, *Chem. Mater.* 30 (24) (2018) 8739–8746.
[35] R. Fetzer, T. Zhang, F. Lindner, A. Weisenburger, G. Müller, How cell design affects the performance of sodium-antimony-bismuth liquid metal batteries, *J. Power Sources.* 591 (2024) 233823.
[36] H. Zhou, H. Li, S. Yan, X. Zhou, W. Zhang, S. Feng, Y. He, K. Jiang, K. Wang, In situ displacement reactions of molten sodium anode and multi-cationic halide electrolytes enabling high-performance liquid metal batteries, *Chem. Eng. J.* 485 (2024) 149786.
[37] D. J. Bradwell, H. Kim, A. H. C. Sirk, D. R. Sadoway, Magnesium-Antimony liquid metal battery for stationary energy storage, *J. Am. Chem. Soc.* 134 (4) (2012) 1895–1897.
[38] P. Leung, S. C. Heck, T. Amietszajew, M. R. Mohamed, M. B. Conde, R. J. Dashwood, R. Bhagat, Performance and polarization studies of the Magnesium–Antimony liquid metal battery with the use of in-situ reference electrode, *RSC Adv.* 5 (2015) 83096–83105.
[39] R. A. Guidotti, P. J. Masset, Thermally activated ('thermal') battery technology: Part IV. anode materials, *J. Power Sources.* 183 (1) (2008) 388–398.
[40] R. P. Clark, K. R. Grothaus, An improved calcium anode for thermal batteries, *J. Electrochem. Soc.* 118 (1971) 1680.
[41] S. Poizeau, H. Kim, J. M. Newhouse, B. L. Spatocco, D. R. Sadoway, Determination and modeling of the thermodynamic properties of liquid calcium–antimony alloys, *Electrochim. Acta*, 76 (2012) 8–15.
[42] T. Ouchi, H. Kim, X. Ning, D. R. Sadoway, Calcium-Antimony alloys as electrodes for liquid metal batteries, *J. Electrochem. Soc.* 161 (2014) A1898–A1904.

12 New Emerging Materials for Highly Efficient Liquid Metal Batteries

Alaa Bedair, Mahmoud Hamed, and Fotouh R. Mansour

12.1 INTRODUCTION

Electrochemical energy storage devices, such as batteries, are pivotal in enabling sustainable energy generation from sources such as solar, wind, and geothermal power [1]. To date, various types of batteries have been utilized for the rapid and efficient storage and discharge of energy, with lithium-ion batteries (LIBs) being the most widely adopted, particularly in portable electronic devices. LIBs are also gaining traction in other sectors, such as stationary energy storage systems and electric vehicles, with the global market projected to reach a valuation of A$200 billion by 2027 [2]. Despite their success, the current energy density at the cell level and pricing at the pack level in practical systems fall short of expectations, with typical values around 350 Wh kg^{-1}, 750 Wh L^{-1}, and A$170 kW h^{-1}, respectively [3]. This has spurred ongoing research over the past few decades to develop more efficient rechargeable batteries capable of storing higher energy densities—targeting around 500 Wh kg^{-1}—at a lower cost [3]. One promising avenue is the exploration of metal-electrode-based battery chemistries, a concept dating back to the 1970s, which has the potential to significantly enhance battery performance. For instance, Li metal anodes offer a high specific capacity of 3860 mA h g^{-1} and a highly electropositive potential of −3.04 V relative to the standard hydrogen electrode (SHE) [4]. However, a major challenge is the propensity of solid metal anodes, including Zn and Li, to form dendrites, which can compromise safety and hinder their application in high-security environments [5]. Therefore, the ongoing research on metal-based batteries continues to prioritize the discovery of alternatives that possess high energy densities, cost-effectiveness, and strong safety characteristics. This discovery sparked curiosity in the field of liquid-to-liquid electrochemistry as a potential replacement for solid-to-solid chemistries over time. It is worth indicating that liquid systems have a competitive advantage over solid-state systems in several aspects. Firstly, they can operate at high rates with high voltage efficiencies owing to the superior kinetics provided by the liquid-liquid interfaces. Secondly, liquid electrodes are resistant to microstructural deformation and dendrite formation. Thirdly, liquid systems allow for flexible energy storage components. Lastly, liquid systems have the potential for

 DOI: 10.1201/9781003489726-12

reduced cost as the active materials used are abundant and cheap [6–10]. Liquid metal batteries (LMBs) are batteries that have at least one electrode made of liquid metal [11].

LMBs hold significant potential to revolutionize large-scale energy storage due to their attractive features, including high power density, long cycle life, low cost, self-healing capabilities, high efficiency, scalability, and the use of readily available materials [12]. In recent decades, substantial progress has been made in understanding battery chemistries and modeling the various physical processes in high-temperature liquid metal batteries (HT-LMBs). Advances in battery chemistry research have allowed for the manipulation of key physical and chemical properties of electrodes and electrolytes, such as electronegativity, phase transition behaviors, ion conductivity, diffusivity, and wettability. These developments are crucial for enhancing the performance of LMBs. Furthermore, modeling efforts provide valuable tools for guiding the design and optimization of LMBs to achieve maximum performance. Figure 12.1 presents a schematic diagram illustrating the relationship between battery chemistry and performance metrics.

The composition and architecture of electrodes and electrolytes significantly influence their physical and chemical properties, which are crucial to the overall performance of the battery. Factors such as macrostructural parameters and boundary conditions during charge and discharge cycles are incorporated into a multiphysics

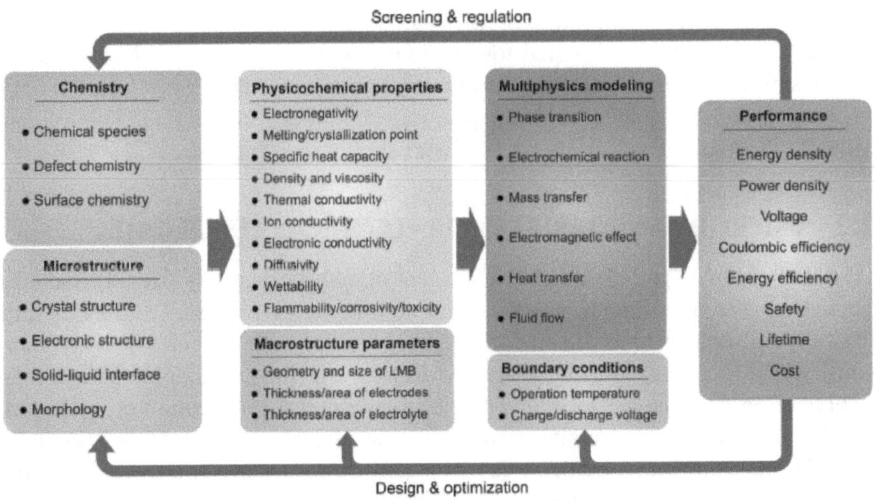

FIGURE 12.1 A logic diagram that illustrates the relationship between battery chemistry and performance metrics. The chemical composition and microstructure of the electrode and electrolyte are critical in shaping their physicochemical properties, which, in turn, influence overall battery performance. These properties, combined with macrostructural parameters and boundary conditions, are integrated into a multiphysics modeling framework that enables precise performance prediction of the battery.

Source: Adapted with permission [11], Copyright (2023), Elsevier.

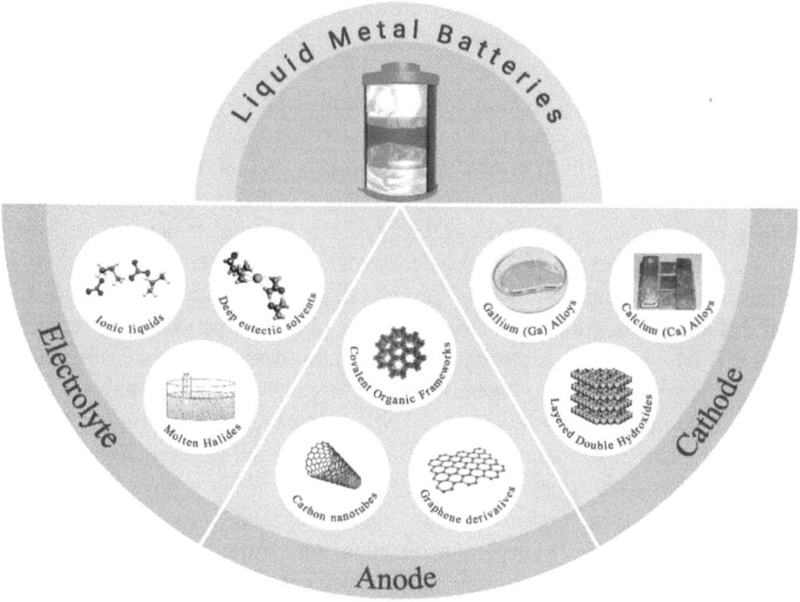

FIGURE 12.2 Emerging materials for liquid metal batteries.

modeling approach to accurately predict battery performance. To enhance battery efficiency, focus must be directed toward four key areas: evaluating battery chemistry, designing optimal electrode and electrolyte microstructures, refining macrostructural parameters, and setting appropriate boundary conditions. This chapter will explore emerging materials for LMBs as shown in Figure 12.2.

12.2 EMERGING MATERIALS FOR ELECTROLYTES OF LMBs

In modern organic synthesis, the development of environmentally friendly and cost-effective solvents is essential, alongside maintaining effectiveness. Traditionally, the use of organic solvents has raised concerns about safety due to their volatility, flammability, and the release of toxic vapors. Moreover, in industrial applications, these solvents often decrease performance because of complex purification processes. As a result, the need for suitable alternatives became apparent. In recent years, ionic liquids (ILs) have gained recognition as viable substitutes for organic solvents, finding applications in various industrial and scientific fields, particularly in electrochemical energy storage. Their non-flammable nature, superior thermal and chemical stability, and significantly higher ionic conductivities make them attractive options. Another emerging class of solvents, deep eutectic solvents (DESs), offers similarly favorable physicochemical properties. A DES is formed from a mixture of a hydrogen bond acceptor (HBA) and a hydrogen bond donor (HBD), which, when combined in a specific molar ratio, form a new phase with a lower freezing point than the individual components due to hydrogen bonding. Despite their advantages, DESs are often unsuitable as substitutes for IL electrolytes in energy storage because they typically have a narrower electrochemical stability window compared to most room-temperature ionic liquids (RTILs). For example, lithium bis(trifluoromethanesulfonyl)imide has been used as an HBA to produce a DES in combination with formamide, resulting in ionic conductivity comparable to RTILs. However, the maximum operating voltage range of this DES was limited to 2.2 V [13].

12.2.1 Ionic Liquids

Metal batteries, whether using nonaqueous electrolytes such as in lithium metal batteries or aqueous electrolytes as in zinc metal batteries (ZMBs), face several internal challenges [14]. During metal deposition, these batteries are prone to dendrite formation, which can penetrate the thin separator and pose serious safety risks [15]. Additionally, metal batteries are subject to side reactions, including the continuous consumption of the metal anode, the repetitive formation and decomposition of a solid-electrolyte interphase (SEI), and in aqueous systems, the generation of hydrogen. These reactions result in reduced Coulombic efficiency (CE). Furthermore, like conventional secondary ion batteries, metal batteries in aqueous or organic electrolyte systems suffer from poor performance at low temperatures and are prone to explosions in organic electrolytes [16]. A crucial factor in overcoming these challenges is optimizing the electrolyte, which plays a vital role in enhancing battery performance.

ILs, which consist of an organic cation and an organic or inorganic anion, are essentially organic salts with melting points below 100 °C [17]. ILs are known for their advantageous properties, including high thermal and chemical stability, excellent ion conductivity, non-flammability, and broad electrochemical stability. Notably, the functional groups in ILs can modify the solvation layer around metal ions, improving their performance at low temperatures. When used in metal batteries, ILs offer a wide range of features that contribute to enhanced electrochemical performance [18].

RTILs also show great promise as electrolytes due to their broad electrochemical stability potential window (ESPW), which ranges from 3 to 6 volts. However, a significant limitation of RTILs is their high viscosity, which results from smaller average void radii relative to the ion radii. To address this issue, organic solvents such as acetonitrile, butyronitrile, and benzonitrile are commonly used to dilute the solution, reducing viscosity and improving ionic conductivity. However, diluting RTILs with organic solvents also reduces the efficiency of their ESPW, presenting a trade-off between viscosity and electrochemical stability.

Although sodium-sulfur (Na-S) batteries offer high energy density, long cycle life, and sustainability, their practical use is hindered by safety and operational challenges due to the requirement to operate at temperatures above 300°C. Reducing the operating temperature of Na-S batteries negatively impacts performance, primarily due to the formation of insulating sulfur/polysulfides, which obstruct the conduction of sodium ions in the β"-alumina solid electrolyte (BASE). Lower temperatures also promote the growth of sodium metal dendrites below its melting point and exacerbate the shuttle effect in the absence of BASE [19]. To address these issues, Wang et al. [19] developed a sodium-sulfur (Na||S) LMB that integrates a dual electrolyte system consisting of BASE and a novel inorganic ionic liquid. This design enables the battery to operate at a lower temperature of 150°C. Investigations revealed that the ionic liquid exhibits high ionic conductivity, a wide electrochemical potential range, and exceptional thermal and chemical stability. Charge-discharge tests demonstrated excellent reversibility at 150°C, with a notable reversible capacity of 795 mAh/g of sulfur at a current density of 0.1 mA, and an average capacity of 381 mAh/g of sulfur maintained over 1,000 cycles at 0.5 mA. These findings highlight the potential of using ionic liquids in dual electrolyte systems to enhance the performance of sodium-sulfur cells.

12.2.2 Deep Eutectic Solvent

DESs are innovative liquid systems formed by hydrogen bonding interactions between two solid components, resulting in a eutectic blend with a substantially reduced melting point compared to each individual constituent. DESs share similar advantageous properties with ILs, yet their production is more cost-effective and straightforward. Furthermore, DESs are highly biodegradable and generally exhibit lower toxicity levels than ILs. The growing interest in DESs stems from their potential to enhance various processes [20], particularly in sustainable applications. Typically, DESs are formed by mixing a quaternary ammonium salt with an HBD such as alcohols, amides, carboxylic acids, or glycols. The hydrogen bond donor interacts with the halide ion of the recipient to form a stable eutectic mixture, with choline chloride being the most commonly used quaternary ammonium salt in the literature. However, DESs based on quaternary ammonium salts are often limited by low ionic conductivity and high viscosity, restricting their practical applications. To address these challenges, alkali metal salts have been introduced as HBAs in DES formulations, resulting in a significant reduction in viscosity and an enhancement in ionic conductivity.

Recent studies have focused on integrating DESs into rechargeable ZMBs as part of a broader effort to improve the performance of aqueous zinc batteries. A major challenge with these batteries is the lack of compatibility between aqueous electrolytes and zinc metal, which can lead to issues such as passivation, corrosion, and water decomposition on the zinc anode. Initial investigations into the chemical stability of zinc in various electrolytes, as depicted in Figure 12.3, demonstrated that zinc foils exposed to traditional aqueous electrolytes (0.25M Li_2SO_4 + 0.5M $ZnSO_4$ and 0.5M LiTFSI + 0.5M $Zn(TFSI)_2$) over 15 days developed tarnish and complex surface coatings associated with non-faradaic processes [21]. In contrast, no noticeable changes were observed in the LZ-DES/2H_2O system. The surface structure of the zinc was further examined using fluorescence microscopy, optical microscopy, and scanning electron microscopy (SEM), which revealed significant differences between aqueous electrolytes and the water-in-DES electrolyte. The aqueous electrolytes promoted the accumulation of micron-scale passive aggregates on zinc, whereas the water-in-DES electrolyte exhibited a dendrite-free zinc deposition.

Moreover, differential electrochemical mass spectrometry (DEMS) was employed to evaluate the compatibility of the electrolyte with zinc by detecting gas release

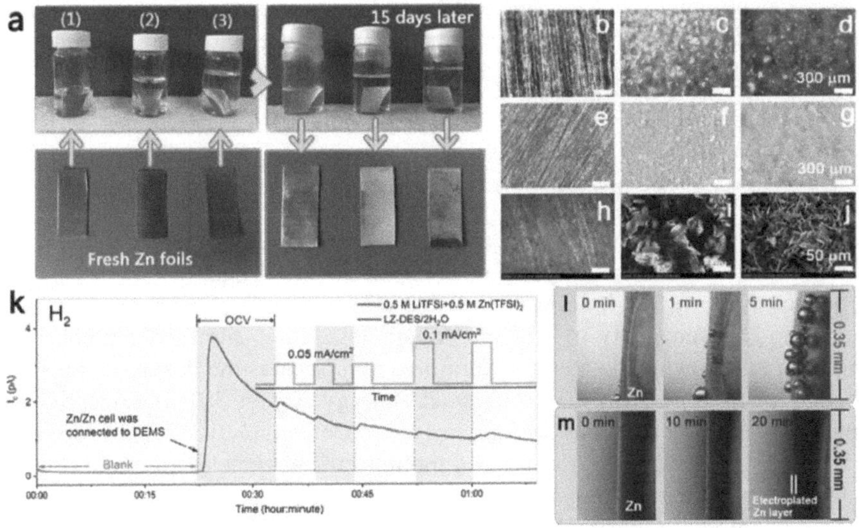

FIGURE 12.3 provides a detailed analysis of the chemical and electrochemical compatibility of aqueous electrolytes with zinc metal. In panel (a), the reactivity of zinc foils was evaluated under three different conditions at room temperature: (1) LZ-DES/2H_2O, (2) 0.25M Li_2SO_4 + 0.5M $ZnSO_4$, and (3) 0.5M LiTFSI + 0.5M $Zn(TFSI)_2$. The images on the left represent the initial state of the zinc foils, while the right-side images depict the state of the foils after 15 days of immersion. To further assess the surface morphology of the zinc foils, they were examined after immersion in the three different electrolyte solutions using a combination of analytical techniques, including fluorescence microscopy (b–d), optical microscopy (e–g), and SEM (h–j). The morphological analysis highlights the distinct differences in surface characteristics across the different electrolyte systems. In panel (k), online DEMS data is presented for Zn||Zn cells using either 0.5M LiTFSI + 0.5M $Zn(TFSI)_2$ or LZ-DES/2H_2O as the electrolyte. The inset of panel (k) illustrates the applied current density profile during the experiments. Prior to the DEMS analysis, the cells underwent a 20-minute stationary observation period to monitor the production of chemical hydrogen (H_2). Following this, current densities of 0.05 and 0.10 mA/cm^2 were applied to assess the electrochemical production of H_2. The zinc electrodeposition process was observed using in situ optical microscopy, as shown in panels (l) and (m). Electrodeposition was conducted at a current density of 0.2 mA/cm^2 in two different electrolyte systems: 0.5M LiTFSI + 0.5M $Zn(TFSI)_2$ (l) and LZ-DES/2H_2O (m). This in situ observation reveals critical insights into the electrodeposition behavior, highlighting the differences between the two electrolytes. The figure emphasizes the superior performance of the LZ-DES/2H_2O electrolyte in minimizing gas formation and promoting smooth zinc deposition.

Source: Adapted with permission [21], Copyright (2019), Elsevier.

during the open-circuit voltage (OCV) phase. In diluted aqueous electrolytes, spontaneous hydrogen (H_2) generation occurred upon contact with metallic zinc, particularly in the 0.5M LiTFSI + 0.5M $Zn(TFSI)_2$ solution. However, no detectable hydrogen gas was produced during the operando testing of LZ-DES/2H_2O. The difference between the electrolytes became more apparent when observing zinc deposition using in situ optical microscopy, where gas bubbles, particularly hydrogen, were seen obstructing zinc deposition in diluted aqueous electrolytes. In contrast, the water-in-DES electrolyte enabled the formation of a smooth, dendrite-free zinc layer at a current density of 0.2 mA/cm^2 over 40 minutes, illustrating its superior chemical and electrochemical stability.

12.2.3 Molten Halides

Molten halides are commonly used as electrolytes in high-temperature LMBs due to their high ionic conductivity and wide electrochemical stability window. These properties make molten halides compatible with a variety of electrode materials, enhancing the versatility of LMBs. Nonetheless, their significant potential to evaporate and flow at elevated temperatures complicates the design and production of the corresponding electrochemical energy storage (EES) devices—both manufacturing expenses and cycling longevity. Conversely, the majority of these EES devices necessitate lithium-based molten salts as the electrolyte to improve performance, which not only escalates costs but also depletes the already scarce lithium resources. Quasi-solidification is an efficient approach in electrolyte design to mitigate the drawbacks of electrolyte leakage and volatilization in room-temperature batteries utilizing liquid electrolytes. The effectively engineered quasi-solid-state (QSS) electrolyte exhibits elevated ionic conductivity and electrochemical stability, reduced combustibility, and enhanced mechanical strength, hence being extensively utilized in sophisticated lithium-based and sodium-based batteries. Ceramic nanopowders have been effectively employed as supplementary fillers to boost mechanical strength and ionic conductivity among various QSS electrolytes. QSS electrolyte, composed of a molten eutectic mixture of Na_2CO_3 and K_2CO_3, combined with nanoparticles of yttrium-stabilized zirconia (YSZ) in a mass ratio of 1:1 was developed by Zhang and coworkers [22]. The QSS electrolyte exhibits reduced volatility compared to the pristine molten Na_2CO_3-K_2CO_3 eutectic, thereby markedly inhibiting the evaporation of molten salts due to robust interactions at the interface between the molten salt and YSZ nanoparticles at elevated temperatures [22]. In this study, the QSS electrolyte, composed of liquid salts and solid oxide nanopowders, had the potential to inhibit the evaporation and flow of molten salts by a robust interaction at the interface between the solid oxide nanopowders and the molten salts. The QSS-molten salt iron-oxygen batteries presented superior cycling charge-discharge capabilities, exhibiting high columbic and energy efficiencies in non-lithium molten salts, along with a liquid-solid-solid 3PI related electrochemical redox reaction that contrasts with the reversible deposition-dissolution mechanism found in lithium-containing molten salts. The results may enhance the advancement of cost-effective and high-performance molten salt metal-air batteries, characterized by notable stability and safety, through the strategic engineering of quasi-solid-oxide electrolyte materials.

However, the high operating temperatures required to maintain molten halides in a liquid state (typically 300°C to 600°C) increase system complexity and limit their application range. Additionally, molten halides can be highly corrosive, necessitating the use of corrosion-resistant materials in battery construction.

12.3 EMERGING MATERIALS FOR ANODE OF LMBs

12.3.1 Covalent Organic Frameworks

Covalent organic frameworks (COFs) are crystalline porous polymers distinguished by their ability to precisely integrate organic units to form predetermined topologies and nanopores. Recently, COFs have garnered significant attention for their potential in creating highly efficient organic materials for diverse applications, including gas storage, catalysis, and optoelectronics. The design and synthesis of COFs hinge on three critical factors: the reversibility of dynamic covalent processes, the diversity of building blocks, and the framework's structural integrity [23].

Carbon-based COFs are characterized by low mass densities, high thermal stability, and permanent porosity, thanks to their composition of lightweight components linked by robust covalent bonds. These frameworks are categorized into two types based on the dimensionality of their structures: two-dimensional (2D) and three-dimensional (3D). Two-dimensional COFs consist of covalent bonds confined to planar sheets, which are stacked to form an ordered layer with aligned columns. This columnar stacking allows for unique p-system structures that are challenging to achieve with conventional methods, facilitating the movement of charge carriers along the stacking direction. Such features render 2D COFs promising for applications in optoelectronics and photovoltaics due to their potential p-electronic and photo-functional properties. In contrast, three-dimensional COFs extend their structure into three dimensions using building blocks with sp^3 carbon or silane atoms. These frameworks are renowned for their exceptionally high specific surface areas, sometimes exceeding 4000 m^2/g, and their low densities, which can be as low as 0.17 g/cm^3.

Recently, COFs and their derivatives have emerged as highly adaptable materials with significant potential for addressing challenges associated with metal anodes, particularly lithium metal anodes. COFs offer several advantages that make them particularly promising for this application. A key feature of COFs is their incorporation of functional groups within their frameworks. These functional groups interact with metal ions to facilitate an even distribution of ion flow during the plating process, resulting in uniform and condensed metal deposition [24]. Additionally, COFs possess versatile structures due to the variety of building blocks available for their synthesis. This versatility allows for the development of various frameworks, including COF-derived conductive porous carbons, which can be employed in metal anode fabrication without requiring a host material. This capability helps address the issue of volume expansion commonly associated with metal anodes. COFs also feature uniform and unobstructed channels that effectively regulate ion migration. This property ensures consistent ion flow during the plating process, reducing the likelihood of dendrite formation. Furthermore, COFs can be engineered to grow on diverse substrates, such as current collectors and metal anodes, thereby enhancing the deposition properties of these anodes. In addition to their structural advantages, COFs exhibit remarkable thermal and electrochemical stability, making them well-suited for stabilizing metal anodes. By modifying the interfacial properties of metal anodes, COFs improve ion movement, serve as nucleation sites, and minimize corrosion. As a result, COFs facilitate the development of metal anodes that are secure, stable, durable, and resistant to dendrite formation [24].

12.3.2 Carbon Nanotubes

Self-healing materials can significantly extend the lifespan of LIBs. One effective strategy involves using self-healing polymers (SHPs) as binding agents for lithium-active materials such as silicon (Si) [25, 26]. SHPs possess the ability to elongate and repair mechanical damage and fractures within the electrode, thereby enhancing the stability of both the mechanical and the electrical connections between the active components. Another approach focuses on utilizing self-healing liquid materials for the electrode itself. Due to their fluidity and surface tension, these materials can be heated to high temperatures to become liquid and then used in the electrode, offering additional advantages for self-healing [27–29].

Carbon nanotubes (CNTs) are cylindrical carbon-based structures with diameters in the nanometer range and lengths extending to the micrometer scale. They exhibit an aspect ratio exceeding 1000 and feature a hexagonal atomic arrangement similar to that of graphite. CNTs consist of a cylindrical layer of graphitic material, known as graphene, formed into a continuous cylinder with a diameter of approximately 1 nm. CNTs can be categorized into two main types based on their structure: single-walled carbon nanotubes (SWCNTs) and multi-walled carbon nanotubes (MWCNTs) [30, 31]. SWCNTs are created by rolling a single layer of graphite (a graphene layer) into a cylindrical form, while MWCNTs are composed of multiple concentric layers of rolled graphite, resulting in a cluster of nested SWCNTs [32, 33].

CNTs are renowned for their exceptional properties, including high optical activity, mechanical strength, electrical conductivity, and thermal stability. Among these attributes, CNTs also demonstrate superior tensile strength. Wu et al. [34] introduced a novel self-healing anode for LIBs by utilizing liquid metal at room temperature to enhance the battery's cycle life. This study presents a new liquid metal alloy, composed of gallium (Ga) and Tin (Sn), which is stabilized within a framework of reduced graphene oxide (RGO) and CNTs. The Ga-Sn alloy serves as a self-regenerating anode for LIBs under normal room-temperature conditions. The incorporation of Sn lowers the melting point of Ga, facilitating self-repair at ambient temperatures. Specifically, Tin exhibits a higher specific capacity of 993 mA h g^{-1} compared to gallium's 769 mA h g^{-1}, thereby enhancing the anode's overall capacity.

The RGO/CNT framework further improves the electrical conductivity of the electrode and prevents aggregation or separation of the Ga-Sn alloy nanoparticles from the current collector during cycling. Consequently, the developed anode demonstrates a remarkable capacity of 775 mA h g^{-1} at a current density of 200 mA g^{-1} and maintains nearly 100% of its capacity over 4000 cycles. The Ga-Sn alloy, with a theoretical capacity of 795 mA h g^{-1}, surpasses the capacity of conventional graphite anodes by more than twice. The alloy, with a Ga:Sn ratio of 88:12 by weight, has a melting point of 20°C, which is lower than the melting points of pure Ga and Sn and below the typical room temperature of 25°C. Additionally, the alloy exhibits notable supercooling capabilities, remaining stable in liquid form even at temperatures as low as −61°C below its melting point. The LM alloy demonstrates excellent stability in its liquid state, making it suitable for applications in LIBs. Figure 12.4 illustrates the transition of Ga and Sn from their solid states to the liquid state of the LM alloy at ambient temperature.

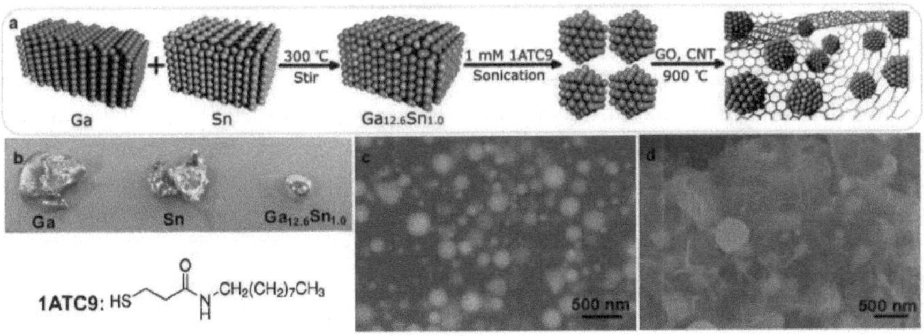

FIGURE 12.4 Synthesis and characterization of the reduced graphene oxide (RGO)/carbon nanotube (CNT)-supported LMNP anode. (a) The synthesis technique of the LMNP anode supported by RGO and CNTs. (b) The phase states of Gallium (Ga) and Tin (Sn), which are in solid form, contrasted with the liquid state of the LM alloy at room temperature (RT). (c) Illustrates the LMNP NPs post-sonication. (d) Demonstrates the incorporation of LMNPs into the carbon framework, resulting in the formation of a three-dimensional structure.

Source: Adapted with permission [34].

12.3.3 Graphene Derivatives for Highly Efficient LMBs

Graphene is considered the foundational structure for all carbon allotropes, including graphite, CNTs, and fullerenes. Discovered in 2004 [35], graphene has garnered significant attention in both industry and academia for its remarkable properties across various domains such as chemistry, mechanics, electricity, optics, and thermodynamics [36]. Its unique characteristics make it an ideal candidate for advancing the development of nanocomposites. The performance of graphene-based nanocomposites is influenced by multiple factors, including the manufacturing methods, the graphene-to-matrix ratio, the distribution of graphene within the matrix, and the type of graphene material used, such as nanosheets, nanoribbons, or chips.

Graphene derivatives, including GO, rGO, and functionalized rGO, often outperform pure graphene due to their enhanced properties [37]. These derivatives address common challenges associated with graphene, such as its hydrophobic nature, tendency to aggregate, and inclination to restack as nanosheets. Functionalizing graphene-based nanocomposites is an effective strategy for tailoring materials for specific applications, driving rapid advancements in the use of graphene-based materials across various industries [38].

Graphene consists of a monolayer of carbon atoms arranged in a hexagonal lattice, exhibiting a thickness of just one or a few atoms, and is recognized as a two-dimensional monolayer. The carbon atoms in graphene exhibit sp^2 hybridization [39]. The primary difference between GO and graphene lies in the presence of oxygen-containing groups in GO, including epoxy and hydroxyl groups on both sides of the two-dimensional plane, with carboxylic groups located near the edges [40]. rGO is produced through the chemical or thermal reduction of GO, which decreases the number of oxygen groups and increases porosity compared to GO, resulting in a higher density of defects [40, 41]. Functionalization of graphene, GO, and rGO enhances their sorption characteristics by forming both covalent and noncovalent bonds [42]. Covalent functionalization introduces oxygen-containing or other functional groups, such as amines, amides, nitro, and thiol, while noncovalent functionalization employs weaker interactions like π-π stacking, van der Waals forces, and electrostatic interactions to bind with metallic ions, organic molecules, and organometallic compounds.

Graphene-based nanocomposites, with their multiple phases, are of considerable interest for their ability to improve stability by reducing π-π stacking between sheets, enhancing dispersibility in water, and decreasing the solubility of GO to minimize material losses [43]. Given the exceptional properties of graphene and its derivatives, they are highly suitable for anode production, as exemplified by the rGO/CNT anode mentioned in the CNT section [34].

Gallium-based liquid metals are notable for their high electrical conductivity and ability to remain in a liquid state at room temperature, making them attractive for various applications, including flexible electromagnetic interference (EMI) shielding. Despite their potential, existing LM composites often fall short of acceptable EMI shielding performance due to the challenge of balancing high shielding efficiency (SE) with minimal thickness. Additionally, there is a growing need for EMI shielding materials that are resilient in diverse environmental conditions.

12.4 EMERGING MATERIALS FOR CATHODES IN LIQUID METAL BATTERIES

12.4.1 Layered Double Hydroxides

Layered double hydroxides (LDHs) are a class of clay minerals characterized by a layered structure similar to that of brucite, incorporating intercalated anions [44]. LDHs exhibit a host-guest architecture, where positively charged host layers are interspersed with guest anions and water molecules occupying the interlayer spaces. The synthesis of LDHs involves the replacement of divalent cations with trivalent cations, resulting in positively charged hydroxide sheets. These sheets are balanced by intercalated hydrated anions, creating a net positive charge that attracts negatively charged species to reside on their surface or within the interlayer spaces. The stability of the guest species is maintained through electrostatic interactions and hydrogen bonding, although surface adsorption rather than intercalation often challenges the incorporation of these species [45]. The unique layered structure of LDHs, combined with their ability to incorporate mixed-valence transition metal ions, provides significant potential for the development of high-performance molecular and nanoscale catalysts. Additionally, LDHs can undergo calcination to form mixed metal oxides (MMOs) with a uniform distribution of M^{2+} and M^{3+} ions and an extensive surface area [46]. An efficient and aesthetically pleasing approach for hydrogen production involves electrolysis, which uses electricity or sunlight to dissociate water molecules. The oxygen evolution reaction (OER), a crucial step in water splitting, consists of four proton-coupled electron transfer (PCET) steps and an oxygen-oxygen bond formation, which are kinetically slow and necessitate the use of catalysts to accelerate the reaction [46].

12.4.2 Calcium (Ca) Alloys

Calcium (Ca) alloys present a promising solution for use as cathodes in high-temperature liquid metal batteries (HT-LMBs). As an earth-abundant and low-cost element, calcium has been explored as an electrode material in batteries, particularly after the concept of LMBs was introduced. Despite its high melting point of 842°C and the high solubility of calcium in molten salt electrolytes, alloying has emerged as a key strategy to overcome these challenges. Calcium's low electronegativity and low cost ($0.14 per mole) make it an attractive candidate for large-scale energy storage systems, particularly for grid applications [47]. However, the main obstacles to the widespread use of Ca-based cathodes have been the high operating temperatures required to maintain the material in a liquid state and the dissolution of calcium in molten salt electrolytes, which can lead to self-discharge and reduced Coulombic efficiency. Alloying calcium with other metals such as bismuth (Bi) and magnesium (Mg) has shown potential in mitigating these issues [47].

Alloying calcium with Mg has proven effective in reducing the melting temperature of the electrode to below 600°C, allowing for more practical operating conditions in calcium-based LMBs. The Ca-Mg system provides a balance between lowering the operating temperature and suppressing calcium's solubility in the electrolyte, which is critical for reducing self-discharge and improving overall battery efficiency. The eutectic behavior of the Ca-Mg alloy ensures that the electrode remains liquid during operation, even as the concentration of calcium varies with the state-of-charge, thus providing stable performance across charge-discharge cycles [48]. Moreover, calcium's solubility in molten salt electrolytes can be further suppressed by alloying with other metals, such as copper (Cu) and bismuth (Bi). For example, alloying calcium with 70 mol % copper at 800 °C significantly reduces calcium's solubility in calcium chloride, from 2.1 to 0.3 mol %. This reduction in solubility minimizes self-discharge and enhances the Coulombic efficiency of the battery, making calcium alloys more suitable for long-term energy storage. Among the most promising calcium-based chemistries for LMBs are Ca-Al, Ca-Pb, and Ca-Sb alloys. Although the low cell voltage of Ca-Al and Ca-Pb presents a limitation, the Ca-Sb couple stands out due to its favorable electrochemical characteristics and compatibility with existing electrolyte systems. The development of Ca-Sb-based LMBs holds potential for achieving high energy densities, long cycle life, and cost-effective energy storage solutions [48]. Calcium alloys represent a significant advancement in the field of liquid metal batteries. By addressing the challenges of high melting points and solubility in electrolytes through alloying strategies, calcium-based LMBs could become a key technology for grid-scale energy storage and other large-scale applications. Further research into optimizing alloy compositions and electrolyte compatibility will be crucial for realizing the full potential of calcium as a cathode material in next-generation energy storage systems.

12.4.3 Gallium (Ga) Alloys

Metals typically exhibit elevated melting temperatures due to the robust electrostatic attraction forces between metal ions and conduction electrons, and only a limited number of elemental metals exist in a liquid state at ambient temperature [49]. Mercury (Hg), possessing a melting point of –39 °C, is one of the most recognized liquid metals at ambient temperature. According to quantum physics, all accessible subshells of Hg are fully occupied. This distinctive electron configuration, which impedes the extraction of electrons from subshells, resembles that of noble gases with completely filled outer shells [49]. Consequently, Hg exhibits characteristics akin to noble gases, which possess weak interatomic interactions and low melting points. However, the use of Hg is limited because of its significant toxicity [49]. Additional liquid metal possibilities comprise cesium (Cs) with a melting point of 28 °C, rubidium (Rb) with a melting point of 39 °C, francium (Fr) with an estimated melting point of 8 °C, and gallium (Ga) with a melting point of 30 °C. Nonetheless, these elemental metals possess constraints, including strong reactivity, radioactivity, and scarcity in nature [49]. Besides pure metals, alloy systems offer enhanced flexibility for the extensive selection and systematic evaluation of fusible metallic electrodes in the design of liquid metal batteries. A fusible alloy is a metallic alloy that may be readily melted at comparatively low temperatures. It is important to note that while many eutectic alloys qualify as fusible alloys, not all fusible alloys are classified as eutectic alloys. Common fusible alloys comprise mercury-based alloys, alkali metal-based alloys, Ga-based alloys, and bismuth (Bi)-based alloys, which have been employed in die casting, fast prototyping, cooling of nuclear reactors, synthesis of nanomaterials, and energy devices [49].

Gallium-based alloys, specifically gallium-tin alloy (EGaSn) nanoparticles undergo significant changes during the lithiation and de-lithiation processes, where the liquid metal transforms into a solid alloy with volume expansion. This phase transition, while beneficial for energy storage, poses challenges like crack formation and particle detachment, which leads to the rupture of the solid-electrolyte interphase (SEI) layer and irreversible capacity loss. To address these issues, a dual-insurance electrode structure is designed to buffer volume expansion and maintain electrical connectivity. The use of a conductive polypyrrole (PPy) layer effectively prevents electrical detachment of solid products and reduces new surface exposure, while also maintaining the liquid metal's self-healing properties. This innovative structure enhances the cycling stability and rate capability of the electrode, making it a robust candidate for high-performance batteries. Electrochemical testing reveals that the PPy coating significantly improves the stability of the cathode during cycling, enhancing the electrode's conductivity and adhesion (Figure 12.5). The incorporation of poly(acrylic acid) (PAA) as a binder further strengthens the electrode's integrity through hydrogen bonding, which mitigates capacity depletion and extends the life of the battery. The CV curves and charge-discharge profiles demonstrate that EGaSn nanoparticles exhibit excellent reversibility and capacity retention, attributed to the self-healing nature of the liquid metal and the protective PPy layer [50].

The development of Ga-based alloy cathodes offers a promising avenue for advanced lithium-ion batteries. Their ability to combine high capacity, self-healing properties, and stability during cycling makes them ideal for applications requiring long-lasting and high-performance energy storage. Further optimization of particle size, surface modifications, and binder materials could enhance the electrochemical performance even more, paving the way for the next generation of lithium-ion batteries with improved efficiency and durability.

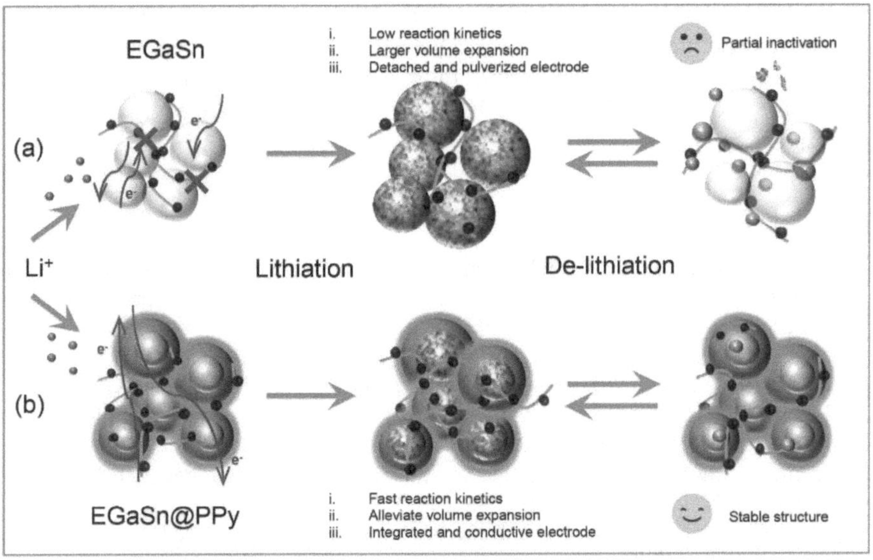

FIGURE 12.5 The electrochemical steps of (a) EGaSn and (b) EGaSn@PPy anodes.
Source: Adapted with permission [50], Copyright (2022), Elsevier.

12.5 CONCLUSION

Upon examining the latest advancements in rechargeable LMBs, it is evident that significant strides have been achieved in both battery chemistry and multiphysics modeling to tackle the crucial concerns. When it comes to HT-LMBs, the cost of materials and fabrication is not a significant issue with liquid metals or alloys and molten salts. However, it is crucial to enhance the performance of the cell by reducing the operating temperature. This can be achieved by investigating new battery chemistries or mechanisms, while still maintaining the three-liquid-layer structure. The absence of dendrites, ease of cell production and material retrieval, and utilization of readily available materials make them economically appealing for large-scale stationary energy storage. More efforts need to be made on the multiphysics modeling to improve the design and operations of HT-LMBs on a greater scale, expediting their commercialization. Regarding MT-LMB and RT-LMB, they provide potential for generating high energy density. However, the interfaces between the liquid electrode and the solid electrolyte present a complex challenge that has hindered their widespread implementation. In addition, the development of a solid electrolyte that possesses both high ionic conductivity and high transference number is a persistent obstacle in the advancement of high-performance lithium metal batteries. The performance of batteries is influenced not only by their inherent physicochemical features but also by the way they are configured and the techniques used for testing. Hence, it is important to consistently investigate and enhance the multiphysics modeling in order to comprehend the interconnection between electrochemical and thermophysical processes inside the LMB system. This understanding is crucial for optimizing battery performance in practical scaling-up scenarios. As new battery technologies and materials continue to advance and our understanding of the interplay between several physical processes improves, it is plausible to expect that LMBs will become strong contenders in the field of grid-scale stationary energy storage and other related applications.

REFERENCES

[1] W. Chen, G. Li, A. Pei, Y. Li, L. Liao, H. Wang, J. Wan, Z. Liang, G. Chen, H. Zhang, J. Wang, Y. Cui, A manganese–hydrogen battery with potential for grid-scale energy storage, *Nat. Energy* 3 (2018) 428–435.

[2] M.E. Arroyo-de Dompablo, A. Ponrouch, P. Johansson, M.R. Palacín, Achievements, challenges, and prospects of calcium batteries, *Chem. Rev.* 120 (2020) 6331–6357.

[3] H. Gao, B.M. Gallant, Advances in the chemistry and applications of alkali-metal–gas batteries, *Nat. Rev. Chem.* 4 (2020) 566–583.

[4] D. Lin, Y. Liu, Y. Cui, Reviving the lithium metal anode for high-energy batteries, *Nat. Nanotechnol.* 12 (2017) 194–206.

[5] D. Chao, W. Zhou, F. Xie, C. Ye, H. Li, M. Jaroniec, S.Z. Qiao, Roadmap for advanced aqueous batteries: From design of materials to applications, *Aqueous Zinc Batter.* (2023) 1–44.

[6] H. Kim, D.A. Boysen, J.M. Newhouse, B.L. Spatocco, B. Chung, P.J. Burke, D.J. Bradwell, K. Jiang, A.A. Tomaszowska, K. Wang, W. Wei, L.A. Ortiz, S.A. Barriga, S.M. Poizeau, D.R. Sadoway, Liquid metal batteries: Past, present, and future, *Chem. Rev.* 113 (2013) 2075–2099.

[7] J. Muldoon, C.B. Bucur, T. Gregory, Quest for nonaqueous multivalent secondary batteries: Magnesium and beyond, *Chem. Rev.* 114 (2014) 11683–11720.

[8] X. Guo, L. Zhang, Y. Ding, J.B. Goodenough, G. Yu, Room-temperature liquid metal and alloy systems for energy storage applications, *Energy Environ. Sci.* 12 (2019) 2605–2619.

[9] Y. Ding, X. Guo, Y. Qian, G. Yu, Low-temperature multielement fusible alloy-based molten sodium batteries for grid-scale energy storage, *ACS Cent. Sci.* 6 (2020) 2287–2293.

[10] X. Guo, Y. Ding, G. Yu, Design principles and applications of next-generation high-energy-density batteries based on liquid metals, *Adv. Mater.* 33 (2021) 1–18.

[11] S. Wu, X. Zhang, R. Wang, T. Li, Progress and perspectives of liquid metal batteries, *Energy Storage Mater.* 57 (2023) 205–227.

[12] S. Qian, C. Xing, M. Zheng, Z. Su, H. Chen, Z. Wu, C. Lai, S. Zhang, $CuCl_2$-modified lithium metal anode via dynamic protection mechanisms for dendrite-free long-life charging/discharge processes, *Adv. Energy Mater.* 12 (2022) 1–11.

[13] S. Choudhury, U. Mahanta, R. Prasanna Venkatesh, T. Banerjee, Ionic liquid derived novel deep eutectic solvents as low viscous electrolytes for energy storage, *J. Mol. Liq.* 366 (2022) 120245.

[14] J. Guan, N. Li, L. Yu, Artificial interphase layers for lithium metal anode, *Acta Phys. Chim. Sin.* 37 (2020) 2009011–0.

[15] Z. Zhang, X. Yang, P. Li, Y. Wang, X. Zhao, J. Safaei, H. Tian, D. Zhou, B. Li, F. Kang, G. Wang, Biomimetic dendrite-free multivalent metal batteries, *Adv. Mater.* 34 (2022).

[16] M. Song, C.L. Zhong, Achieving both high reversible and stable Zn anode by a practical glucose electrolyte additive toward high-performance Zn-ion batteries, *Rare Met.* 41 (2022) 356–360.

[17] N. Nasirpour, M. Mohammadpourfard, S. Zeinali Heris, Ionic liquids: Promising compounds for sustainable chemical processes and applications, *Chem. Eng. Res. Des.* 160 (2020) 264–300.
[18] J. Chen, W. Zhou, Y. Quan, B. Liu, M. Yang, M. Chen, X. Han, X. Xu, P. Zhang, S. Shi, Ionic liquid additive enabling anti-freezing aqueous electrolyte and dendrite-free Zn metal electrode with organic/inorganic hybrid solid electrolyte interphase layer, *Energy Storage Mater.* 53 (2022) 629–637.
[19] D. Wang, J. Hwang, C. Chen, K. Kubota, K. Matsumoto, R. Hagiwara, A β"-alumina/inorganic ionic liquid dual electrolyte for intermediate-temperature sodium–sulfur batteries, *Adv. Funct. Mater.* 31 (2021) 1–9.
[20] E.L. Smith, A.P. Abbott, K.S. Ryder, Deep Eutectic Solvents (DESs) and their applications, *Chem. Rev.* 114 (2014) 11060–11082.
[21] J. Zhao, J. Zhang, W. Yang, B. Chen, Z. Zhao, H. Qiu, S. Dong, X. Zhou, G. Cui, L. Chen, "Water-in-deep eutectic solvent" electrolytes enable zinc metal anodes for rechargeable aqueous batteries, *Nano Energy* 57 (2019) 625–634.
[22] S. Zhang, Y. Yang, L. Cheng, J. Sun, X. Wang, P. Nan, C. Xie, H. Yu, Y. Xia, B. Ge, J. Lin, L. Zhang, C. Guan, G. Xiao, C. Peng, G.Z. Chen, J.-Q. Wang, Quasi-solid-state electrolyte for rechargeable high-temperature molten salt iron-air battery, *Energy Storage Mater.* 35 (2021) 142–147.
[23] X. Feng, X. Ding, D. Jiang, Covalent organic frameworks, *Chem. Soc. Rev.* 41 (2012) 6010.
[24] C. Wei, L. Tan, Y. Zhang, K. Zhang, B. Xi, S. Xiong, J. Feng, Y. Qian, Covalent organic frameworks and their derivatives for better metal anodes in rechargeable batteries, *ACS Nano.* 15 (2021) 12741–12767.
[25] Y. Sun, J. Lopez, H. Lee, N. Liu, G. Zheng, C. Wu, J. Sun, W. Liu, J.W. Chung, Z. Bao, Y. Cui, A stretchable graphitic carbon/si anode enabled by conformal coating of a self-healing elastic polymer, *Adv. Mater.* 28 (2016) 2455–2461.
[26] C. Wang, H. Wu, Z. Chen, M.T. McDowell, Y. Cui, Z. Bao, Self-healing chemistry enables the stable operation of silicon microparticle anodes for high-energy lithium-ion batteries, *Nat. Chem.* 5 (2013) 1042–1048.
[27] J.W. Boley, E.L. White, R.K. Kramer, Mechanically sintered gallium-indium nanoparticles, *Adv. Mater.* 27 (2015) 2355–2360.
[28] M.D. Bartlett, A. Fassler, N. Kazem, E.J. Markvicka, P. Mandal, C. Majidi, Stretchable, high-k dielectric elastomers through liquid-metal inclusions, *Adv. Mater.* 28 (2016) 3726–3731.
[29] B.J. Blaiszik, S.L.B. Kramer, M.E. Grady, D.A. McIlroy, J.S. Moore, N.R. Sottos, S.R. White, Autonomic restoration of electrical conductivity, *Adv. Mater.* 24 (2012) 398–401.
[30] F. Ghaemi, A. Amiri, R. Yunus, Methods for coating solid-phase microextraction fibers with carbon nanotubes, *TrAC—Trends Anal. Chem.* 59 (2014) 133–143.
[31] X.-Y. Song, J. Chen, Y.-P. Shi, Different configurations of carbon nanotubes reinforced solid-phase microextraction techniques and their applications in the environmental analysis, *TrAC Trends Anal. Chem.* 86 (2017) 263–275.
[32] S. Iljima, T. Ichihashi, Single-shell carbon nanotubes of 1-nm, *Nature* (1993) 603–605.
[33] M. He, S. Zhang, J. Zhang, Horizontal single-walled carbon nanotube arrays: Controlled synthesis, characterizations, and applications, *Chem. Rev.* 120 (2020) 12592–12684.
[34] Y. Wu, L. Huang, X. Huang, X. Guo, D. Liu, D. Zheng, X. Zhang, R. Ren, D. Qu, J. Chen, A room-temperature liquid metal-based self-healing anode for lithium-ion batteries with an ultra-long cycle life, *Energy Environ. Sci.* 10 (2017) 1854–1861.

[35] K.S. Novoselov, A.K. Geim, S. V. Morozov, D. Jiang, Y. Zhang, S. V. Dubonos, I.V. Grigorieva, A.A. Firsov, Electric field effect in atomically thin carbon films, *Science* 306 (2004) 666–669.
[36] V. Dhand, K.Y. Rhee, H. Ju Kim, D. Ho Jung, A comprehensive review of graphene nanocomposites: Research status and trends, *J. Nanomater.* 2013 (2013) 1–14.
[37] A.A. Iqbal, N. Sakib, A.K.M.P. Iqbal, D.M. Nuruzzaman, Graphene-based nanocomposites and their fabrication, mechanical properties and applications, *Materialia* 12 (2020) 100815.
[38] X. Zhao, M. Yang, Graphene Nanocomposites, *Molecules* 24 (2019) 2440.
[39] F. Pena-Pereira, V. Romero, I. de la Calle, I. Lavilla, C. Bendicho, Graphene-based nanocomposites in analytical extraction processes, *TrAC—Trends Anal. Chem.* 142 (2021) 116303.
[40] A.T. Smith, A.M. LaChance, S. Zeng, B. Liu, L. Sun, Synthesis, properties, and applications of graphene oxide/reduced graphene oxide and their nanocomposites, *Nano Mater. Sci.* 1 (2019) 31–47.
[41] N. Manousi, E. Rosenberg, E.A. Deliyanni, G.A. Zachariadis, Sample preparation using graphene-oxide-derived nanomaterials for the extraction of metals, *Molecules* 25 (2020) 2411.
[42] T.J.M. Fraga, M.N. Carvalho, M.G. Ghislandi, M.A. da Motta Sobrinho, Functionalized graphene-based materials as innovative adsorbents of organic pollutants: A concise overview, *Brazilian J. Chem. Eng.* 36 (2019) 1–31.
[43] M. Raji, N. Zari, A. el kacem Qaiss, R. Bouhfid, Chemical preparation and functionalization techniques of graphene and graphene oxide, in: *Functionalized Graphene Nanocomposites Their Derivatives*, Elsevier, Amsterdam, the Netherlands, 2019, pp. 1–20.
[44] I.A. Abdallah, S.F. Hammad, A. Bedair, R.M. Abdelhameed, M. Locatelli, F.R. Mansour, Applications of layered double hydroxides in sample preparation: A review, *Microchem. J.* 192 (2023) 108916.
[45] S. Yu, G. Choi, J.-H. Choy, Multifunctional layered double hydroxides for drug delivery and imaging, *Nanomaterials* 13 (2023) 1102.
[46] X. Long, Z. Wang, S. Xiao, Y. An, S. Yang, Transition metal based layered double hydroxides tailored for energy conversion and storage, *Mater. Today* 19 (2016) 213–226.
[47] S. Wu, X. Zhang, R. Wang, T. Li, Progress and perspectives of liquid metal batteries, *Energy Storage Mater.* 57 (2023) 205–227.
[48] H. Kim, D.A. Boysen, J.M. Newhouse, B.L. Spatocco, B. Chung, P.J. Burke, D.J. Bradwell, K. Jiang, A.A. Tomaszowska, K. Wang, W. Wei, L.A. Ortiz, S.A. Barriga, S.M. Poizeau, D.R. Sadoway, Liquid metal batteries: Past, present, and future, *Chem. Rev.* 113 (2013) 2075–2099.
[49] Y. Ding, X. Guo, G. Yu, Next-generation liquid metal batteries based on the chemistry of fusible alloys, *ACS Cent. Sci.* 6 (2020) 1355–1366.
[50] Y. Qi, C. Shen, Q. Hou, Z. Ren, T. Jin, K. Xie, A self-healing liquid metal anode for lithium-ion batteries, *J. Energy Chem.* 72 (2022) 522–531.

13 Liquid Metal Batteries for Large-Scale Energy Storage

Vikas Yadav and Rajendra Kumar Singh

13.1 INTRODUCTION

The rapid initiation of nations towards the zero-carbon economy has significantly changed the whole scenario of global energy consumption, compelling us to expedite the transition from conventional fossil-based fuels to eco-friendly and renewable energy options [1–3]. Among these technologies, electrochemical energy storage systems (EESS), primarily batteries, play an important role in energy harvesting from intermittent sources such as solar, wind, and geothermal [4]. Particularly lithium-ion batteries (LIBs) have fulfilled the demand in portable electronic devices and are being established in other niche markets such as stationary units and electric vehicles. Despite the success of LIBs, there is still a need for energy storage devices that offer higher energy densities at lower costs. In particular, alkali metals and alkaline-earth metals such as lithium (Li), sodium (Na), potassium (K), magnesium (Mg), and calcium (Ca) have great potential in building high energy density (HED), metal-based rechargeable batteries [5]. However, challenges like uncontrolled growth of dendrites, volume expansion, and decomposition of electrolyte have hindered the direct employment of these earth metals in solid-state batteries (SSBs), raising safety worries and reducing cyclability [6]. To address this, an effective approach is to replace solid-state chemistries with liquid-state electrochemistry. In various reported electrochemistries of batteries, liquid metal batteries (LMBs) have shown the potential to fulfill the needs of grid-scale energy storage (GSES) technology. LMBs consist of at least one liquid electrode out of two electrodes. Liquid systems provide several advantages, including high voltage capabilities at high rates, immunity to microstructural deformation and dendrite formation, flexibility in energy storage components, and reduced costs due to the use of abundant and inexpensive materials [7–10]. These advantages are directly beneficial for the development of large GSES technology [11]. The concept of LMBs was first introduced in the early twentieth century with the development of the Hoopes cell, used for high-purity aluminum. In the 1960s, the tri-layered cell popped-up as the bimetallic cell, enjoying over a decade of rigorous research into HED batteries for mobile applications [12]. The demand for GSES in the early twenty-first century motivated a revival of LMBs. Initially, all types of LMBs used a molten salt electrolyte and two different metal

DOI: 10.1201/9781003489726-13

electrodes with high melting points. These components naturally separated into three layers due to differences in density and their inability to mix with each other. Operating temperatures of these LMBs were typically higher than 350 °C [13–15]. Despite challenges like the need for rigorous thermal management and high corrosion of active cell components, high-temperature LMBs have demonstrated potential for practical use, particularly for GSES. Another type of LMB uses solid electrolytes instead of molten salts, such as the early Na-S and ZEBRA batteries developed since the 1960s [16]. These batteries use molten sodium as the anode, along with an Na^+ selective ceramic conductor as the solid-state electrolyte, and are typically operated at temperatures of around 300–350 °C [17]. To address the challenges associated with high operating temperatures, recent efforts have focused on lowering the operational temperature by incorporating fusible materials such as Ga-based and liquid Na-K alloys, which remain liquid at or near room temperature. Ga-based liquid metals are particularly promising due to their safety profile and moderate voltage range, while Na-K liquid alloys offer the potential to replace traditional Li anodes as charge carrier reservoirs [18, 19]. Different types of LMBs are now being developed based on their operating temperature ranges. High-temperature LMBs (HT-LMBs) operate above 350 °C using molten salt electrolytes and liquid metal electrodes. Medium-temperature LMBs (MT-LMBs) function in the 100–350 °C range with solid or liquid electrolytes, while room-temperature LMBs (RT-LMBs) operate close to ambient conditions using at least one liquid metal electrode. Each type of LMB faces distinct challenges, particularly in terms of battery chemistry and system design, which researchers are addressing through innovative strategies. For HT-LMBs, research has emphasized achieving high reversible cell potentials and low costs, exemplified by the Mg-Sb cell. For MT-LMBs, the focus has been on developing solid electrolytes that enable the use of lower-melting-point materials, enhancing the practicality of these systems. For RT-LMBs, efforts have been directed toward finding fusible materials that can operate at room temperature, with Ga-based and Na-K liquid alloys showing significant promise. In addition to battery chemistry, understanding the underlying physical mechanisms of LMBs is crucial. Unlike solid-state batteries, LMBs involve many chemical reactions and various complex physics phenomena, including electric potential distribution, electro-vortex flow, thermal convection flow, mass transport, and heat transfer [20]. In-depth understanding of these processes is vital for optimizing battery performance and accelerating the deployment of LMBs for GSES applications. Despite the challenges associated with LMBs, such as high operating temperatures and material reactivity, recent findings have broadened the future possibilities of this developing technology. The working temperature of LMBs is influenced by various factors, including electrode materials, electrolyte solubility, wettability, energy density, and efficiency. Comprehensive research efforts are focused on optimizing these factors to improve the performance and practicality of LMBs. In summary, the transition to a net-zero carbon economy has driven significant advancements in battery technologies, with LMBs developing as a suitable solution for a high-efficiency energy storage device. By addressing challenges related to operating temperature, material reactivity, and multi-physical processes, researchers are working toward the large-scale deployment of LMBs for GSES technology, contributing to a sustainable energy future.

13.2 SELECTION OF LIQUID METAL ELECTRODES

Screening liquid metal electrodes involves selecting materials that can effectively function in LMBs by meeting specific criteria. These materials are highlighted in Figure 13.1 and are selected based on three main requirements:

13.2.1 Suitable Phase at Operating Temperatures

The metal must remain liquid within a practical temperature range. This means having a melting point below 1000 °C and a boiling point above 25 °C at normal pressure. Ensuring the metal stays in a liquid state is crucial for forming stable liquid-liquid electrode-electrolyte interfaces.

13.2.2 High Electrical Conductivity

The metal should exhibit high electrical conductivity, with a minimum conductivity exceeding 1 $S.cm^{-1}$. This is necessary for efficient electron transport, which is vital for minimizing resistance and enhancing battery performance.

13.2.3 Nonradioactive and Stable Isotopes

The metal must be nonradioactive and available as a stable, naturally occurring isotope to ensure safety and viability for long-term energy storage applications.

To determine suitable candidates, metals are initially categorized as positive or negative electrodes based on their deposition potentials from aqueous

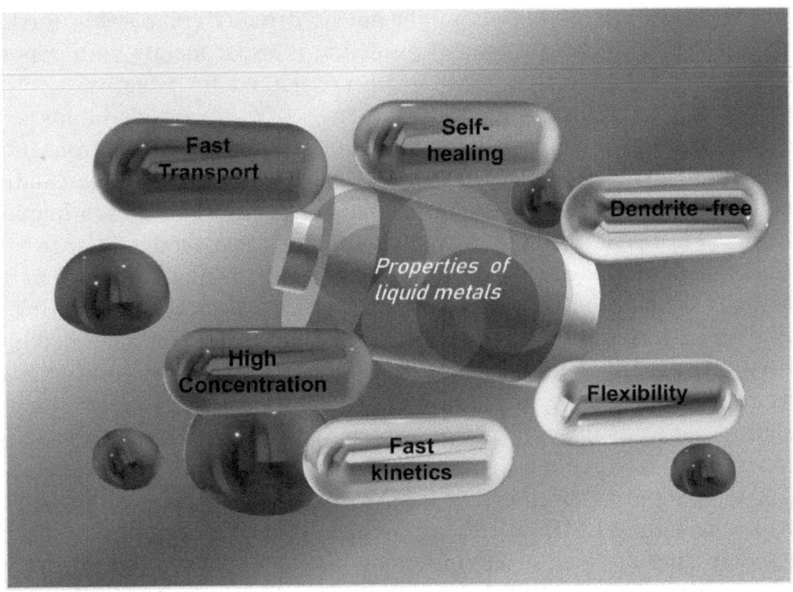

FIGURE 13.1 Properties of liquid metals.

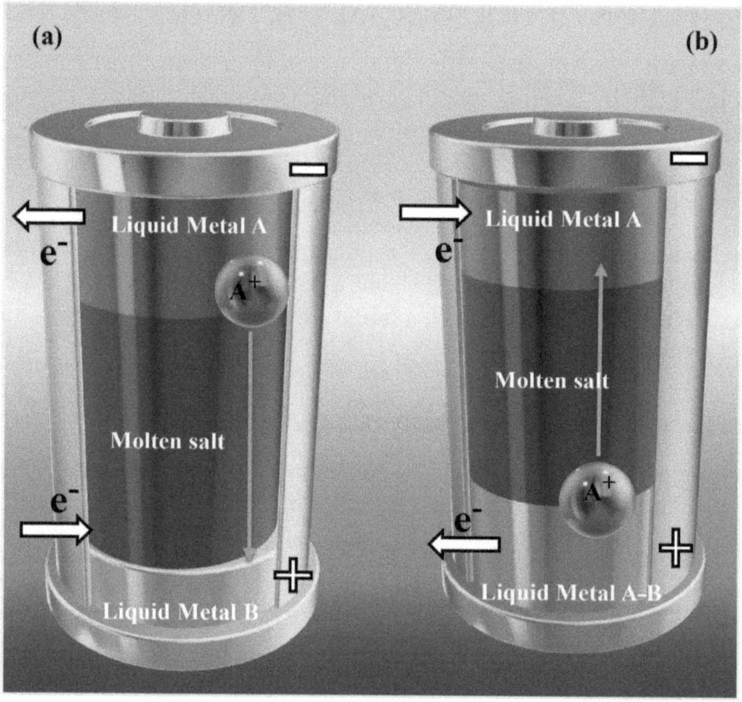

FIGURE 13.2 Discharging (a) and charging (b) process of LMBs.

solutions. While these potentials might not be directly comparable for LMBs with different electrolytes, a general guideline is used: metals with deposition potentials more negative than –2.0 V are considered for negative electrodes, while those with potentials more positive than –1.0 V are considered for positive electrodes. Additionally, metals are categorized on the basis of electronegativity: more electropositive metals such as Li, K, Na, Ca, Mg are suitable candidates for anodes, and similarly more electronegative metals such as semimetals, or chalcogens (Pb, Bi, Te, Sb, Ga, S) are candidates for cathodes [21]. In a standard bimetallic three-layer LMB, the anode shrinks during discharge as metal A undergoes oxidation, releasing electrons ($A \rightarrow A^{n+} + ne^-$). The resulting positive ions travel through the molten salt electrolyte to the cathode, driving the discharge process. At the same time, the cathode layer grows thicker as these cations are reduced and form a liquid alloy (A in B). This interaction between the electrodes creates the cell voltage, which drives the LMB. During charging, the process reverses. The melting points of pure metals and alloys cover a wide range of temperatures, from below 0 °C to over 800 °C, providing numerous options for designing LMBs with varying operating temperatures. Alkali metal alloys, possessing the lowest melting points and abundant in the Earth's crust, are particularly promising candidates. However, Ga based alloys demonstrate low melting points, but Bi-based alloys despite being less abundant and having higher melting points, are suitable for alkali-ion storage at moderately elevated temperatures. A thorough understanding of the phase change behavior of these alloys is essential, as it has a significant impact on the operating temperatures of LMBs. In bimetallic cells, the charging and discharging processes are similar to dealloying and alloying reactions. The melting point of a metal, influenced by factors such as bond strength, valence electron count, ionization degree, and crystal structure, is generally higher for pure substances compared to impure ones—a phenomenon known as "melting point depression."

Eutectic systems demonstrate this, and mixing energy is given as $\Delta G_{mix} = \Delta H_{mix} - T\Delta S_{mix}$, where ΔG_{mix} represents the Gibbs free energy, $\Delta H_{mix,}$ represents the enthalpy, T represents the temperature, and ΔS_{mix} stands for entropy. The Gibbs free energy of mixing drives the fusion of bimetallic alloys, while a change in the Gibbs free energy, which can be positive or negative, determines the stable phases of the mixture at different temperatures. In designing LMBs, the thermal properties of the electrolyte are also crucial [22]. Typically, for high-temperature (HT)-LMBs, the melting temperature is greater than 350 °C and molten halide salts are used as electrolytes. Medium-temperature (MT)-LMBs (100 °C $<$ T $<$ 350 °C) use solid ceramics as electrolytes, and room temperature (RT)-LMBs (T ~ 25 °C) use solid or liquid electrolytes [23]. Research in LMBs has resumed since 2010 due to the increased demand for GSES technology, highlighting the ongoing importance of these criteria in developing efficient and reliable LMBs.

13.3 FLUID DYNAMICS IN LMBs

Fluid motion in LMBs is driven by magneto-hydrodynamic effects, due to the interaction between a magnetic field (either external or generated by the battery current) and the cell current. Based on magneto hydrodynamics and fluid dynamics, LMBs show fluid flow phenomena such as Tayler Instability, Electro-vortex Flow, Rayleigh-Bénard convection, Marangoni Convection, etc.

13.3.1 Tayler Instability

The Tayler instability (TI) is a kink-type instability caused by the interaction of a current passing through a conducting fluid with its own magnetic field, similar to the instability observed in Z-pinches and tokamak phenomena in plasma physics. In LMBs, resistivity and viscosity help to stabilize the system, but despite this, a non-compressible liquid destabilizes the plasma [20].

Current density $J_{0,}$ flowing along the circular cross section of cylindrical conductor generates a rotating magnetic field B_0 with a purely azimuthal component B_{Φ}. The resulting Lorentz force from J_0 and B_0 creates an inwardly directed pinch force, which is prone to both of them an axisymmetric "sausage" and non-axisymmetric "kink" instabilities. In liquid metals, which have significantly higher kinematic viscosity ν compared to plasmas, due to fast interface kinetics and finite electrical conductivity **σ**, the onset of these instabilities can be characterized by a critical value of the Hartmann number (equation (13.1)) [24].

$$H_a = B_0 R \sqrt{\frac{\sigma}{\rho\nu}} = \frac{I u_0}{2\pi} \sqrt{\frac{\sigma}{\rho\nu}} \tag{13.1}$$

where $B_0 = B_\Phi$ (R) is the azimuthal magnetic field at the cylinder radius R, the liquid's density, I, is the total current, and the vacuum permeability, respectively.

The onset of a non-axis symmetric TI occurs with a dominant azimuthal mode m = 1 when the critical value of Hartmann number (Ha) exceeds ~ 20. The velocity of the saturated TI scales with Ha^2. The critical Ha translates directly into a critical current dependent on the liquid metal's properties. Anodic materials such as K, Li, Na, and Mg, with critical currents between 1 and 2 kA, are more prone to TI than denser, less conductive cathode metals. This suggests that TI is less likely to affect cathode mixing but could threaten the anodic compartment. Note that total current, not current density, determines TI criticality. As can be seen from the stability criterion derived by Vandakurov and Tayler, the kink-type instability depends on the radial (r) dependence of B_Φ [24].

$$\frac{\partial\left(rB_\phi^2\left(r\right)\right)}{\partial r} > 0; \tag{13.2}$$

By changing B_Φ (r), in eqs. (13.2), there is a probability of stabilizing against the TI.

13.3.2 Electro-Vortex Flows

The Tayler instability represents a current-driven instability under the ideal condition of a perfectly uniform current, acting as an upper bound for maintaining a current-carrying fluid at rest. In practical applications, non-uniform current distributions are more common, leading to rotational Lorentz forces that cannot be balanced by pressure and hence generate electro-vortex flows (EVFs) [20, 24]. Such inhomogeneities in current density can arise from the finite thickness of current collectors. An extreme case, as described by Shercliff et al., involves a point current source in an insulating plate, where the current spreads radially, creating a Lorentz force with a rotational component that drives a flow balanced only by friction. EVFs are important in industrial processes such as electroslag remelting, weld pools, and aluminum refining.

Here, we examine a milder case compared to Shercliff et al., with more gradual electric field gradients. The computational setup sketched consists of a cylinder filled with liquid metal, bounded by current collectors of the same diameter and conductivity. A constant potential is applied over a circular area on the outer surfaces of the current collectors. This finding aligns with Vlasyuk et el., who observed that for EVF in a cylinder, the relationship between total current and axial flow velocity transitions from quadratic to linear depending on a dimensionless parameter that characterizes the ratio of electromagnetic force to fluid resistance due to viscosity and density. The mean velocities are consistent with expected values for confined EVFs.

13.4 HIGH-TEMPERATURE (HT)-LMBs

High-temperature liquid metal batteries (HT-LMBs) represent a significant advancement in electrochemical energy storage technologies. These systems operate at elevated temperatures, typically above 350 °C, to maintain all components—anode (Li, Na, K, etc.), electrolyte, and cathode (Sb, Hg, etc.)—in a liquid state. This high operating temperature is essential for establishing liquid-liquid (L-L) and electrode-electrolyte (E-E) interfaces, which expedite the mass transport of reactant elements and products and it also helps to suppress the ohmic losses raised by highly conductive molten electrolytes. As a result, HT-LMBs can achieve relatively high voltage capability and fast charge-transfer kinetics, making them suitable for high-rate applications. However, the high operating temperature of HT-LMBs give birth to several challenges that hinder the widespread application of HT-LMBs. Extreme temperatures can weaken insulating seals, requiring careful thermal management and creating a highly reactive environment. This increased reactivity can cause electrode materials to dissolve in the molten salt electrolyte, which raises self-discharge rates and makes it difficult for the cell to recover its performance. Additionally, the all-liquid components limit battery mobilization and pose potential hazards, such as thermal runaway and short circuits in the event of liquid phase leakage.

Moreover, the high temperature also incurs multiple cost-based challenges, as the external energy input and requirements for hermetic sealing significantly increase the overall cost, making HT-LMBs less competitive compared to existing technologies in both grid-scale and portable markets. Reducing the operating temperature, while detrimental to the reaction kinetics and performance, could enable the use of polymeric sealing materials with enhanced mechanical strength, reduce heat loss, improve resistance to corrosion, and simplify the packing process. To balance these factors, approaches such as electrode engineering through alloying and optimization of molten electrolytes with multiple components are being investigated. These strategies aim to maintain high performance while mitigating the drawbacks associated with high operating temperatures. HT-LMBs, with their unique three-liquid-layer cell structure, hold promise for large-scale energy storage applications, provided these challenges can be effectively addressed [25].

13.4.1 Lithium-Based HT-LMBs

Lithium-based HT-LMBs have shown significant potential due to their HED and performance. Additionally, Li is the smallest element as compared to other metals such as K, Na, etc., and shows an intermediate melting point of 180 °C. Because in its own molten electrolyte, Li has significantly lower solubility because its valence electron is closer to the nucleus than in Na or K. One notable system is the Li||Te cell, where the metalloid Tellurium (Te) acts as a positive electrode, providing a high open-circuit voltage (OCV) of up to 1.75 V. Initially aimed at electric vehicle applications, the Li||Te system, as first assembled by Cairns et al., operates at 480 °C and demonstrates excellent rate performance and better cyclability test [26]. To prevent short circuits and safety concerns, a paste of molten salt electrolyte and inert filler materials is used. However, challenges such as poor conductivity and high solubility of Te in molten salts can deteriorate the Coulombic efficiency of LMBs. Recent advancements include alloying Te with Sn to enhance conductivity and reduce solubility, resulting in high-performance Li||Te-Sn cells with a discharge voltage of 1.6 V and energy density of 495 Wh kg^{-1} at 500 °C. Another promising system is the Li||Bi cell, which utilizes Bismuth (Bi) as the positive electrode. Despite Bi's lower melting point (271.3 °C) and favorable properties, issues such as solid discharge product formation and high polarization voltage need to be addressed. Innovations such as the self-healing Li||Bi cell and the Bi-Ga alloy electrode, which offer improved performance and reduced operating temperatures, highlight the potential for lithium-based HT-LMBs in specific applications like military and large-scale energy storage, despite challenges related to material cost and abundance [27].

13.4.2 Sodium and Magnesium-Based HT-LMBs

Sodium-based HT-LMBs are attractive due to sodium's low cost, high abundance in the Earth's crust, and environmental friendliness. Although lithium-based batteries dominate the market, the cost escalation, limited availability, and random distribution of lithium make sodium a viable alternative for large-scale grid storage. Early examples like the Na||Sn cell, developed by General Motors in the 1960s, operated at 700 °C but faced high self-discharge rates due to sodium's solubility in molten salts. Recent advancements involve optimizing electrolyte compositions and reducing operating temperatures. For instance, the NaI-LiI-KI system serves as an effective electrolyte for Na-based LMBs, allowing operation below 350 °C with low sodium solubility and sufficient Na^+ conductivity [28]. Moreover, ternary molten chloride salt mixtures have been developed to inhibit sodium dissolution, enabling stable operation and high Coulombic efficiency in Na||Bi cells. These innovations demonstrate the potential of sodium-based HT-LMBs for cost-effective and efficient GSES.

Magnesium-based HT-LMBs offer another promising alternative due to magnesium's abundance on Earth and high gravimetric capacity. Additionally, due to the low solubility of Mg in its molten electrolyte, Mg-based LMBs offer a more attractive perspective since only half as many ions would need to achieve the same capacity as Li. With a melting point of 650 °C, Mg releases two electrons per atom when oxidized (Mg^{2+}), providing higher capacity than Li, Na, or K. Mg's low solubility in molten salts ensures low self-discharge rates and high Coulombic efficiency. The Mg|NaCl-KCl-$MgCl_2$|Sb cell, featuring a liquid Mg anode and Sb cathode, exhibits high performance at 700 °C, with Coulombic and energy efficiencies of 94% and 73%, respectively. Despite these advantages, the high operating temperature and limited rate capability pose challenges for real-world applications [29].

Efforts have been made to lower the operating temperature through alloying, such as using eutectic Ca-Mg alloys. Ca is the fifth most abundant element on Earth and has some good features such as nontoxic nature, SHE potential (–2.87 V), and high theocratical capacity (332060 mAh.cm^{-3}). Besides this, Ca-based LMBs face two main problems: a high melting point (842 °C) and high solubility of Ca in molten electrolyte.

Nevertheless Ca-Mg-based alloys have shown promise in reducing solubility and improving cell performance. Mg-based HT-LMBs thus represent a viable option for GSES, provided the technical challenges, same as those mentioned above for Ca-based LMBs, can be addressed.,

13.4.3 Why High Temperature and Its Effects?

The systems discussed earlier operate at high temperatures above their respective melting points to enable the formation of liquid-liquid interfaces between the electrodes and electrolytes. This allows for rapid mass transport of reactants and products, as well as low ohmic losses due to the high conductivity of molten electrolytes, resulting in high voltage efficiencies even at high current rates. The fast charge-transfer kinetics, a key feature of HT-LMBs, is also reliant on these high operating temperatures.

However, the requirement for high operating temperatures is a significant obstacle to the broader adoption of HT-LMBs. It can degrade insulating seals, necessitate stringent thermal management, and create an environment with intense chemical reactivity. Specifically, this high reactivity can lead to the solubility of metallic electrode species in the molten salt electrolyte, causing high self-discharge rates and irreversible performance degradation. Moreover, the use of all-liquid components restricts battery mobility and poses a safety risk in the event of liquid phase leakage, which could lead to rapid heat generation and short-circuiting within the cell. Additionally, the high operating temperatures bring about various cost-related challenges. The need for external energy input and hermetic sealing significantly increases the overall cost, making HT-LMBs less competitive compared to existing technologies in both grid-scale and portable markets.

Lowering the operating temperature of HT-LMBs can hinder reaction kinetics and performance but allows for the use of stronger polymeric seals, reduced heat loss, improved corrosion resistance, and easier packaging. To balance these factors, strategies like electrode engineering through alloying and optimizing molten electrolytes with multiple components have been explored. These approaches are discussed in the following sections.

13.4.4 Alternative to Conventional Liquid Metals

Alloying strategies were initially explored to address the high melting point of calcium in Ca-based LMBs. Alloys like Ca-Bi, Ca-Sb, Ca-Mg, and Ca-Ge were investigated, with a Ca-Mg (90:10 mol %) alloy showing 99% Coulombic efficiency at temperatures below 550 °C. While decreasing Ca concentration in these alloys can improve cyclability, it may lead to voltage loss.

Eutectic alloys, such as Li-Bi, not only reduce melting points but also hold potential for lowering the melting points of cathodes. However, challenges like polarization and solid layer formation can impede reaction kinetics. An optimized Bi-Ga alloy has shown improved performance by reducing operating temperature to 500 °C and enhancing charge transfer.

Antimony (Sb) offers a cost-effective alternative to bismuth, with Sb-based alloys such as Pb-Sb demonstrating good performance at lower temperatures. However, environmental concerns with lead-based alloys have led to the development of more benign options such as Sb-Sn and Sb-Bi-Sn. While these alloys improve melting points and solubility, they can decrease energy density and complicate Li^+ transfer kinetics due to the formation of multiple intermetallic layers.

The eutectic alloy approach effectively lowers melting points and electrode solubility in electrolytes, but balancing reduced operating temperatures with maintaining energy output remains a challenge.

13.4.5 Electrolyte Optimization and Alternatives

Molten halides are commonly used as electrolytes in HT-LMBs due to their high ionic conductivity, cost-effectiveness, and safety. These electrolytes enable fast charge-transfer kinetics through liquid–liquid interfaces but face challenges such as the formation of subvalent ions at high temperatures, leading to high self-discharge rates and reduced energy efficiency. Alloying strategies can help suppress metal solubility in molten salts and lower the melting point of the electrolyte, as seen in $LiCl\text{-}CaCl_2$ systems.

To avoid issues with liquid electrolytes, solid electrolytes like Na-based beta-alumina (Na-BASE) have been explored. Na-BASE offers a less reactive interface but faces challenges like reduced wettability and mechanical stability at lower temperatures. Techniques such as alloying, doping, and adding oxygen getters have been used to improve performance, but cost and material stability remain concerns.

Alternative strategies include using K-based electrolytes, which show promise due to potassium's lower melting point and abundance, although research is still in its early stages. Overall, optimizing the balance between operating temperature and ionic conductivity is crucial for advancing HT-LMBs.

13.5 MEDIUM-TEMPERATURE LMBs

The operating temperature of maximum HT-LMBs is primarily restricted by the elevated melting points of molten salt electrolytes, while the metal or alloy electrodes, such as Li, Na, Sn-Pb, and Si-Pb, tend to have smaller melting points. To overcome the challenge of developing an electrolyte with both a low melting point and a high ionic conductivity, solid electrolytes have emerged as a promising solution. In this context, the melting points of the metal electrodes play a critical role in defining the operating temperature of LMBs. Sodium-based medium-temperature liquid metal batteries (MT-LMBs), which function below 350 °C using solid electrolytes, were initially demonstrated in systems like NaIIS and ZEBRA, which have been the subject of extensive research over the years. More recently, lithium-based MT-LMBs have gained attention as a potential candidate for large-scale energy storage applications.

13.5.1 Lithium-Based MT-LMBs

The development of solid electrolytes over the years has breathed new life into lithium-based LMBs by eliminating lithium molten salt electrolytes, much like the progress seen in sodium-based MT-LMBs. For example, by using a garnet-type $Li_{6.4}La_3Zr_{1.4}Ta_{0.6}O_{12}$ (LLZTO) solid ceramic electrolyte, researchers have successfully developed Li||LLZTO||Sn-Pb and Li||LLZTO||Bi-Pb cells capable of stable cycling at 240 °C. These cells deliver a high volumetric capacity of 896 mAh.cm^{-3} with minimal capacity degradation over 30 cycles. In a puncture test, the surface temperature only rose by 80 °C, indicating a significant safety margin. The theoretical volumetric energy densities for the Li||LLZTO||Sn-Pb and Li||LLZTO||Bi-Pb cells are 570 Wh. L^{-1}and 940 Wh. L^{-1}, respectively, making them strong candidates for GSES applications [30, 31].

In 2020, new medium-temperature Li||S and Li||Se batteries were introduced, utilizing the LLZTO ceramic electrolyte. Operating at 240 °C and 300 °C, these batteries achieved a stable Coulombic efficiency of 99.99%, with a power density of 180 mW.cm^{-2} and an energy efficiency exceeding 80%. The Li||LLZTO||S cell-maintained stability over 50 cycles, and its performance could be further enhanced by improving the sealing to prevent sulfur vapor loss. The theoretical energy densities for Li||LLZTO||S and Li||LLZTO||Se batteries are approximately 1850 Wh.kg^{-1} and 920 Wh.kg^{-1}, respectively, outperforming most existing GSES systems [32].

In addition to sodium and lithium, potassium is also employed in MT-LMBs. Lu et al. designed a novel K||S battery using a potassium conducting solid electrolyte (BASE) consisting of 91 wt % K-β"-Al_2O_3, 8.4 wt % K-β-Al_2O_3, and 0.6 wt % ZrO_2. This BASE is effectively wetted by molten potassium at 150 °C and remains compatible with K without cracking. With a tetraglyme-based catholyte containing polysulfides and K salts, the K||S LMB demonstrated minimal capacity degradation over 1000 cycles at 150 °C. An in-situ Raman study revealed the discharge reactions from K_2S_5 to K_2S_4, K_2S_3, and K_2S_2. However, the capacity is very small as compared to Na||S batteries. The K||S battery achieves a similar energy density due to its higher discharge voltage. Following a slight capacity increase in the first 70 cycles, the K||S cell demonstrated stable performance over 1000 cycles with minimal degradation [33].

13.5.2 Sodium-Based MT-LMBs

The Na||S battery operates as a LMB consisting of a molten sodium as negative electrode, a molten sulfur as positive electrode, and β-Al_2O_3 as solid electrolyte. Originally developed by Ford in 1967, this system offers several advantages such as high gravimetric energy density of ~760 Wh.kg^{-1}, excellent Coulombic efficiency, and low-cost materials. However, it requires high operating temperatures of 300–350 °C to achieve sufficient ionic conductivity and maintain proper wetting between molten sodium and the electrolyte [21]. These conditions, well above the melting points of sodium (98 °C) and sulfur (120 °C), present safety challenges due to the corrosive sodium polysulfides (e.g., Na_2S_4), which degrade components and increase the risk of thermal hazards.

To address these challenges, researchers have explored intermediate-temperature Na||S batteries, which operate around 150 °C using a denser β''-Al_2O_3 membrane as the separator and tetraglyme as the solvent for the catholyte. This approach has demonstrated promising performance, especially due to the high solubility of Na_2S_4 in tetraglyme, though issues such as poor wetting of molten sodium on the ceramic membrane remain a concern [34].

In a recent advancement, Wang et al. introduced a Na||S battery that also operates at 150 °C, utilizing a BASE ceramic separator combined with an innovative inorganic ionic liquid. This system showed high ionic conductivity, improved thermal stability, and a reversible capacity of 795 $mAh.g^{-1}$ with long-term cycling stability, exceeding 1000 cycles at a current of 0.1 mA.

Another type of Na-based MT-LMB is the Na||metal halide battery. The Na||Ni/$NiCl_2$ battery, developed by the ZEBRA group in the 1980s, uses a BASE membrane and $NaAlCl_4$ electrolyte. Although the energy density is lower than Na||S cells, the Ni/$NiCl_2$ cathode is safer and less corrosive [16]. In 2015, a Na||$FeCl_2$-based cell was developed, showing stable cycling performance at 190 °C with S-based additives improving the cathode's performance. Recently, advancements in Na||metal chloride cells have improved specific power and energy by refining cathode design. Although other cathode materials, including ionic liquid- and aqueous-based systems, have been explored, challenges like high cost and safety risks persist. A fully molten, inorganic sodium battery with a NaI-$AlBr_3$ catholyte operating at 110 °C was also reported. Efforts to further lower the operating temperature have led to the development of a Na||Bi-Pb-Sn LMB using a liquid organic electrolyte (1 M NaI in tetraglyme), capable of operating at ~100 °C. Although the rate capability is limited by the lower ionic conductivity of organic electrolytes, there is potential for improvement with advanced electrolyte development [35].

13.6 ROOM-TEMPERATURE LMBs

The development of LMBs that operate at room temperatures (0–40 °C) is an intriguing goal, as it can acquire a similar configuration to lithium-ion batteries. It would enable broader applications beyond stationary energy storage. At these temperatures, solid-state battery chemistries and technologies, along with common solutions from LIBs, can be applied to RT-LMBs. Various electrolyte options are available, including solid, organic, and inorganic liquid electrolytes. Metals and alloys with low melting points, such as mercury (Hg), gallium (Ga) and its alloys, rubidium (Rb), cesium (Cs), francium (Fr), and sodium-potassium (Na-K) alloys, have been investigated. However, the commercial feasibility of Hg is hindered by its toxicity, and the radiation risks posed by Rb, Cs, and Fr further limit their applications. Ga emerges as a safer and more favorable option for RT-LMBs due to its moderate potential range relative to lithium. Another promising candidate is the Na-K alloy, which remains in liquid form down to −12.6 °C and benefits from cost-effective raw materials and straightforward synthesis [9].

13.6.1 Sodium and Potassium Alloy-Based Room-Temperature LMBs

The Na-K alloy, known for its low reduction potentials (−2.71 V for Na and −2.92 V for K against the standard hydrogen electrode), is a promising alternative to lithium anodes in addressing dendrite growth issues. With specific capacities of 629 $mAh.g^{-1}$ for Na and 579 $mAh.g^{-1}$ for K, the Na-K alloy is a strong candidate for room-temperature liquid metal batteries (RT-LMBs). However, key challenges lie in enhancing the wettability of the liquid metal on various substrates and ensuring its compatibility with different electrolytes. To tackle the wettability issue, Yu et al. developed a liquid metal anode using fibrous carbon by immersing the carbon in the Na-K alloy. This technique secures the liquid metal in place, enabling stable cycling across multiple cathode types [36, 37].

While Na-β" alumina solid electrolyte (Na-BASE) is typically used for molten sodium at high temperatures, the Na-K alloy can cause damage to Na-BASE due to ion exchange, resulting in pulverization. In contrast, K-BASE, which has a lower free energy, remains stable in contact with molten K and the Na-K alloy, making it a more suitable electrolyte for this system [33].

The electrochemical behavior of liquid alkali alloys, like Na-K, is complex, particularly with the simultaneous presence of both Na and K in the anode. This raises the question of whether the alloy acts primarily as a Na or K anode. Research by Goodenough et al. revealed that the type of alkali metal stripped from the Na-K anode depends on the ion preference of the cathode material. This dual-anode capability allows the Na-K alloy to function as a dendrite-free anode in both Na-metal and K-metal LMBs. Moreover, the electrolyte can affect ion transport by forming a solid-electrolyte interphase (SEI) that is either Na-rich or K-rich on the anode surface, further influencing charge selection.

Inspired by this mechanism, Guo et al. created a ternary hybrid-cation battery by pairing a $Li_2[Ni_{0.33}Mn_{0.33}Co_{0.33}]\ O_2$ cathode with an Na-K anode in a Li/K hybrid electrolyte [38]. In this system, only Li ions intercalate into the cathode, while K ions undergo stripping and deposition on the anode. The selective charge process is determined by the ion distribution in the SEI layer. Additionally, Ding et al. investigated the in-situ formation of Na-K alloy on the electrode, demonstrating that a carefully chosen electrolyte system can enhance this process and protect the alkali metal electrode through a galvanic replacement reaction.

This approach offers promising strategies for improving performance and stability in room-temperature and low-temperature LMB systems.

13.6.2 Ga-Based RT-LMBs

Beyond traditional alkali metals, alloys such as Ga and its combinations with elements like Indium (In) have garnered attention due to their low melting points and beneficial electrochemical traits when alloyed with alkali ions. Early explorations of Ga as an anode material in Li-ion batteries at around 40 °C revealed that Ga solidifies rapidly upon lithiation, forming phases such as Li_2Ga_7, LiGa, and Li_2Ga. During cycling, Ga exhibits multiple voltage plateaus. Notably, Ga displays self-healing properties upon de-lithiation, though its solid form tends to show reduced cyclability at room temperature due to slower kinetics [39]. Alloys such as Ga-In, with a melting point of 15.3 °C, and Ga-Sn, which melts at 10.7 °C, have also been investigated as self-healing anodes for Li-ion batteries. For instance, Wu et al. achieved superior cycling performance, with a Ga-Sn alloy embedded in a 3D carbon framework, capable of sustaining up to 4000 cycles. This was further enhanced by Yu et al., who dispersed bulk liquid metal into nanoparticles via ultrasonication, improving stability and preventing structural degradation.

Ga-based liquid metals have also shown potential in addressing interface issues in non-lithium-based systems such as zinc-ion and aluminum-ion batteries. A notable development in the field is the room-temperature LMB designed by Ding et al., which utilized Na-K and Ga-In alloys. In this battery, the immiscibility and density differences of the negative electrode, electrolyte, and positive electrode allow them to self-segregate into three distinct layers. Ga-In alloys, in particular, have been found to adhere well to gold surfaces, with theoretical calculations indicating a much higher adsorption energy on gold compared to materials like aluminum or stainless steel. This finding opens doors for improved surface and interface design in LMBs.

Despite significant advancements in LMBs, challenges remain in making them viable for grid-scale energy storage. Optimizing metal alloying strategies can help reduce melting points and expand battery chemistries, as demonstrated by research into alloys like Ca-Bi, Ca-Sb, and Na-based alloys. Such approaches not only help lower operating temperatures but also provide potential for improving both anode and cathode materials. Moreover, exploring multi-element fusible alloys offers cost-effective solutions for reducing operating temperatures and enhancing performance.

Reducing the internal resistance of liquid metal or alloy electrodes is another key challenge, with strategies focusing on enhancing the wettability and electrical conductivity of these materials. In high-temperature liquid metal batteries (HT-LMBs), molten salt electrolytes offer high ionic conductivity and are cost-effective, though the high solubility of metals can lead to self-discharge, reducing energy efficiency. For medium- and room-temperature liquid metal batteries (MT-LMBs and RT-LMBs), solid ceramic electrolytes present challenges such as poor wettability at lower temperatures. Effective interface engineering, particularly in systems involving Na-K and Ga-based alloys, is essential for optimizing performance.

Sealing and corrosion remain critical concerns, particularly at high temperatures. Polymers may suffice at temperatures below 200 °C, but for higher temperatures, glass and ceramics must be considered, despite the difficulties they present in terms of thermal expansion and chemical stability. Corrosion of reactive metals like Li and Na can also shorten battery life, making it essential to develop materials and sealing techniques that can withstand these harsh conditions [40].

While research continues to push the boundaries of what is possible with LMBs, it is clear that temperature alone should not be the sole criterion for evaluating their potential. A balanced consideration of factors such as energy density, efficiency, cycle life, safety, cost, and environmental impact is crucial to making these technologies viable for large-scale energy storage. Although HT-LMBs still hold promise for their simplicity and cost-effectiveness, future research must adopt a more holistic approach to fully realize the potential of LMB technology.

13.7 CONCLUSIONS

Reviewing the advancements in rechargeable LMBs reveals significant progress in battery chemistry and multi-physics modeling aimed at overcoming key challenges.

In the case of high-temperature LMBs (HT-LMBs), materials such as liquid metals, alloys, and molten salts do not present significant cost barriers. However, improving cell performance requires reducing operating temperatures through the exploration of new battery chemistries or mechanisms while preserving the three-liquid-layer structure. These batteries exhibit advantages such as being free from dendrite formation, being easy to fabricate, and offering cost-effective material recovery, making them attractive for large-scale, stationary energy storage applications. Nevertheless, further work on multi-physics modeling is crucial to optimize design and scale-up operations, pushing these batteries closer to commercialization. Medium- and room-temperature LMBs (MT-LMB and RT-LMB) offer potential for higher energy density but face challenges related to the interfaces between liquid electrodes and solid electrolytes. Addressing these issues and developing solid electrolytes with high ionic conductivity and transference numbers remain significant hurdles for future high-performance LMBs. Moreover, battery performance is affected not only by the intrinsic properties of the chemistries but also by the cell configurations and testing protocols. Thus, continued efforts in multi-physics modeling are essential for understanding the interactions between electrochemical and thermophysical processes, thereby optimizing LMB performance for practical large-scale applications. With further advancements in battery chemistry and multi-physics modeling, LMBs show considerable promise for future energy storage solutions.

13.8 ACKNOWLEDGMENTS

One of the authors (RKS) thankfully acknowledges the financial assistance by SERB, New Delhi, India. While VY acknowledge financial support from CSIR, New Delhi, in the form of JRF.

REFERENCES

[1] M. Park, J. Ryu, W. Wang, J. Cho, *Nat Rev Mater* 2 (2016).
[2] Y. Liu, Y. Zhu, Y. Cui, *Nat Energy* 4 (2019) 540–550.
[3] H. Li, H. Yin, K. Wang, S. Cheng, K. Jiang, D.R. Sadoway, *Adv Energy Mater* 6 (2016).
[4] W. Chen, G. Li, A. Pei, Y. Li, L. Liao, H. Wang, J. Wan, Z. Liang, G. Chen, H. Zhang, J. Wang, Y. Cui, *Nat Energy* 3 (2018) 428–435.
[5] M.M. Thackeray, C. Wolverton, E.D. Isaacs, *Energy Environ Sci* 5 (2012) 7854–7863.
[6] X.B. Cheng, R. Zhang, C.Z. Zhao, Q. Zhang, *Chem Rev* 117 (2017) 10403–10473.
[7] H. Kim, D.A. Boysen, J.M. Newhouse, B.L. Spatocco, B. Chung, P.J. Burke, D.J. Bradwell, K. Jiang, A.A. Tomaszowska, K. Wang, W. Wei, L.A. Ortiz, S.A. Barriga, S.M. Poizeau, D.R. Sadoway, *Chem Rev* 113 (2013) 2075–2099.
[8] J. Muldoon, C.B. Bucur, T. Gregory, *Chem Rev* 114 (2014) 11683–11720.
[9] X. Guo, L. Zhang, Y. Ding, J.B. Goodenough, G. Yu, *Energy Environ Sci* 12 (2019) 2605–2619.
[10] X. Guo, Y. Ding, G. Yu, *Adv Mater* 33 (2021).
[11] B. Dunn, H. Kamath, J.-M. Tarascon, *Electrical Energy Storage for the Grid: A Battery of Choices System Power Ratings*, Module Size, n.d.
[12] S. Wu, X. Zhang, R. Wang, T. Li, *Energy Storage Mater* 57 (2023) 205–227.
[13] J.M. Newhouse, S. Poizeau, H. Kim, B.L. Spatocco, D.R. Sadoway, *Electrochim Acta* 91 (2013) 293–301.

[14] J.G. Simpson, G. Hanrahan, E. Loth, G.M. Koenig, D.R. Sadoway, *Renew Sustain Energy Rev* 149 (2021).
[15] R. Chen, Q. Li, X. Yu, L. Chen, H. Li, *Chem Rev* 120 (2020) 6820–6877.
[16] J. Coetzer, Short communication a new high energy density battery system, *J Power Sources* 18 (1986).
[17] H. Pan, Y.S. Hu, L. Chen, *Energy Environ Sci* 6 (2013) 2338–2360.
[18] K. Wang, K. Jiang, B. Chung, T. Ouchi, P.J. Burke, D.A. Boysen, D.J. Bradwell, H. Kim, U. Muecke, D.R. Sadoway, *Nature* 514 (2014) 348–350.
[19] F. Zhu, J. Ge, Y. Gao, B. Cai, Z. Zhang, F. Jia, S. Jiao, *Electrochem Commun* 155 (2023).
[20] P. Personnettaz, P. Beckstein, S. Landgraf, T. Köllner, M. Nimtz, N. Weber, T. Weier, *J Power Sources* 401 (2018) 362–374.
[21] Y. Ding, X. Guo, G. Yu, *ACS Cent Sci* 6 (2020) 1355–1366.
[22] J.H. Cho, M. Aykol, S. Kim, J.H. Ha, C. Wolverton, K.Y. Chung, K.B. Kim, B.W. Cho, *J Am Chem Soc* 136 (2014) 16116–16119.
[23] K. Wang, K. Jiang, B. Chung, T. Ouchi, P.J. Burke, D.A. Boysen, D.J. Bradwell, H. Kim, U. Muecke, D.R. Sadoway, *Nature* 514 (2014) 348–350.
[24] T. Weier, A. Bund, W. El-Mofid, G.M. Horstmann, C.C. Lalau, S. Landgraf, M. Nimtz, M. Starace, F. Stefani, N. Weber, in: *IOP Conf Ser Mater Sci Eng*, Institute of Physics Publishing, 2017.
[25] ANL-6725 Chemical Separations Processes for Plutonium and Uranium (TID-4500, 28th Ed.) AEC Research and Development Report, n.d.
[26] M.R. Gao, Y.F. Xu, J. Jiang, S.H. Yu, *Chem Soc Rev* 42 (2013) 2986–3017.
[27] H. Li, K. Wang, H. Zhou, X. Guo, S. Cheng, K. Jiang, *Energy Storage Mater* 14 (2018) 267–271.
[28] Q. Gong, W. Ding, A. Bonk, H. Li, K. Wang, A. Jianu, A. Weisenburger, A. Bund, T. Bauer, *J Power Sources* 475 (2020).
[29] D.J. Bradwell, H. Kim, A.H.C. Sirk, D.R. Sadoway, *J Am Chem Soc* 134 (2012) 1895–1897.
[30] J. Popovic, *J Electrochem Soc* 169 (2022) 030510.
[31] Y. Ding, G. Yu, *Angew Chem Int Ed* 56 (2017) 8614–8616.
[32] L. Huang, J. Li, B. Liu, Y. Li, S. Shen, S. Deng, C. Lu, W. Zhang, Y. Xia, G. Pan, X. Wang, Q. Xiong, X. Xia, J. Tu, *Adv Funct Mater* 30 (2020).
[33] X. Lu, M.E. Bowden, V.L. Sprenkle, J. Liu, *Adv Mater* 27 (2015) 5915–5922.
[34] X. Lu, B.W. Kirby, W. Xu, G. Li, J.Y. Kim, J.P. Lemmon, V.L. Sprenkle, Z. Yang, *Energy Environ Sci* 6 (2013) 299–306.
[35] Y. Ding, X. Guo, Y. Qian, G. Yu, *ACS Cent Sci* 6 (2020) 2287–2293.
[36] L. Xue, H. Gao, W. Zhou, S. Xin, K. Park, Y. Li, J.B. Goodenough, *Adv Mater* 28 (2016) 9608–9612.
[37] L. Xue, W. Zhou, S. Xin, H. Gao, Y. Li, A. Zhou, J.B. Goodenough, *Angew Chem* 130 (2018) 14380–14383.
[38] X. Guo, Y. Ding, H. Gao, J.B. Goodenough, G. Yu, *Adv Mater* 32 (2020).
[39] M. Zabrocki, W. Gąsior, A. Dębski, *J Mol Liq* 332 (2021).
[40] M.G. Nicholas, R.M. Crispin, Diffusion bonding stainless steel to alumina using aluminium interlayers, *J Mater Sci* 17 (1982) 3347–3360.

14 Liquid Metal-Based Flexible Devices

Paola N. García-Hernández, Perla J. Pérez-Díaz, Ana K. Cuentas-Gallegos, and Daniella E. Pacheco-Catalán

The development of flexible LM-based batteries is gaining importance due to the need for combining the inherent advantages of LM as electrodes with the flexibility of substrates, allowing for innovative applications across various sectors. Traditional battery designs are often constrained by the rigidity of electrode infrastructure, limiting their adaptability in flexible and portable devices. The fabrication of flexible batteries using LM represents a promising direction, leveraging the unique properties of these metals to create more versatile and efficient energy storage solutions.

Among the various elements considered for use in LM, Ga stands out due to its low toxicity, good biosafety, intrinsic fluidity, and excellent electrical and thermal conductivities. This has made Ga the most used LM. Additionally, materials such as EGaIn (75.5% Ga, 24.5% In) and Galinstan (68.5% Ga, 21.5% In, 10% Sn) have been developed to further optimize the performance of these materials. EGaIn, for example, has a lower viscosity compared to pure liquid Ga, making it highly suitable for flexible battery applications [1]. Table 14.1 shows some of the characteristics of the Ga alloys mentioned above; thanks to these characteristics, their use in flexible device applications has been encouraged. One of the most important characteristics is the lower melting point of the Ga alloys compared to Ga as an element, ensuring the liquid phase at room temperature. Furthermore, the viscosity remains virtually unchanged, but the boiling point decreases, which means that Ga alloys should not be used in systems with a potential temperature increase.

In the field of flexible LM-based batteries, Liu et al. [5] fabricated for the first time a soft and stretchable battery using EGaIn as the anode, carbon paste coated with MnO_2 slurry as a cathode, KOH/PAAm hydrogel as anode electrolyte, LiOH/KOH/PAAm hydrogel as cathodic electrolyte, and Ecoflex as package, with a basic configuration as shown in Figure 14.1. Since the electrolyte is alkaline, Ga has more electrochemical activity compared to In. A portion of Ga in the EGaIn alloy participates in the discharge reaction, generating GaO_3^{3-} trough as per the following reaction: $Ga{-}3e^- + 6OH^- \leftrightarrow GaO_3^{3-} + 3H_2O$.

In this study [5], the flexible EGaIn-MnO_2 battery shows excellent electrochemical stability under different mechanical stress conditions, including stretching and bending. It was observed that as stretching increases, the discharge voltage levels also increase, especially at higher current densities, with the voltage difference

DOI: 10.1201/9781003489726-14

TABLE 14.1
The Physical Properties of Common Liquid Metals [2–4]

	Ga	EGaIn	Galistan	$Ga_{61}In_{25}Sn_{13}Zn_1$
Melting point (°C)	29.8	15.7	10.5	7.6
Boiling point (°C)	2204	2000	1300	900
Density (10^3 kgm^{-3})	6.08	6.28	6.36	6.5
Viscosity (10^{-7} m^{2}s^{-1})	3.24	2.7	2.98	0.71
Surface tension (Nm^{-1})	0.72	0.624	0.533	0.5
Conductivity (10^6 Sm^{-1})	3.7	3.4	3.1	2.8
Thermal conductivity (W(m°C)$^{-1}$)	29.4	42.2	44.8	48.2

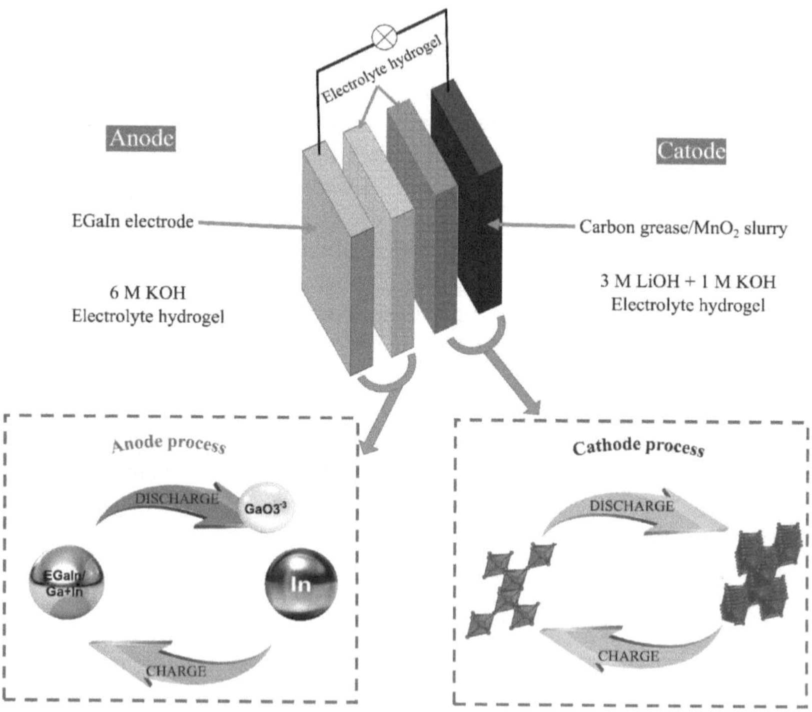

FIGURE 14.1 Configuration of a battery with liquid EGaIn in the anode, and the schematic representation of electrochemical reaction of the battery.

between charging and discharging decreasing from 0.51V to 0.13V as the stretch increases from 0% to 100% (Figure 14.2a). The battery's impedance is low, with R_{int} increasing from 8.4 to 51.3 Ω and R_{ct} decreasing from 85.6 to 18.5 Ω as the stretch increases (Figure 14.2b). Additionally, the battery maintains stable discharges under tension and bending, showing that its electrochemical performance remains almost unchanged even under extreme mechanical stress (Figures 14.2c and 2d). The

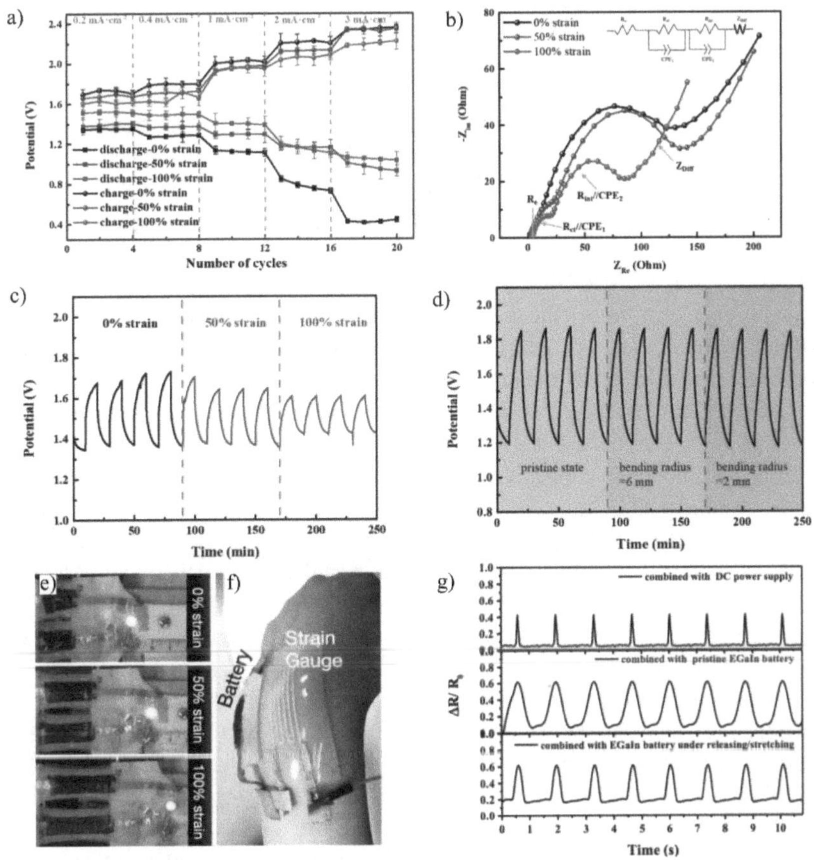

FIGURE 14.2 (a) Rate performance of discharge–charge cycle of the EGaIn-MnO_2 battery with increasing static strain from 0 to 100% at different discharge current densities. (b) Corresponding Nyquist plots of the impedance of the array with increasing tensile strain of 0, 50, and 100% at open circuit voltage (OCV). (Inset is equivalent circuit model.) (c) Full discharge curves of the battery at various tensile strains of 0, 50, and 100%. (d) Cycle curve of the battery in the bending mode with radii = 6 mm and 2 mm. (e) Photographs of the stretchable EGaIn-MnO_2 battery array in series of two stretched under 0, 50, and 100% strain integrated with LEDs. (f) Photograph of battery-powered strain sensor that is mounted on the wrist. (g) Strain sensor measurements during stretching with a frequency of ≈1 Hz when powered with a DC power supply, pristine EGaIn battery, and releasing/stretching EGaIn battery, respectively.

Source: Adapted with permission [5]. Copyright (2019) Advanced Energy Materials.

battery's ability to continuously power an LED and a strain gauge under dynamic load is demonstrated. The LED remains bright even when the battery is stretched up to 150% (Figure 14.2e). When mounted on the wrist (Figure 14.2f), the stretchable battery successfully powers a strain gauge, showing consistent measurements of $\Delta R/R \approx 0.6$ (Figure 14.2g). This consistency is observed regardless of whether the power is supplied by a direct current (DC) power source, an unstretched EGaIn-MnO_2 battery, or a stretched EGaIn-MnO_2 battery alongside the strain gauge. These results confirm that the flexible EGaIn-MnO_2 battery can work effectively under various external mechanical conditions.

More recently, in 2024, Naman Shi et al. [6] developed a self-healing flexible anode based on LM (GaInSn) for lithium-ion batteries. Their research leverages the flexibility of carbon cloth and the fluidity of LM to fabricate anodes composed of LM nanoparticles, carbon nanotubes, and carbon cloth. Carbon nanotubes serve to prevent the agglomeration of LM nanoparticles. This study establishes that the amount of LM is a critical factor, as excessive amounts negatively impact specific capacity, as do insufficient amounts. The optimal proportion determined in this research was 0.46 $mgcm^{-2}$ named LM/CNTs. The material exhibited a charge retention of 95.8% at 2 Ag^{-1}, with a minimum charge transfer resistance of 90 ohms.

The lithium charge storage mechanisms in the electrode reveal a predominantly capacitive behavior, as indicated by the parameter b in the equation $i = av^b$, which takes values of 0.91 and 0.93 for the cathodic and anodic peak currents, respectively (Figure 14.3 A,ii). In Figure 14.3 A.i, the redox peaks observed in the voltammogram at various scan rates are shown. Specifically, a capacitive contribution of 82% was determined for the curve at 0.8 mVs^{-1} (Figure 14.3 A,iii), with an increase in the capacitive contribution observed as the scan rate increased (Figure 14.3 A,iv), attributed to the enhanced conductivity provided by the presence of carbon nanotubes. This excellent electrochemical performance is also attributed to the self-healing property of the LM. During the discharge process, the GaInSn nanoparticles fracture due to the insertion of Li^+, but during the charge process, as Li^+ is extracted, these fractures automatically repair themselves.

Another area of interest for the application of LMs in flexible devices has been triboelectric nanogenerators (TENGs), which harvest energy by converting mechanical energy to electrical energy. An example of this is the work done by Nayak Suryakanta et al. [7], where they created a triboelectric foam using the LM alloy (LMA) Galinstan ($Ga_{62}In_{22}Sn_{16}$) and Ecoflex (Figure 14.3 B). In order to achieve a porous structure in the elastomer, they used a soluble sacrificial template of NaCl particles. When characterizing the foam material, they observed that both the porosity and the conductivity of the elastomer matrix are key to the triboelectric properties of the foam, and they also confirmed that the presence of LMA and the pores in the foam are essential for charge generation. After evaluating the output short circuit current (ISC) and the OCV with respect to the deformation cycles in samples with different proportions, they found that the optimum is 3 parts by weight of LMA in the Ecoflex matrix (LMA:NaCl:Ecoflex = 3:5:10), since a higher concentration of LMA causes a degradation in the load as a consequence of an overload of the LMA in the elastomer. They were also able to evaluate that a percentage close to 28% of porosity in the foam composite is adequate, since when there is an increase in the

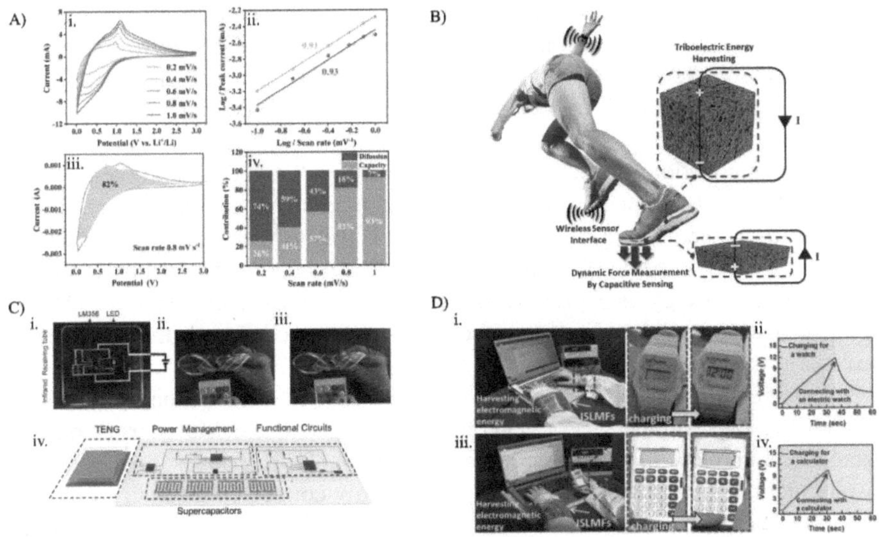

FIGURE 14.3 (a) Electrochemical characterization of LM/CNTs, (i) voltammograms at different scanning rates; (ii) the relationship between peak current and scanning rate; (iii) the contribution of capacity behavior to charge storage at a scanning rate of 0.8 mVs^{-1}; and (iv) contribution of capacitance and diffusion behavior to charge storage at different scanning rates. (b) Concept of smart footwear with energy harvesting, force sensing capabilities, and wireless data connectivity. (c) Infrared pair tube circuit, (i) photo image of the circuit; (ii) photo image of the stretched infrared pair tube circuit before receiving signal, and (iii) after receiving signal; (iv) schematic illustration of multifunctional self-powered system with Cu-EGaIn electrodes. (d) Performance of ISLMFs to harvest electromagnetic energy from ambient electrical appliances, (i) demonstration on harvesting electromagnetic energy from a laptop to drive and electronic watch and (ii) real-time charge/discharge curve, (iii) Demonstration of harvesting electromagnetic energy from a laptop to drive a calculator and (iv) real-time charge/discharge curve.

Sources: (a) Adapted with permission [6]. Copyright (2024) Applied Materials Today. (b) Adapted with permission [7]. Copyright (2019) Nano Energy. (c) Adapted with permission [8]. Copyright (2022) Nano Energy. (d) Adapted with permission [9]. Copyright (2021) Advanced Energy Materials.

pore density of the foam, a degradation in the generation of triboelectric charge is caused by the structural collapse of the foam.

In order to perform a practical evaluation of the foam, they prepared a 10 mm thick porous 3D shoe insole. They molded support structures in the form of hexagonal columns to reduce the effective stiffness of the foam, resulting in an insole with a base thickness of 3.5 mm and column thickness of 6.5 mm. As in the evaluation of the foam material, they placed the insole between copper foil electrodes. They tested a 73 kg person, producing an instantaneous peak power of 2.6 mW (instantaneous power density of 13 μWcm^{-2}) while jogging. In this study, they demonstrated that the insole can be used for both triboelectric generation and force and weight measurements in motion.

Another application for these LM TENGs was made by Yang and collaborators in 2022 [8], who developed a self-powered system featuring stretchable and flexible electrodes based on LM Cu-EGaIn. The system consists of an energy harvester, specifically a TENG, capable of converting mechanical energy into electrical energy. This electrical energy is subsequently stored in micro-supercapacitors (MSCs), which utilize Cu-EGaIn/MXene nanocomposite electrodes and are employed to power an infrared pair tube circuit.

When pressure is applied to the Cu-EGaIn-based TENG, physical contact between the two friction layers generates opposing triboelectric charges on their surfaces. Upon releasing the pressure, the separation of the surfaces induces an electrostatic field. As the two electrodes are electrically connected through an external load, free charges from one electrode flow to the other to establish an opposing electric potential, balancing the induced electrostatic field. When the two friction layers make contact again, the triboelectric potential vanishes, and electrons flow back to their original positions. This cyclical process enables energy to be stored in the Cu-EGaIn/MXene electrodes, which demonstrated a capacitance of up to 1.72 mFcm^{-2} with a charge retention of 92.4% after 2,160 cycles.

To demonstrate functional application, an infrared pair tube circuit was designed using printed Cu-EGaIn electrodes. This functional circuit integrates patterned Cu-EGaIn electrodes with packaged electronic components (Figure 14.3 C,i). When exposed to an external infrared signal, the LED in the circuit lights up, demonstrating functionality under both stretched and bent conditions (Figure 14.3 C,ii and iii). Figure 14.3 C,iv illustrates the complete self-powered, multifunctional system, which integrates the LM-based TENG (Cu-EGaIn), the supercapacitors (Cu-EGaIn/MXene), and the infrared pair tube circuit.

Moreover, in the research of Lai Ying-Chih et al. [9] they ventured into the development of intrinsically elastic liquid metal (ISLM) fibers that act as TENGs and also capture electromagnetic energy (like from laptops or cell phones), thus having the ability to capture two different types of energy sources (Figure 14.3). For the fabrication of the multifunctional ISLM, they used a Styrene-Ethylene/Butylene-Styrene (SEBS) elastomeric hollow fiber, which was filled with EGaIn (75.5 Ga/24.5 In) through an injection process. During the application as a TENG, the SEBS fulfills the function of triboelectric material, while the EGaIn is the conductive electrode. For its triboelectric characterization, while taking advantage of biomechanical energy, they proceeded to sew 2 m of ISLMF in an elastic fabric. The textile was evaluated in 3 parts of the body: the elbow, the knee, and the wrist. The latter produced electricity when touched by hand under conditions of 0.1 N and 4 Hz, and by storing the energy in a 3.3 μF capacitor, 5 V of charge was obtained in approximately 180 s. With the energy obtained, they were able to power small electronic devices such as a digital watch and an electric calculator. This research also evaluated the feasibility of harvesting electromagnetic energy using the multifunctional ISLMF. In this case, the metallic property of EGaIn is used to carry out the process of induced electrification in metals. They determined that the electrostatic potentials depend directly on the distance between the electromagnetic sources and the LMs. To carry out the practical application, they used the same procedure of baking the ISLMF to an elastic fabric; with this they were able to charge a 4.7 μF capacitor to 5 V in 17 s by extracting electromagnetic energy from a laptop. With these results, they determined that the charging time of the capacitor through electromagnetic energy was shorter than when it was charged through biomechanical energy resulting from human movement. They observed that the use of this energy radiated by the laptop did not modify its original energy consumption for its normal operation. In Figure 14.3D, the practical results obtained by powering electrical devices, such as an electric watch (Figure 14.3 D,i) and a calculator (Figure 14.3 D,iii) can be observed. The charging/discharging process of both the watch and the calculator can be observed in Figures 14.3 D,ii and 3 D,iv, respectively.

Within the broader realm of flexible LM-based devices, flexible sensors stand out as a particularly promising application. Sensors are essential for gathering information by detecting and responding to various stimuli from the physical environment or living organisms (e.g. humans and plants), such as light, heat, motion, moisture, pressure, or electrical pulses [10], in order to detect, evaluate, and control them. Traditional sensors, however, are typically rigid, which limits their use in applications that require portability, biocompatibility, and the ability to endure stress from movements or actions involving flexibility (Figure 14.4a and 14.4b).

Flexible sensors based on LMs are emerging as a critical innovation in this area. LMs, such as Ga and its alloys, play a vital role in the development of these sensors. Its properties enable the creation of sensors that are not only flexible but also highly efficient and responsive. This makes them ideal for use in soft robotics, health monitoring (Figure 14.4c), and other fields where flexibility and performance are vital [11].

Due to its low Young's modulus, Ga and its alloys are considered excellent candidates for use as high-response sensors. This characteristic causes rapid deformations in LMs, which is an advantage when obtaining electrical signals that will later be used to calculate the mechanical signal corresponding to the deformation caused [12]. Among the research carried out in the area of biosensors is the work of T. F. Kong et al. [13], where they used a 3D multilayer LM (Ga) microcoil to detect hematocrit levels in the blood, which provides essential information on blood oxygen saturation, in addition to helping to identify other medical conditions. They used the microcoil to detect changes in the transverse relaxation rate of the blood cells and their hematocrit level. This was done with the aim of developing a portable device based on the developed microcoil in the future.

Given the current medical conditions in the population, it is of utmost importance to be able to constantly monitor people's health. Blood glucose is an important parameter to consider in monitoring. For this reason, various sensors have been investigated for noninvasive applications. Kim et al. [14] developed a self-powered saliva glucose biosensor fabricated by n-InGaZnO/p-Si photodetector covered with poly (ethylene glycol) film containing enzymes, which react with glucose and cause the film to change color. Another study was carried out by Lei Mou et al. [15] in which they developed a LM-polymer conductor patch using LM particles and a

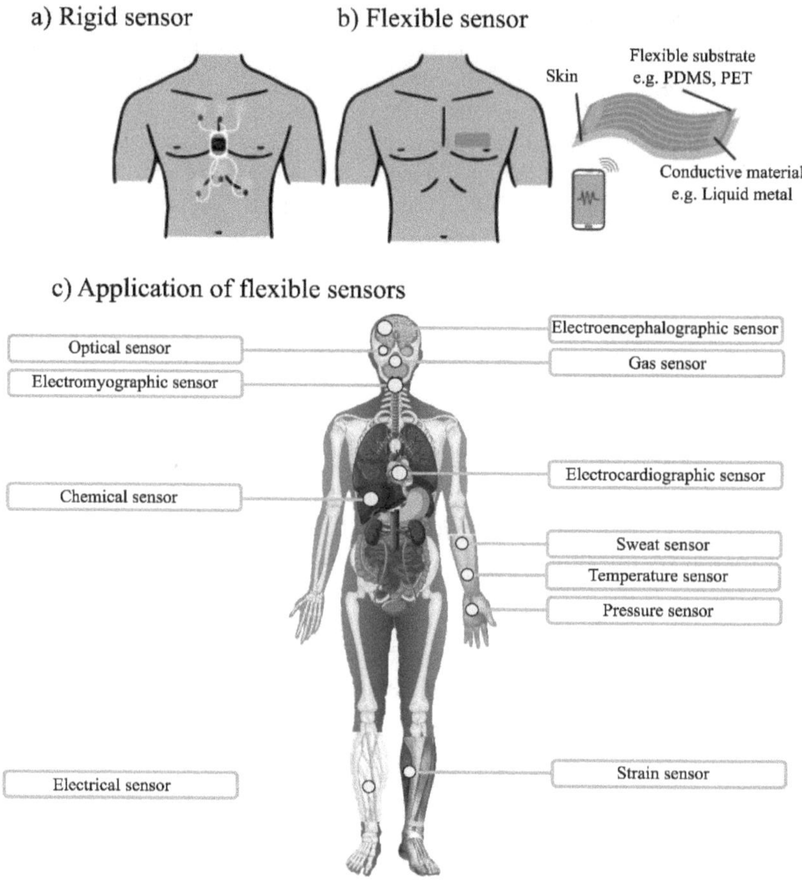

FIGURE 14.4 Comparison between (a) rigid cardiac sensor and (b) flexible cardiac sensor. (c) Application areas of flexible sensors for human health monitoring.

polyethylene terephthalate substrate. With this patch, they were able to detect metabolites in human sweat (glucose, Na^+, K^+, etc.).

Additionally, advances in wearable technology have led to the creation of humidity sensors that track skin moisture and respiration, offering real-time data for healthcare applications [16]. LMs have also found use in neural interfaces, enabling precise monitoring and stimulation of neural activity [17]. These sensors' flexibility and softness reduce tissue damage and improve the potential for therapeutic uses, such as in nerve repair and prosthetics.

Furthermore, since temperature sensors are vital for monitoring the life characteristics of organisms, recent advancements have led to the development of stretchable temperature sensors using LM circuits, capable of maintaining performance under strain. These sensors show high flexibility, stability, and quick response, making them ideal for wearable or micro-scale applications, where precise and durable measurements are crucial [18, 19].

Moreover, optical sensors have seen advancements through the use of LM nanoparticles, which exhibit unique UV plasmonic properties [20]. LM microdroplets have been applied to create optical sensors with linear responses, useful in solar energy harvesting and UV sensing [21]. The flexibility, high conductivity, and low melting point of LMs make it ideal for wearable, foldable devices. Recent developments include stretchable UV sensors integrated with energy storage systems, and LM-based coils in wireless biomedical devices, such as artificial retinas and eye movement tracking systems [22]. These innovations have opened doors to continuous health monitoring and improved biomedical technologies.

As mentioned, there are various applications for LMs, which also leads to the development of a variety of manufacturing methods for these devices, where these methods are adapted to the needs of their final use, with printing and microfluidic technologies being the most commonly used (Figure 14.5a). Other techniques include selective wetting, laser engraving, and wiping. Printing and spraying of LM are particularly favored for their low-cost and rapid prototyping capabilities, allowing for the creation of conductive patterns on flexible substrates (Figure 14.5f). These substrates typically include soft thermoplastic polymers and silicone elastomers such as polyethylene terephthalate (PET), polyurethane (PU), polydimethylsiloxane (PDMS), Ecoflex, and electroconductive polymers [23].

Another prominent technique involves injecting LM into pre-fabricated molds (Figure 14.5d **and** e). These molds can be printed onto substrates, such as PDMS, to achieve specific patterns for various applications [24]. This process generally starts with creating patterns on flexible substrates, like PVA, PET, or PDMS, using 2D machines. A thin film of the same flexible substrate is then placed over the patterns. The LM is injected into the channels, and a vacuum is applied to seal the surfaces, effectively encapsulating the LM [25–27].

For instance, screen-printing technology has been used to create a stretchable radiation shield from LM (Figure 14.5c). This involves uniformly coating a silicon film to a thickness of 0.3 mm [28]. Additionally, nano-ink of LM has been fabricated using ultrasonic baths with PVP as a stabilizer. This nano-ink is then injected into a ball-point pen and tested on various substrates. Once the desired pattern is created, it is fixed with a sealer. A similar technique, known as soft lithography (Figure 14.5g), has been employed to fabricate electronic passive components and wearable sensors on PDMS substrates [29–31].

Laser radiation techniques have also been explored (Figure 14.5b). For example, nano-ink of LM was filtered and transferred to a flexible, transparent substrate using laser patterning. This method demonstrated effectiveness for creating antennas and capacitors [29]. Further advancements include the synthesis of a humidity sensor using Ga_2O_3/LM. This process involves sonicating the LM, spraying it onto a flexible substrate, and employing UV laser sintering. The single-step direct laser writing technique leverages the photothermal effect of the laser to selectively convert insulating traces into conductive ones [16].

The future development of flexible LM devices holds vast potential, driven by continuous advancements in materials science, fabrication techniques, and applications. These devices are expected to revolutionize sectors like healthcare, sports science, wearable technology, soft robotics, and renewable energy, where flexibility,

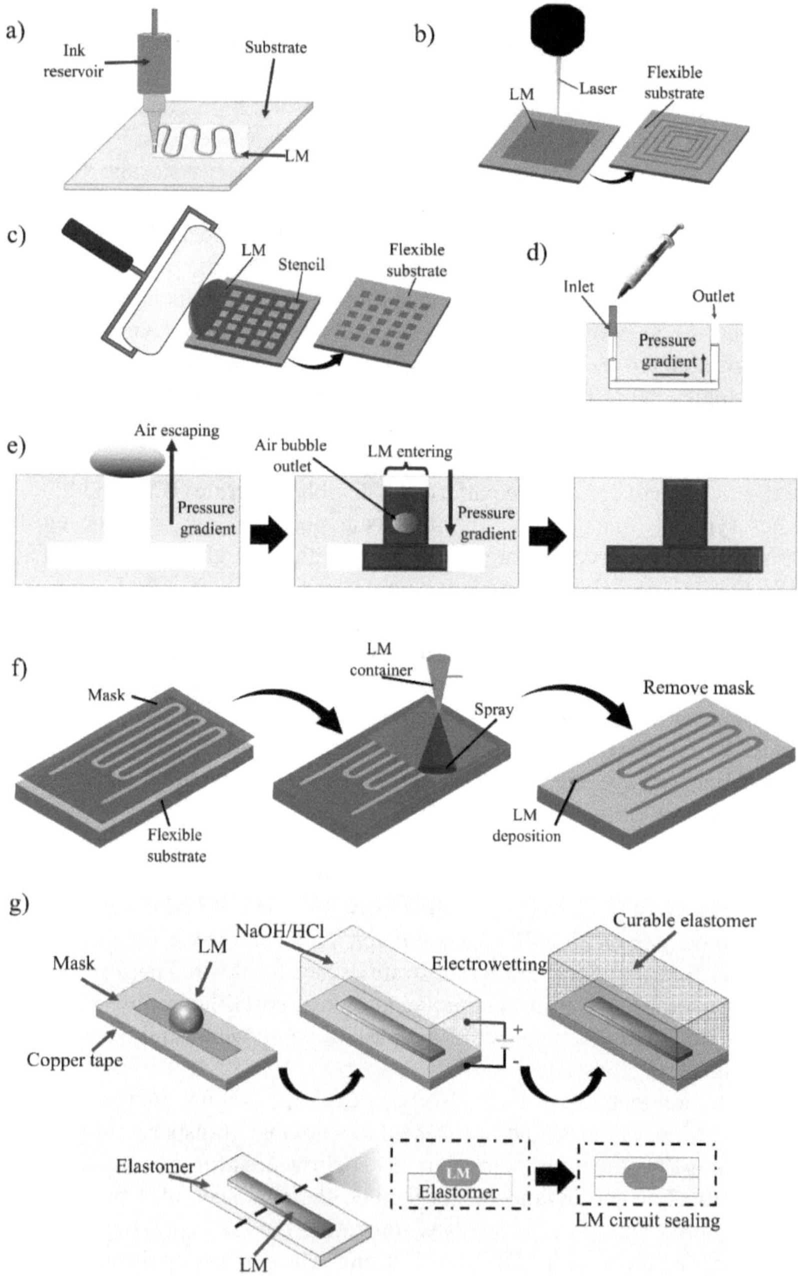

FIGURE 14.5 Patterning methods of LMs. (a) direct printing, (b) laser ablation of LM films, (c) screen-printing method using stencil mask which is made by photolithography, (d) and (e) Injecting and vacuum filling of LMs into a microfluidic channel, (f) spray printing of LM onto a stencil, and (g) soft-lithographic method.

biocompatibility, and durability are essential. The adsorption of ultraviolet and electromagnetic radiation by LMAs, mainly based on Ga and indium (In), can be widely used for the development of protective materials that reduce the harmful effects of electronic devices. In addition, they can be a viable option for the reduction of heavy metal pollutants in the environment. Further research into optimizing LMA, working temperature, leakage, robustness, improving production methods like microfluidics and printing, and integrating energy storage systems with sensors will be critical [32]. As LM technology matures, it is poised to enable more efficient, adaptable, and multifunctional devices, paving the way for innovations in portable diagnostics, health monitoring, batteries, supercapacitors, and beyond.

REFERENCES

[1] M. Yang, G. Li, Y. Gu, J. Song, H. Li, X. Zhao, Y. Huo, Liquid metal-based flexible and wearable thermoelectric cooling structure and cooling performance optimization, *Sci China Mater* 66 (2023) 4001–4011.

[2] S. Chen, J. Liu, Pervasive liquid metal printed electronics: From concept incubation to industry, *IScience* 24 (2021) 102026.

[3] J. Yan, Y. Lu, G. Chen, M. Yang, Z. Gu, Advances in liquid metals for biomedical applications, *Chem Soc Rev* 47 (2018) 2518–2533.

[4] S. Chen, H.Z. Wang, R.Q. Zhao, W. Rao, J. Liu, Liquid metal composites, *Matter* 2 (2020) 1446–1480.

[5] D. Liu, L. Su, J. Liao, B. Reeja-Jayan, C. Majidi, Rechargeable soft-matter EGaIn-MnO2 battery for stretchable electronics, *Adv Energy Mater* 9 (2019) 1902798.

[6] N. Shi, L. Luo, S. Li, C. Xu, J. Liu, S. Sha, J. Zhu, Y. Zhou, G. Cai, R. Zhang, A self-healing liquid metal-based flexible anode for next-generation lithium-ion batteries, *Appl Mater Today* 41 (2024) 102484.

[7] S. Nayak, Y. Li, W. Tay, E. Zamburg, D. Singh, C. Lee, S.J.A. Koh, P. Chia, A.V.Y. Thean, Liquid-metal-elastomer foam for moldable multi-functional triboelectric energy harvesting and force sensing, *Nano Energy* 64 (2019) 103912.

[8] J. Yang, J. Cao, J. Han, Y. Xiong, L. Luo, X. Dan, Y. Yang, L. Li, J. Sun, Q. Sun, Stretcheble multifunctional self-powered systems with Cu-EGaIn liquid metal electrodes, *Nano Energy* (2022) 107582.

[9] Y.C. Lai, H.W. Lu, H.M. Wu, D. Zhang, J. Yang, J. Ma, M. Shamsi, V. Vallem, M.D. Dickey, Elastic multifunctional liquid–metal fibers for harvesting mechanical and electromagnetic energy and as self-powered sensors, *Adv Energy Mater* 11 (2021) 2100411.

[10] Y. Ren, X. Sun, J. Liu, Advances in liquid metal-enabled flexible and wearable sensors, *Micromachines (Basel)* 11 (2020) 200.

[11] S. Jamalzadegan, S. Kim, N. Mohammad, H. Koduri, Z. Hetzler, G. Lee, M.D. Dickey, Q. Wei, Liquid metal-based biosensors: Fundamentals and applications, *Adv Funct Mater* (2024) 2308173.

[12] M. Zhang, X. Wang, Z. Huang, W. Rao, Liquid metal based flexible and implantable biosensors, *Biosensors (Basel)* 10 (2020) 170.

[13] T.F. Kong, N.T. Nguyen, Liquid metal microcoils for sensing and actuation in lab-on-a-chip applications, *Microsyst Technol* 21 (2015) 519–526.

[14] K. Kim, H. Kim, E.J. Jo, H. Jang, J. Park, G.Y. Jung, M.G. Kim, Reactant/polymer hybrid films on p-n junction photodetectors for self-powered, non-invasive glucose biosensors, *Biosens Bioelectron* 175 (2021) 112855.

[15] L. Mou, Y. Xia, X. Jiang, Liquid metal-polymer conductor-based wireless, battery-free epidermal patch, *Biosens Bioelectron* 197 (2022) 113765.

[16] S. Cui, Y. Lu, D. Kong, H. Luo, L. Peng, G. Yang, H. Yang, K. Xu, Laser direct writing of Ga2O3/liquid metal-based flexible humidity sensors, *Opto-Electron Adv* 6 (2023) 190042.

[17] Y.G. Park, G.Y. Lee, J. Jang, S.M. Yun, E. Kim, J.U. Park, Liquid metal-based soft electronics for wearable healthcare, *Adv Healthc Mater* 10 (2021) 2002280.

[18] Q. Wang, M. Gao, L. Zhang, Z. Deng, L. Gui, A handy flexible micro-thermocouple using low-melting-point metal alloys, *Sensors (Switzerland)* 19 (2019) 314.

[19] S.Y. Hong, Y.H. Lee, H. Park, S.W. Jin, Y.R. Jeong, J. Yun, I. You, G. Zi, J.S. Ha, Stretchable active matrix temperature sensor array of polyaniline nanofibers for electronic skin, *Adv Mater* 28 (2016) 930–935.

[20] I.S. Maksymov, A.D. Greentree, Dynamically reconfigurable plasmon resonances enabled by capillary oscillations of liquid-metal nanodroplets, *Phys Rev A (Coll Park)* 96 (2017) 0438291–0438296.

[21] P. Lin, Z. Wei, Q. Yan, J. Xie, Y. Fan, M. Wu, Y. Chen, Z. Cheng, Capillary-based microfluidic fabrication of liquid metal microspheres toward functional microelectrodes and photothermal medium, *ACS Appl Mater Interfaces* 11 (2019) 25295–25305.

[22] Z.N. Zhao, J. Lin, J. Zhang, Y. Yu, B. Yuan, C.C. Fan, L. Wang, J. Liu, Liquid metal enabled flexible electronic system for eye movement tracking, *IEEE Sens J* 18 (2018) 2592–2598.

[23] Z. Guo, G. Liao, L. Ren, H. Qiao, Z. Huang, Z. Wang, X. Qi, Non-invasive flexible sensor based on liquid metal for human physiological detection, *Next Nanotechnol* 5 (2024) 100042.

[24] S. Ali, D. Maddipatla, B.B. Narakathu, A.A. Chlaihawi, S. Emamian, F. Janabi, B.J. Bazuin, M.Z. Atashbar, flexible capacitive pressure sensor based on PDMS substrate and ga-in liquid metal, *IEEE Sens J* 19 (2019) 97–104.

[25] R. Guo, J. Liu, Implantable liquid metal-based flexible neural microelectrode array and its application in recovering animal locomotion functions, *J Micromech Microeng* 27 (2017) 104002.

[26] Y. Wu, Y. Zhou, W. Asghar, Y. Liu, F. Li, D. Sun, C. Hu, Z. Wu, J. Shang, Z. Yu, R.-W. Li, H. Yang, Liquid metal-based strain sensor with ultralow detection limit for human–machine interface applications, *Adv Intell Syst* 3 (2021) 2000235.

[27] J. Cao, X. Li, Y. Liu, G. Zhu, R.W. Li, Liquid metal-based electronics for on-skin healthcare, *Biosensors (Basel)* 13 (2023) 84.

[28] Y. Deng, J. Liu, Liquid metal based stretchable radiation-shielding film, *J Med Devices Trans ASME* 9 (2015) 0145021–0145024.

[29] M. Zhang, G. Li, L. Huang, P. Ran, J. Huang, M. Yu, H. Yuqian, J. Guo, Z. Liu, X. Ma, Versatile fabrication of liquid metal nano-ink based flexible electronic devices, *Appl Mater Today* 22 (2021) 100903.

[30] M.G. Kim, H. Alrowais, S. Pavlidis, O. Brand, Size-scalable and high-density liquid-metal-based soft electronic passive components and circuits using soft lithography, *Adv Funct Mater* 27 (2017) 1604466.

[31] Y. Tao, F. Han, C. Shi, R. Yang, Y. Chen, Y. Ren, Liquid metal-based flexible and wearable sensor for functional human–machine interface, *Micromachines (Basel)* 13 (2022) 1429.

[32] Z. Lin, T. Li, S. Yang, B. Ji, Z. Wang, Revolutionizing flexible electronics with liquid metal innovations, *Device* 2 (2024) 100331.

Index

S

T

W

Z

Kolibri 360 GmbH
Ferdinand-Jühlke-Straße 7
99095 Erfurt, Deutschland
produktsicherheit@kolibri360.de